Essential Skills

for Human Services

Essential Skills for Human Services

Cynthia Cannon Poindexter
Boston University

Deborah Valentine
University of Wyoming

Patricia Conway
University of Wyoming

Brooks/Cole • Wadsworth

I(T)P®An International Thomson Publishing Company

Belmont • Albany • Bonn • Boston • Cincinnati • Johannesburg
London • Madrid • Melbourne • Mexico City • New York • Pacific Grove • Paris
Singapore • Tokyo • Toronto • Washington

Sponsoring Editor: *Lisa Gebo*
Editorial Assistant: *Susan Wilson*
Marketing Team: *Steve Catalano, Aaron Eden*
Project Editor: *Janet Hill*
Marketing Communications: *Kyrrha Seven*
Copy Editor: *Margaret Pinette*
Indexer: *Meg McDonnell*

Design Editor: *E. Kelly Shoemaker*
Cover Illustration: *Judith Harkness*
Interior and Cover Design: *Lisa Henry*
Manufacturing Buyer: *Vena Dyer*
Composition: *Joan Mueller Cochrane*
Cover Printer: *Webcom*
Printer & Binder: *Webcom*

For more information, contact

WADSWORTH PUBLISHING COMPANY
10 Davis Drive
Belmont, CA 94002
USA

International Thomson Editores
Seneca 53
Col. Polanco
11560 México, D. F., México

International Thomson Publishing Europe
Berkshire House 168-173
High Holborn
London, WC1V 7AA
England

International Thomson Publishing GmbH
Königswinterer Strasse 418
53227 Bonn
Germany

International Thomson Publishing Asia
60 Albert Street #15-01
Albert Complex
Singapore 189969

Thomas Nelson Australia
102 Dodds Street
South Melbourne, 3205
Victoria, Australia

International Thomson Publishing Japan
Hirakawacho Kyowa Building, 3F
2-2-1 Hirakawacho
Chiyoda-ku, Tokyo 102
Japan

Nelson Canada
1120 Birchmount road
Scarborough, Ontario
Canada M1K 5G4

Printed in Canada

10   9   8   7   6   5

Library of Congress Cataloging-in-Publication Data

Poindexter, Cynthia Cannon, 1998
    Essential skills for human services / Cynthia Cannon Poindexter,
Deborah Valentine, Patricia Conway.
        p.   cm.
    Includes bibliographical references and index.
    ISBN 0-534-34690-1
    1. Social service.   I. Valentine, Deborah P., 1998.
II. Conway, Patricia Gail.   III. Title.
HV31.P65   1998
361'..0023'73—dc21   98-37157

*We dedicate this book to all of the people whose stories
and experiences appear throughout these chapters.
We are grateful to these teachers.*

CONTENTS

CHAPTER TWO
Attitudes and Values   14

CHAPTER THREE
Awareness of Others   28

CHAPTER SEVEN
Crisis Intervention 108

CHAPTER EIGHT
The Helping Interview and the Problem-Solving Process 122

PART THREE

Models for Helping 137

CHAPTER NINE

Case Management 139

CHAPTER TEN
Class Advocacy and Community Organization   159

PART FOUR
Applications to Special Areas   173

CHAPTER ELEVEN
Violence in Relationships and Families   175

CHAPTER TWELVE
Gerontology and Aging 201

CHAPTER THIRTEEN
HIV and AIDS   223

CHAPTER EIGHTEEN
Putting It All Together

316

THINK ABOUT IT EXERCISES

The Purpose and Approach of This Book

This book is an introductory text for beginning helpers; that is, human services workers, students, and volunteers who wish to develop the skills, values, and knowledge to effectively support and guide persons who approach social services systems for assistance. The text also can by used by professional helpers who are serving as trainers, educators, or teachers with human services workers, students, or volunteers.

The overarching goal of the textbook in general parallels the purpose of "helping" to enhance the quality of life for individuals, couples, families, groups, and communities. The authors present an experiential, skills-building approach that emphasizes the importance of strengths and helping relationships.

Human services workers, students, and volunteers are part of a multidisciplinary group of helpers who provide valuable support in a variety of settings. Human services helpers interface with other professionals who may have more specialized degrees, such as social workers, nurses, clergy persons, psychologists, physicians, music and art therapists, criminal justice personnel, counselors, and educators. Cooperation, collaboration, and communication with specialists therefore are becoming increasingly important for human services helpers. Human services workers, students, and volunteers are essential components of diverse service areas, such as child welfare or development, health care, corrections, crisis intervention, gerontology, mental health, addictions, recreation, economic aid, nutrition, support for persons with disabilities, community organization, and intervention for relationship and community violence.

Human services helpers often are the first or only line of defense for those seeking assistance, and as such are extremely important players in maintaining physical and mental health for many individuals, couples, families, groups, or communities. Human services helpers fulfill a variety of roles, such as peer counselors, case managers, advocates, residential managers, caregivers, group facilitators, and teachers. Because human services helpers often work as aides in diverse disciplines, a broad base of knowledge is necessary.

Regardless of where human services workers are employed or volunteering, what discipline or population interests them the most, or which type of licensed processionals are working with them, certain essential, basic, or universal skills of helping are necessary. Helpers must be skilled in relationship building, active listening, supporting someone's informed choices, crisis intervention, problem solving, interviewing, record maintenance, accessing and using community resources, managing service and treatment plans, advocating for persons who are having difficulty obtaining access to the support they desire, obtaining consultation and supervision, and assisting individuals, couples, families, groups, or communities in meeting their goals. This textbook introduces these concepts and techniques. Whether you are a student in a human services Associates of Arts or Bachelor's degree program, a volunteer in a human services organization, a peer-support person or human services worker in an agency, or someone considering a professional degree program in any of the helping professions, this text introduces you to ways of relating to persons with whom you work in order to facilitate the helping process.

The authors have been involved for decades in the education and training of community-based volunteers and human services students and workers. Out of those experiences comes our commitment to providing beginning helpers with the tools and perspectives that will facilitate their success as effective and appropriate supports to individuals, couples, families, groups, and communities.

Effective helping is a blend of knowledge, skills, personal qualities, and values. In this textbook you will be introduced to basic knowledge about certain populations, ethical guidelines, essential helping skills, situations in which these helping skills might be applied, circumstances in which supervision and consultation should be used, and guidelines for making decisions about your role in the helping relationship. Examples and illustrations are used throughout to illuminate the helping techniques and to highlight difficulties or dilemmas in service provision. In addition, there are exercises that you may do at your own pace, as part of class assignments, or in a group of learners.

This book can be used in the following ways:

1. As a textbook for "Introduction to Human Services," "Interviewing," "Counseling Skills," or "Introduction to Helping" college classes in two- or four-year human services programs.
2. As a training guide for professional facilitators who are teaching groups of paid or volunteer helpers.
3. As a self-help guide for beginning helpers who wish to be better prepared to work with others. If you are using this book for this purpose, we recommend that you keep a journal in which you record your reactions, responses, and questions, and that you work with a consultant who can help you clarify and apply the material.

Although reading and studying are important tools in acquiring a new perspective of any subject, no human services helper can learn appropriate techniques simply by reading about them. This book is best used as one part of a comprehensive training program that includes experiential learning, group discussions, reinforcement of content, supervision and consultation, and on-the-job training.

This textbook certainly cannot cover all of the concerns, methods, or topics in human services, not even for beginning helpers. Even for the subjects we included, we have just skimmed the surface. Each reader will be able to identify important issues that were omitted. We encourage teachers and learners to supplement the use of this text with other materials, especially if you wish to explore a topic or population in depth. This textbook is written for audiences who are practicing in the human services network in the United States. International contexts are not addressed in this material.

Organization of This Book

The first section, "Laying the Groundwork," introduces the beginning helper to the concept of help, to underlying values or help-giving, and to an awareness of self and others. The second section, "Foundations for Helping," explores the helping relationship, use of self, communication and interviewing, and principles and techniques of active listening, crisis intervention, and problem solving. Two common models for helping (case management and community organizing) are presented in the third section. Specific situations are addressed in the fourth section, such as relationship violence and grief, as well as special populations, such as persons who are elderly and persons with HIV, developmental disabilities, or mental illnesses. The fifth and final section assists helpers with information regarding consultation and stress management. Throughout each section illustrations, exercises, and questions prompt thought. The authors hope that this book can serve as some illumination on a journey toward becoming a more effective human services helper. For educators and trainers, a brief "Instructors' Guide" is available from Brooks/Cole Publishing Company.

Several chapters received the attention of friends, and we acknowledge and appreciate their input and care: Lana Ka'opua for Chapter 3, Mick Smyer for Chapter 12, and Eric Noel for Chapter 13. We also thank Douglas Chung, Grand Valley State University; Frank Clark, University of Montana; Mary Davidson, Columbia Greene Community College; Sharon Eisen, Mott Community College; Kathy Elpers, University of Southern Indiana; Alan Kemp, Pierce College; Blanchita Porter, Howard University; Guy Wylie, Western Nebraska Community College; and John Heapes, Harrisburg Community College, all of whom provided thoughtful and useful feedback on the entire text.

We are indebted to our fabulous Brooks/Cole representative, Lisa Gebo, who never lost patience with us or faith in the project.

Cynthia Cannon Poindexter
Deborah Valentine
Patricia Conway

Essential Skills

for Human Services

PART ONE

Laying the Groundwork

The first three chapters of this textbook introduce the student to the world of human services and of helping. Chapter 1 explores, among other topics, what help is, why it might be needed, and where it might be offered. Chapter 2 begins to examine the important attitudes, philosophies, and values that undergird the human services field, such as focusing on strengths, suspending judgment, supporting choice, being trustworthy, maintaining privacy, respecting differences, and being aware of oneself. Chapter 3 asks the reader to be aware of others, including the concepts of culture and developmental stages, the importance of language usage, and striving for cultural competence. Although this section is by no means a comprehensive treatment of the basics of human services, no beginning helper should embark on any social service job without considering these fundamental issues.

CHAPTER ONE

Introduction:
What Does It Mean to Be a Helper?

This chapter introduces you to the reasons persons choose to become helpers, the definition of helping, the philosophy of this textbook, types and models of helping, and settings where human services workers and volunteers may be found.

Why Be a Helper?

People choose to offer assistance to others for many reasons. You may have had experiences in your own life that brought you into contact with a helper and increased your appreciation for being able to find appropriate and timely support. Possibly you have recently endured a tragedy or crisis and now feel that you have wisdom and comfort to share with those in similar situations. Perhaps you were raised in a family, culture, religion, neighborhood, or community in which helping others was valued and encouraged. Maybe you have always been someone whom others sought when they needed a good listener. Many people are attracted to the human services field because they have a capacity to be helpful, sometimes because of their personal life experiences, and they have a desire to be supportive and useful to persons who are in need or in turmoil.

Whatever the origins of your desire to help others, it is not enough to have good intentions. A genuine desire to help and a real sense of caring are absolutely necessary, but they must be supplemented with the tools, skills, and techniques of helping. Just as accomplished ballet dancers require the love of their art, and basketball players their sport, as well as years of hard work on form and technique, good helpers must use both their hearts and their heads to be useful to persons experiencing difficulties.

Think About It #1

1. Think about a time in your life when you benefited from contact with a helper. In what ways did this person help you? What did he or she do or not do? In what ways did this person have an impact on your life?

3

2. Think about some of the helping roles that you have played in your life. In what ways were they rewarding? In what ways were they difficult?

What Is Helping and Why Is It Needed?

Help is defined in *The American Heritage College Dictionary* (1997) in this way: "to give assistance to; aid; to contribute to the furtherance of; promote; to give relief to; to ease; relieve; to change for the better; improve" (p. 631). At times individuals, couples, families, groups, and communities request help from an outside source. Sometimes the weight of social and personal problems makes it difficult to solve all of one's problems without the input of trusted helpers.

Our society is now facing many problems. The reporting of sexual assault, child and elder abuse, and relationship violence has risen over the past few decades. The number of older persons living in their communities has increased dramatically in recent years. HIV, a life-threatening and infectious disease, surfaced less than two decades ago and has taken a tremendous toll on human life. At a time when U.S. residents with mental, physical, or developmental disabilities, as well as chronic diseases, are receiving care in their homes and communities, more family members are working outside the home and caregivers are feeling more overwhelmed. Poverty, crime, gang membership, and single-parent households are occurring more frequently, at a time when federal and state governments are decreasing financial support. The need for "help" is all too obvious.

The nature, amount, and frequency of help that is provided by mental health, social services, and counseling professionals have dramatically increased over the past three decades. It is now estimated that one-third of all Americans consult a mental health professional or seek therapy sometime in their lifetimes (Meredith, 1987). Typically, persons in trouble do not seek social services or counseling first; rather, asking for assistance is the last choice, after they have tried everything else to relieve the stress of the situation (Meredith, 1987).

Why do many persons seek out others when they are in need of consoling, care, and new solutions? Humans are social animals and often find comfort in appropriate, invited, and accepting "social support," which has consistently been found to lessen the effects of emotional stress (Berkman, 1984; Cohen, 1983; Cohen and Wills, 1985; Gottlieb, 1985; Pearlin, 1985; Sandler and Barrera, 1984). Parents caring for children with special needs, for example, list emotional support as their single most needed personal resource (Friesan, 1989; McCubbin, McCubbin, Patterson, Cauble, Wilson and Warwick, 1983). "People embedded in secure social networks suffer less physical, emotional, and social dysfunction than their counterparts who are not members of support networks" (Germain and Patterson, 1988, p. 79). In fact, in one study, older men and women who reported interacting consistently with two or fewer people over a three-year period died at a much higher rate over the next eight years than those elders who regularly interacted with a larger social network (Bower, 1997). Furthermore, African-American adults who reported few social contacts had significantly higher blood pressure readings than those with more social contacts (Bower, 1997). There also is convincing evidence that social support can act as a protective factor in preventing and reducing the risk of child abuse and neglect (Thompson, 1997). Therefore, helping is based on the notion that support invited and welcomed from another person can be nurturing, healing, and transforming.

Think About It #2

1. What is your response to the information above regarding the importance of support from others? Does this surprise you? Why or why not?
2. Speculate on why and how social support might have an effect on physical and mental health. What is your theory about this?
3. Has there ever been a time in your life when you were hungry for social support? What happened?

Note that in the last sentence regarding social support we specified that help should be "invited and welcomed." If individuals, couples, families, groups, or communities do not willingly accept the input of a human services helper or agency, then the "help" is not real. Remember the old joke about the Boy Scout who said he had done his good deed for the day but it had taken a long time to do so? The Scoutmaster said, "What did you do?" The boy answered, "Well, I helped an old lady across the street." The Scoutmaster said, "So, why did it take so long?" The Scout said, "Sir, she didn't want to go." Even though this illustration is silly, it makes the point that true "help" can only happen within a participatory partnership. It is not something that can be done "to" someone. This is a very important concept. Forgetting this guideline is the basis for many mistakes and false steps for human services workers, students, and volunteers. Sometimes people are sent to our agencies against their wishes, as when someone is ordered by a court of law to report for counseling, pay a fine, or go to jail. Often we refer to this type of service recipient as "involuntary." Even in these situations the person must on some level "let you in" and accept your input. You still cannot force someone to gratefully accept what you have to offer.

Illustration

Michelle broke a bone in her foot while she was moving into a new apartment. With a cast on her foot, she was a little discouraged but still felt able to take care of the unpacking and minor repairs herself. She was looking forward to having the pleasure of fixing up her new place herself. One morning two of her friends rang the doorbell and greeted her with big grins on their faces. "We're here to help you!" they announced gleefully. Not wanting to turn away her kind friends, and thinking "I could sure use some help with a few things," Michelle welcomed them. However, throughout the day Michelle's friends unpacked boxes that she had not asked them to unpack, and made decisions about where to place furniture and knickknacks without consulting her. Michelle was hurt and sometimes tearful about their leaving her out but was unwilling to confront them because she knew that they meant well.

What Are Human Services Helpers?

Human services workers are trained to provide a wide range of emotional and practical support to a wide variety of persons who are seeking assistance with life's difficulties and challenges. Human services workers and volunteers can be found in a variety of settings and activities. Examples are: working as a house parent in a residential facility for persons with retardation or mental illness; acting as an advocate for sexual assault victims who must go to the police station or to court; arranging for and managing services for persons who are physically impaired or ill; providing parenting and household management training to family members who have been violent; assisting persons in a nursing home or adult day care center; offering counseling to the survivors of relationship violence; facilitating parent-child communication in schools; staffing a crisis or suicide hotline; or providing support for parents caring for children with special needs.

Some human services workers go on to obtain a bachelor's and/or master's degree and align themselves with a particular profession, such as social work, counseling, pastoral care, or psychology. No matter what plans you make for the future, what will serve you and the persons seeking help the most is your genuine desire to help, your respect for the rights and dignity of others, and your nonjudgmental, gentle presence. Human services workers and volunteers are often very good at speaking "the language of the heart" (Borkman, 1990). At this level, you will be able to work effectively side by side with licensed specialists, supporting individuals, couples, and families, as well as supplementing and augmenting the treatment plan developed by the licensed specialist.

A Note on Terms and the Philosophy of Helping
Helpers and Those Seeking Help

Throughout this book the reader will see a distinction between helpers (human services workers, students, and volunteers) and persons who are seeking help. Although the use of these labels is a necessary mechanism to talk about the process of helping, the authors believe that these distinctions can blur the important fact that the "helper" and the "help seeker" are really *partners* in helping and problem-solving. As we have stated previously, true helping must be a mutual endeavor. There are two major reasons for this.

First, the person who is requesting assistance from you or your organization is the expert on his or her own needs, goals, and life. The role of the helper is to assist someone in finding his or her own solutions; this requires much more skill than merely telling him or her what to do. Helpers do possess authority and skills because they have training, knowledge, experience, and an agency backing them. However, helpers do not have the right to take over someone else's decision making. Each of us has the right to choose for ourselves how we will live our lives, within legal and ethical limits. The helper must acknowledge the idea that others have the right to make choices based on personal and cultural perspectives and needs. The choices of other persons may not always be the first choice of the helper.

Second, every person deserves to be treated with dignity and respect, as a person who is in charge of his or her own life, no matter what the current situation is or how difficult it is to manage. Sometimes persons who are occupying the helper role forget that the help seekers deserve to be treated kindly and politely. Think of times when you were sad or in crisis and needed emotional support or practical assistance from another person. You probably did not want to be treated as an object, a "case," a bad person, or merely a collection of symptoms. You wanted to be treated as a valuable, functioning, worthwhile person who was temporarily in need of a helpful connection with another person. You probably also wanted that person to be caring, compassionate, considerate, concerned, warm, welcoming, and not critical. Even when someone seems angry, hesitant, or unmotivated, a human services helper should treat that person with respect.

Person-First Language

It is important that whenever referring to an individual or group, the humanity or personhood should come before the condition, issues, problem, or illness. You will notice that this textbook uses terms such as "woman with mental retardation," "person with hemophilia," or "children living with HIV," rather than terms such as "mentally retarded woman," "hemophiliac," or "AIDS babies." Labels have powerful social and political results, and we feel strongly about using terms that remind ourselves and others that people are people *first*, with characteristics being secondary. This is an important value for us, and at times we will remind readers of our belief about the power of language.

Use of Genuine Self

This book emphasizes the helping relationship, communication, respect, and rapport as much as techniques and tasks. As you have already read, real helping occurs only between people who are partners and with a helper who is genuinely concerned and can appropriately communicate that caring. In the human services field, "caring" is just as important as "curing."

Throughout the reading of this textbook (indeed, throughout life) it will be useful to remember how you would want to be treated and how you would want your loved ones treated. Use that awareness of what it feels like to be troubled, and translate that into the way that you offer help. This will assist you to value and nurture the helping relationship and to be genuinely yourself within that relationship.

Respect for Diversity

In addition, there will be a dual theme throughout this textbook: commonalities between human beings who are seeking help, and diversity and differences in persons and communities. Identifying common patterns and needs among individuals is useful to helpers. However, no person, group, or culture can be stereotyped and understood based only on generalities. Diversity should be identified, understood, and affirmed. This book seeks to point out what is similar and what is different in persons who may request help.

An important aspect of respecting diversity and becoming culturally competent is demonstrating an acceptance of whatever definition of "family" we are offered by the people who approach us for assistance. Too often in human services work we find a discounting of someone's significant support persons because they do not fit with traditional or mainstream definitions of "kin," "relatives," or "family." Throughout this book, when we use the word "family," we mean "family of choice."

Finding Strengths

Focusing first on the strengths and resources that individuals, couples, families, groups, and communities possess, rather than concentrating only on deficits, barriers, or weaknesses, is vital in human services work. Taking a strengths perspective enhances the chance that the helping efforts will succeed. As people realize their strengths, they are frequently better able to build on those strengths. In addition, once strengths are identified, acknowledged, and talked about, people tend to feel better about themselves and are much more likely to be able to work on deficits and weaknesses. Helpers should concentrate on identifying inner and outer resources and helping those who ask for assistance to recognize their own strengths. Persons who are troubled or in crisis usually have no difficulty recognizing their barriers and failures; it is the helper's job to assist them with identifying the successes and strengths as well. Frequently a helper can identify a strength that the person never realized he or she possessed. In these instances helpers should try to point out some of the strengths that they see, offering their perceptions for the other person to consider. Focusing on strengths does not mean ignoring, discounting, or invalidating the real challenges that people are facing. But, as Saleebey (1996) states, "It is as wrong to deny the possible as it is to deny the problem" (p. 297). Relying on this strengths perspective in the helping professions is an attempt to correct the previously overwhelming focus on the negative aspects in the others' lives: "emphasis on what is wrong, what is missing, and what is abnormal" (Saleebey, 1996, p. 297).

What Types of Help Are Requested?

People may ask for practical help and/or emotional support. Practical help may include obtaining food, clothing, shelter, day care, or transportation. Someone who is sick or has a disability may require more personal practical activities, such as assistance with bathing, feeding, toileting, dressing, and taking medicines. Human services helpers can provide practical help directly or arrange for it from another source. For example, you may drive someone to the doctor yourself or may arrange for the local Council on Aging to provide transportation services to that person.

In addition to practical help, persons receive much benefit from emotional support. Emotional support consists of listening actively and nonjudgmentally and being fully "present." Parents who are caring for children with special needs, for example, list emotional support as their single most needed resource (Friesan, 1989; McCubbin et al., 1983). Take a moment to think about why the care and concern of others would make a difference to caregivers who are worried or frustrated. Again, beginning helpers provide emotional support directly (by listening nonjudgmentally to the concerns of the individual who asks for help) and/or indirectly (by referring him or her to other service providers or agencies).

Illustration

Mrs. Tell is a 26-year-old married Caucasian woman who is receiving prenatal care from the local county public health department. She has two preschool children: Amy, age five, and Loren, age three. Mrs. Tell has been taking medication for a mental disorder called "manic depression," which can cause unexpected and uncontrollable mood swings. She has stopped taking the medication during her pregnancy and does not intend to take it while she is breast-feeding because she is concerned about the possible impact on the baby. Mrs. Tell reports to the nurse at the prenatal care clinic that she is experiencing reduced energy, increased sadness, difficulty thinking clearly, and decreased patience with Amy and Loren. Because of this mild depression, Mrs. Tell has quit her job at K-mart. Her husband, a 37-year-old Caucasian man with a history of illegal drug use, is not working. The Tells consequently have moved in with Mrs. Tell's parents, and have found that, as a result, they argue frequently. Mrs. Tell's mother receives disability payments and does not work outside the home; her father owns a small business.

Think About It #3

1. What are the strengths and resources that Mrs. Tell demonstrates, possesses, or has access to?
2. What are some of the difficulties or challenges that the Tell family faces?

3. What might you as the helper provide to the Tell family directly?

4. What other services might be put into place for various members of the Tell family?

Helping Whom?

Helping may occur with (1) individuals, (2) couples or families, (3) groups, and (4) neighborhoods or communities. Each of these formats will be briefly discussed below. Note that this text does not directly address work with couples, families, or groups. However, illustrations throughout the book highlight human services helpers' involvement in these areas.

One-on-One (Individual) Support

In addition to emotional support and practical assistance, human services workers, students, and volunteers offer help to individuals through case management (also called service coordination). These three categories are not mutually exclusive; each contains elements of the other. They are briefly discussed below.

Emotional support. Frequently, helping happens in a private, one-on-one relationship with a person who is in need of a connection to a caring person. In this case, helping is likely to consist of listening, emotional support, and counseling. For example, Don is a retired businessman who is working part time for a community-based hospice that provides home care for persons who are dying. He is assigned to support Shirley, an elderly woman who has end-stage leukemia. Several times a week he drops by Shirley's home to chat with her, hold her hand, listen to music with her, or listen to her talk about her life, depending on how she feels and what she wants to do. At times Shirley talks to Don about her feelings about her impending death.

Practical assistance. Human services helpers often work as aides or assistants in institutional settings, such as hospitals; day-care centers; shelters; facilities for persons with mental retardation, developmental disabilities, or mental illnesses; or homes for the elderly, children, or adolescents. For example, Gretchen is an aide at a day-care center for persons with dementia. Her job entails helping the participants with personal needs and with recreational activities. She helps persons with eating and getting to the bathroom, and she also strives to keep them occupied with card games, music, television, or crafts, depending on their interests, attention, and abilities.

Case management or service coordination. Another service delivery form common for beginning helpers is case management, or service coordination, which will be discussed in further detail in Chapter 9. In this mode, the helper's task is to determine what someone needs, arrange for those services, coordinate the agencies involved, and monitor the quality of service delivery. For example, Doris is a service coordinator (often called "case manager" or "case worker") with the local Council on Aging. After the agency nurses assess the elderly persons to determine their needs for practical support and personal care, Doris contacts workers in her own agency, as well as in other agencies, to arrange for the services the person wants and needs. She may arrange for home-delivered meals, home nursing, transportation, home visitors, housekeeping help, home repair, telephone reassurance, or help with bathing, dressing, or toileting. She calls the individual or family once a month to make sure that all of the arranged services are in place and working well.

Couple or Family Support

Human services workers, students, or volunteers may also be asked to provide support to a couple or family who is in crisis or in need of assistance. For example, couples may be experiencing stressors in their relationship, parents may be frustrated or confused about a child's behavior or development, or a group of people who identify as "family" may be struggling with how to get along with each other. Sometimes couples, parents, or families are experiencing extreme difficulties, like unemployment or violence, and can benefit from the intervention of a concerned helper or agency. Couple or family support is illustrated by the experience of David, who is a parent aide in an agency that assists parents who have been neglectful of their children, with the goal of keeping the family together. David does home visits with each of the families in his case load twice a week, in the day or evening, depending on their work schedules, in order to give them tips on stress reduction, anger management, discipline techniques, and communication strategies.

Group Support

Support, assurance, and help can be given in a group setting, such as a support group, self-help group, or psychoeducational group. A group is simply three or more people who have gathered for a common purpose or goal. Not everyone feels comfortable sharing personal information in a group, but for many persons, groups are the best way to find support and encouragement from persons who are experiencing similar difficulties. Frequently the members of groups resemble each other with regard to their "type of problem circumstances surrounding the onset of their problem, the severity and expected duration of their problem, and their emotional reactions to their problems" (Medvene, 1990, p. 133).

Groups are organized for many reasons, such as for persons who have a problem with alcohol consumption, parents who have abused their children,

children whose parents are separating, persons with cancer or HIV, adults or children who are bereaved, men who have battered their partners, adults who have experienced incest as children, couples who are infertile, adoptive families, or caregivers of persons with disabilities or who are frail, elderly, or ill. Often groups such as the above are organized and led (or cofacilitated) by peer facilitators; that is, paid or unpaid persons who have recovered from the problem sufficiently to be of use to others who are struggling. For instance, Sally is a woman who was battered by her former husband and is now working for an agency addressing relationship violence, where one of her jobs is to colead a support group in the shelter for women who have been battered.

Community Organization and Advocacy

In addition, helpers may work to effect change within an organization, neighborhood, or community. Often problems are widespread and cannot be sufficiently addressed unless changes are made in larger systems. For example, LeRoy is a Boy Scout leader who realizes that the young boys in his troop are at great risk for recruitment by gangs and for being the victims of random violence in their neighborhood. He organizes meetings of the parents and guardians in the area to talk about their concerns for the safety of their children. Their discussions led to a plan to institute a neighborhood watch program, an after-school activity program, and a "latchkey kids" program.

Where Might You Work?

In addition to having many choices regarding the type of human services work you do, there are many settings and roles available to human services workers, students, and volunteers. Many social or medical services agencies employ human service helpers in a variety of capacities, including providing practical and emotional support to individuals, couples, families, groups, or communities.

Human services helpers work in addictions counseling and mental health settings, both inpatient and outpatient. Human services helpers often work as house parents at children's shelters, or as house managers or peer counselors in shelters for persons who are homeless, or for women who are experiencing relationship violence. They also work as legal assistants, court advocates, house managers, and parent supporters. They also might work as aides in programs for persons with developmental or physical disabilities, literacy projects, community-based grassroots agencies, or corrections facilities. They may be parole or probation officers, intake interviewers, eligibility workers, or crisis intervention counselors.

Think About It #4

1. The authors state: "Each of us has the right to choose for ourselves how we will live our lives, within legal and ethical limits." Brainstorm what some of those limits might be. Think individually about what your feelings and values are regarding some of those limits.

2. Jot down a stressful or difficult event that changed forever the way you function or view your life. Also note what helped you at the time or what you would have wanted that you did not receive. What did this teach you about helping?

3. *Help* is defined as:"to give assistance to; aid; to contribute to the furtherance of; promote; to give relief to; to ease; relieve; to change for the better; improve." Does this definition fit with your idea of helping?

4. Think about what drew you to helping. What made you want to be a human services worker or volunteer? What do you hope to receive from your experiences as a helper? What do you wish to gain from being a helper?

5. What are the fears, hesitancies, or anxieties that you are experiencing about the textbook, the course, or the process of becoming a human services worker?

6. What are the strengths and gifts that you bring to the task of being a helper? What relationship skills and helping qualities do you possess?

7. What relationship skills or helping qualities would you like to acquire or improve over the next few months?

CHAPTER TWO

Attitudes and Values

Some attitudes, beliefs, values, and behaviors seem to increase the likelihood that effective and constructive help will occur. Being a responsible and effective helper means being aware of our own needs and feelings, cultivating the qualities that are known to be helpful when talking with and listening to others, and adhering to ethical standards. This chapter will address those helper attitudes, beliefs, values, and ethics that promote well-being in others and develop self-awareness. This chapter is rich with illustrations to help learners understand the applications of these basic helping qualities.

Helper Attitudes and Beliefs

The following attitudes and beliefs associated with effective helping will be discussed in this section: seeing strengths first; suspending judgment; supporting choice; and being honest and trustworthy.

Seeing Strengths First

Help giving will be more effective if helpers assume a positive stance toward others. Instead of focusing on weaknesses, focus on strengths. Instead of trying to catch people doing something wrong, strive to catch them doing something right. Helpers who assume that others have the capacity to manage events in their lives and build upon resources and strengths are more effective than helpers who label individuals, couples, and families "deficient," "sick," or in need of our correction (Dunst, Trivette, Davis and Cornwall, 1988). Being aware of the abilities and strengths in all human beings is an essential feature of an effective helper. Emphasizing the weaknesses and the shortcomings of individuals experiencing struggles and stresses further convinces them that they are inadequate, lowers their self-esteem, and further robs them of self-confidence and a feeling of personal competence.

Illustration

Nikisha is a 14-year-old African-American girl with visual impairments. The teacher and the teacher's aide at the junior high school Nikisha attends were concerned about Nikisha's aggressive and violent behavior toward male classmates, male teachers, and male administrators. They reported that Nikisha scratched, kicked, hit, and fought in what appeared to be unprovoked attacks at unpredictable times. After a home visit and a lengthy discussion with her 31-year-old-mother, it was discovered that Nikisha's family lived on a very limited income in a very dangerous and unsafe neighborhood. Nikisha's mother had taught her daughter self-defense skills so that she could protect herself against the ever-present possibility of attack or rape.

In the above situation, a person assuming a deficit perspective may have concluded that Nikisha had a behavior problem or was emotionally disturbed. A deficit perspective may have led to the conclusion that the family was uncaring and not appropriately disciplining or setting limits with Nikisha. Utilizing a strengths perspective, however, it was possible to understand Nikisha's behavior in a different way. A loving mother is trying to keep her daughter safe from harm in the only way that she knows how. It would certainly be appropriate for school personnel to help both Nikisha and her mother explore other behavioral alternatives to aggressive behavior while in the school setting. Approaching the mother in a way that appreciates and respects her love and concern for her daughter will increase the likelihood of cooperation, rather than approaching the mother in ways that criticize or assume the worst.

Seeing the best in people is a skill that will be useful to you in all your helping work. Practice identifying your own and other's strengths regularly. Discuss your strengths and the strengths of your friends and family members with others. If identifying and describing strengths is difficult for you, think about why that is the case. The following exercise asks you to list some of your strengths and those of significant persons in your life in order to practice taking a strengths perspective.

Think About It #5

Divide a blank page in three sections. At the top of the first section write, "My Strengths." At the top of the second section of the page, write "Strengths of My Current Family." On the top of the third section, write "Strengths of My Family of Origin" (the family I was raised in). List at least five strengths in each column.

Suspending Judgment

It is also important that helpers not judge negatively the way in which an individual, couple, or family manages decisions and crises. People cope and manage in different ways. Effective helpers provide a safe, nonjudgmental space where others are led gently through feelings and thoughts. This does not mean that you do not and will not have judgments, opinions, values, beliefs, biases, and thoughts. It simply means that you "suspend" (put them out of the way) for the duration of your contact with or on behalf of the persons seeking help.

For example, one mother of a child diagnosed at birth as having mental retardation said that her way of managing this news was to appear to be detached. She got down to business, collected information about the particular disability her child had, and proceeded to locate medical and educational resources in a very matter-of-fact way. She was aware that others were criticizing her for appearing so unemotional, cold, and unfeeling. But this was simply her way of coping and adjusting to a major life crisis.

Human services helpers should listen with compassion to many different ways of coping with stress without making judgments. It is important to honor and validate all feelings, even when they are difficult to hear or accept. (Addressing feelings and ways to listen to feelings in a helpful way are covered in later chapters.)

Illustration

For the past two months, Liz McKinney, a 34-year-old single mother of three school-age children, has been participating in a support group for women who are adult survivors of sexual abuse. Liz is the only woman in the group who receives welfare. The other participants either work outside the home or have partners who provide financially for the family. At one point, Liz mentioned that she was going home to make thick T-bone steaks for dinner with all the fixings, including asparagus spears and an artichoke salad. Two of the other women clearly resented the fact that Liz, receiving food stamps, was able to purchase such expensive food, when they, hard workers and self-sufficient, could never afford to purchase such luxuries. It was possible that their resentment could interfere with their ability to provide Liz with the kind of support, respect, and positive regard she needed from them. Susan, the group facilitator, said, "Liz, you know I love steak, but I never can afford it. Help me understand how you decide to use your food stamps to buy such expensive food." Liz replied in a very nondefensive way. She said, "Well, you know, I don't have money to take my kids to a movie. I certainly don't have money for a baby-sitter. I can't go shopping or really do anything, so when I feel like I really need to treat myself and my kids, the only thing I can do is to buy special food, because I can budget my food stamps. I might have to scrimp at the end of the month. We may be eating rice and beans, but sometimes we just need to remind ourselves that we can do something nice and normal."

Think About It #6

1. What was your initial reaction to Liz's statement that she was going to prepare a very expensive meal bought with food stamps? Did you feel somewhat critical of her decision to use her food stamps in this way?

2. What was your reaction to Liz's explanation in response to Sue's question? Can you better understand Liz's reasoning?

3. Have there been times when you have felt that you were unable to offer support and respect to others because of decisions they have made? Think of ways that these decisions might make sense if you were in the other's position.

Support of Choice

One of the most difficult philosophies of helping to put into practice relates to supporting the informed choice of those who ask for our guidance. This is often called "self-determination," meaning that all of us, unless we hurt ourselves or others, have the right to determine for ourselves how we live our lives.

Helpers may not always agree with what others choose for themselves but nevertheless have a responsibility to educate them about all possible consequences, resources, options, and alternatives and then support their right to make an informed choice. It is often painful for us to see someone continue to use alcohol or other drugs, to refuse traditional medical treatment, or to choose to have an abortion or to have a baby under difficult circumstances. However, ultimately helpers must recognize that, although one can express care and concern to those who ask for services, one cannot force them to behave in certain ways.

Illustration

Lana is a home visitor with a home health agency. Mark is a 32-year-old Asian-American man who has been through several biopsies, two surgical procedures, and six difficult months on chemotherapy to treat a very serious case of lymphoma. Mark's physician has recently recommended another round of chemotherapy. Mark tells Lana during a home visit that he is very tired of fighting the cancer and has decided not to undergo any more traditional medical treatments of any kind. He wants to continue acupuncture and herbal remedies but wants nothing more to do with Western medicine. Mark says that he realizes that he is not recovering and will probably die sooner than if he participates in taking the powerful drugs but that he would rather live out his days without the terrible side effects that the treatments have caused. Both Lana and Mark cry a little as he explains his decision to her. Lana is thinking to herself, "No! He can't do this!" But Lana hugs Mark and says that she respects his decision, cares about him, and wishes him well.

Think About It #7

1. How do you think you will feel as a helper when someone chooses an action with which you do not agree? How do you think you will act when this happens to you?

2. Have you ever made a decision with which some significant persons in your life did not agree? How did they respond to you? What types of responses were you wishing for?

Being Honest and Trustworthy

The need for a helper to possess the qualities of honesty and trustworthiness may seem so obvious as to be not worth mentioning. Most of us consider ourselves honest and trustworthy in all our relationships with others, and most relationships do not survive without a basic sense of trust. Helpers should, however, take special care with trust and honesty in the helping relationship for several important reasons.

First, many individuals are experiencing extreme emotional vulnerability. Helpers may be tempted to reassure or comfort them by leaving out bad news or exaggerating the positives. The intentions may be honorable, but the information is not totally accurate. Being overly optimistic is as dishonest as being overly pessimistic. It is important that helpers share what they know and have experienced. It is also important to balance reality with hope, to balance the bad news with the pledge of our support. When someone discovers that a helper has not been honest, the relationship may be jeopardized.

It is almost never useful to try to persuade persons that their situation is better than they think. Remember a time when you were feeling sad, angry, sick, or overwhelmed. Would it have been helpful if someone told you to look at the bright side? Would it have been helpful if someone reminded you that it could be worse? Usually not. It is important for the helper to acknowledge the feelings, even if they are painful, in a sensitive and caring way. Letting someone express his or her feelings sends a message that the feelings are not too bad, too scary, or too overwhelming to hear, and that makes the person in turmoil feel safer and more able to confront the situation.

Second, when individuals, couples, or families seem to need much, it is sometimes tempting to agree to do more or spend more time than is realistic. Helpers may make promises that they initially intend to honor but regrettably must cancel. It is important to evaluate one's own capabilities and limitations and communicate these honestly and straightforwardly. Once again, the effect is to lose credibility and jeopardize the helping relationship if helpers fail to honor a promise. (See the section on assertiveness in Chapter 6 for help in saying "no" when it is necessary to do so.)

Illustration

Tamara Wilkins works for an organization called "Parents Anonymous," which provides services to parents who have been neglectful or abusive to their children. Tamara's job is to facilitate a recreational/support group for children who have experienced neglect or abuse, whose parents attend a support group. One 8-year-old girl, Angie, grew very attached to Tamara and asked Tamara if she would be caring for her forever. Tamara knew that Angie had experienced several major losses in her life. She had lived in five different foster homes in the past three years. Tamara wanted Angie to feel secure and to be able to count on her. Tamara also knew that she would be moving in a few months and would no longer be working for Parents Anonymous. Tamara thought that it was important to be honest with Angie, even if Angie reacted with anger or sadness. Tamara said, "Angie, I know that you wish I could be here forever with you. That would be nice in a lot of ways. But, you know, we'll both be growing up and having other adventures. Next year, I'll be moving to Arizona. Maybe we can start talking now about ways we can remember each other and ways to say goodbye. I like you very much and will miss you. What are you feeling and thinking when I tell you I have to move away?"

Third, although persons in crisis frequently need information and direction, helpers should not try to represent themselves as experts. It is both unethical and illegal to provide information or services that you have not been trained to provide or to diagnose someone when you do not have the credentials to do so. It is more important to respond with an "I don't know" than to pretend to have information, even if the person seems desperate for an answer. If you do not know, say so, and then work hard to find the information for them or refer them to someone who can help.

Think About It #8

1. Were you surprised to read that there might be situations in which you would be tempted to be less than completely honest? Can you think of similar situations which might be difficult or challenging for you?
2. How would you *feel* if a 38-year-old woman with terminal cancer asked you, "Am I going to die? Will I be able to see my two little children grow up?" What would be your first (or "gut") response to her question? How might you be fully honest without being unnecessarily harsh?

Values and Ethics

The term *values* refers to beliefs or standards that people hold dear or to highly regarded ways of living. *Ethics* are sets of guidelines or principles regarding moral duties or the nature of good and evil. Countless situations in our roles as human services helpers require ethical decision making and/or judgments about values. These experiences are frequently tough on helpers, who (like everyone else) have very strong opinions, beliefs, thoughts, and feelings about behavior and standards. The difference is that human services helpers must allow others the freedom and autonomy to make their own decisions.

Characteristics of effective helpers include the ability not only to suspend judgment but also to set aside their own values and beliefs so that they only minimally interfere with the helping relationship. Some examples of questions involving ethical decisions are as follows:

1. What is the individual's right to choose about life and death?

   - Does an individual with AIDS or Alzheimer's disease have the right to take his or her own life? Does a friend, professional, or acquaintance have the right to assist with the suicide?
   - Is it "moral" for a woman to terminate a pregnancy because of a disability detected in the fetus?
   - Does a person with a life-threatening illness that might be transmitted to a fetus or infant, genetically or through infection, have the right to choose to become pregnant and deliver a child?

2. How much should we intervene and treat people with terminal illness?

   - Should parents have the right to determine the extent of medical interventions used on a child?
   - Who should make ethical decisions about whether to prolong life? Physicians? Family members? The persons who are ill? Philosophers? Clergy persons?
   - Should individuals be required to endure endless painful medical procedures with little hope of recovery?
   - Should the financial cost of medical intervention ever be considered in decisions about treatment?

3. If resources are limited (livers and kidney for organ donations, for example, or time on a kidney machine), who should receive these services and scarce resources? How should selections be made?

4. Who is responsible for the financial, emotional, and physical care of individuals with special needs? Are families solely responsible? Should the community and government help? Or are the government and community primarily responsible?

Think About It #9

1. What are some other moral and ethical dilemmas that you might encounter or have already encountered in your role as human services helper?
2. What do you anticipate would be the most difficult or problematic ethical situations for you to face as a helper? How will you handle these situations?

Among the most important tools for helping are personal characteristics that relate to the helper's understanding of his or her own values. Brammer (1979) maintains that an awareness of our value positions is imperative in helping. This awareness aids helpers in being honest with themselves and others and assists them in avoiding unethical behavior.

For example, it is inappropriate for a helper to use the helping role to push his or her personal religious or political views. Although these values may be strongly and dearly held by the helper and his or her behavior may be guided by them, the helping role should not be used to try to convert vulnerable individuals into another way of thinking. Of course, if an individual, couple, or family desires religious or spiritual assistance, helpers can refer them to the appropriate pastoral counselors. However, if you are working as a paid or volunteer helper for a helping agency, it would be unethical to provide spiritual guidance yourself.

At other times personal values may restrict or limit the behaviors of others. In these instances these values may get in the way of allowing another person to reach his or her potential because of stereotypes regarding behavior. These stereotypes can be based on a person's gender, religion, race, ethnicity, age, ability, or sexual orientation. For example, helpers may limit the options of individuals and families if they assume that it is always the role of females to assume primary caregiving responsibilities for someone who is ill or frail. Helpers might also be limiting the options of children if they think that only financially successful families should be able to adopt.

Illustration

Mrs. Edith Barrow is a 78-year-old African-American woman who is recovering from a series of strokes she suffered 18 months ago. Prior to that time, she had chronically poor health. She developed diabetes approximately 17 years ago and has been losing her eyesight steadily since that time. She has been legally blind for the past eight years. At this time, the stroke has left her nonambulatory, but she has slowly recovered some of her speech and the use of her right hand and arm. Mrs. Barrow lives with her 45-year-old son, Charles, who has been her caretaker for the past 15 years. Charles keeps the house clean, shops, and cooks for himself and his mother. He has not worked outside the home since he was 32 years old. Prior to that he was a construction worker, made a reasonable living, enjoyed his work, and was quite successful. Six months ago, Mrs. Barrow's 39-year-old daughter, Barbara Wheeler, returned to live in the household. Barbara has mild mental retardation; recently her husband ended their eight-year marriage. Barbara has sporadically held part-time jobs in cafeterias. Mrs. Barrow is now distraught because she says her small social security check cannot care for the three of them. Her health is not returning, and she complains that Barbara and Charles bicker with each other constantly.

Think About It #10

1. Identify and examine some of the ways in which the decisions of this family may conflict with your own values.
2. How might you practice putting your judgments aside while you worked with this family?

When ethical principles and values are shared widely among helpers, they are written down into codes. Professional groups such as the National Organization for Human Service Educators, the National Association of Social Workers, the American Psychological Association, and the American Medical Association have professional codes of ethics that serve to guide action and reflect common values regarding the helping relationship and responsibilities.

In general, a helping relationship needs to be handled in such a way that it protects the well-being of individuals, couples, families, groups, and communities and does not exploit or abuse the relationship or the persons seeking help in any way. The following list of ethical guidelines is essential to any helping relationship. This list is certainly not complete but includes these principles: maintaining confidentiality and privacy; respecting differences; being self-aware; recognizing conflicts of interests; and maintaining boundaries.

Confidentiality

Keeping confidential and private certain information received from persons seeking assistance is vital. People who are needing help will share secrets and talk about very personal aspects of their lives. They need to be able to trust helpers with this information. Helpers should always respect the privacy of those they help and hold in confidence all information obtained. "An unjustified breach of confidentiality is a violation of justice and is tantamount to theft of a secret with which one has been entrusted" (Hepworth and Larsen, 1990, p. 72). Maintaining strict confidentiality requires a strong commitment, for sometimes individuals reveal information that is shocking, humorous, or bizarre. To fulfill your responsibility in maintaining confidentiality, it is vital to guard against disclosing information in inappropriate situations, the most common of which include the following: to friends or family members, in hallways or elevators, with supervisors or coworkers within listening range of others, at parties, on public transportation, or with colleagues or coworkers who have no need to know the information.

At times it may be appropriate to share confidences with others, either in your agency or in other organizations that can help the individual, couple, or family. For example, you will need to share information with your supervisor, consultant, another helper, or team leader if you wonder how to be more helpful or if you are worried and feel in over your head.

Illustration

Amy and David Goldhammer have been married for two years. They are both Jewish and 28 years old. They live in a three-bedroom apartment in Chicago. Amy is an attorney with Legal Aid Services. David is a Licensed Practical Nurse at a hospital. Six months ago, their infant daughter, Suzanne, died in a car accident. Amy was driving Suzanne to day care when another car failed to stop at a traffic light and crushed the passenger side of the car where Suzanne was riding in her car seat. After an agonizing three days, Suzanne died from multiple internal injuries and brain damage. Both Amy and David have been living in an emotional fog since the accident. They recently began attending a local support group for parents who have experienced the death of a child. Amy and David have not been able to be especially supportive of each other since their daughter's death. They seem to be handling their grief separately and privately. Amy says that she feels guilty and responsible for Suzanne's death because she was driving and should have been more careful. She also believes that if Suzanne were in the back seat of the car instead of in the front seat, she would not have been badly hurt. Amy confides in Faye, the group facilitator, that she has recently been using cocaine, which she purchases from a colleague at work at least three times a week to deaden her pain.

Think About It #11

1. What is your responsibility to share or not share the information that Amy told you about her cocaine use?
2. With whom would you think about sharing it? Her husband? Her employer? Her doctor? Why or why not?

Several exceptions to the rule about maintaining confidentiality exist. You are required to share confidences, as mandated by law, in the following circumstances: if you suspect child abuse or neglect; if you suspect the exploitation, neglect, or abuse of an incapacitated adult; or if you worry that homicide or suicide is likely to occur. In all of these cases, you should immediately inform your supervisor of your suspicions or worries, as well as what has led you to your conclusions.

A ground-breaking legal case in 1974 illuminated the helper's responsibility to protect the safety of others. The case is Tarasoff versus Regents of the University of California (1974). A university student informed his psychiatrist at the university of his intent to kill his lover. Although the psychiatrist notified the campus police, he did not warn the potential victim of the danger of homicide. The young man did indeed murder her. The dead woman's parents successfully sued the psychiatrist. The legal ruling is that when someone is potentially in danger, the helper has an obligation to give warning not only to the authorities but also to the intended victim. This case states, "the protective privilege ends where the public peril begins" (Tarasoff versus Regents of the University of California. 13 Cal. 3d 177, 1974).

Respecting Differences

Chapter 3 will be devoted entirely to the concept of understanding and respecting diversity (individual and cultural differences). Effective human services workers, students, and volunteers appreciate and celebrate the fact that people differ from each other in a variety of ways, including how they make choices about their lives. Gender, race, ethnic group, age, ability, religion, sexual orientation, and socioeconomic status are just a few of the ways in which human beings differ from each other. That list does not take into account the different backgrounds, experiences, values, beliefs, and political views that make us distinct as well. Helpers respect all people as worthwhile simply because they are human, even when they break the law or make decisions that differ from those we would make.

Being Aware of Yourself

Being aware of one's own strengths and limitations is very important in becoming a helper. Be aware of personal growth and personal blind spots so that the impact on others is as positive and as helpful as possible. As discussed earlier in this chapter, being aware of one's attitudes and beliefs is important. Awareness of our feelings, and having a commitment to meeting our own needs and taking care of ourselves are also crucial.

Sometimes helpers feel judgmental, angry, uninterested, helpless, afraid, or bored when talking with a person seeking help. Why might that happen? The following reasons are possible explanations:

- Sometimes a person's situation reminds helpers too much of personal problems, causing confusion or overinvolvement.
- Sometimes helpers may think that another's problem or situation is not interesting or of little consequence compared to their own acquaintances' problems, and consequently unintentionally trivialize or discount its importance.
- Sometimes helpers are too preoccupied with personal issues, and cannot give another full attention.
- Sometimes helpers are so obsessed with helping and saving another that they act out of personal needs rather than the needs of the other. At times the "need to be needed" outweighs the ability to truly listen to what another person wants.
- Sometimes helpers encounter a person whom they simply do not like or cannot respect, or who behaves in ways that run counter to the helper's deeply held values.

Sometimes helpers can tell personal feelings are getting in the way if they respond in any of the following ways:

- **Preaching.** Are we acting moralistic and judgmental? Do we want people to be "ashamed of themselves?" Do we find ourselves saying: "You should . . . " or "You shouldn't . . . ?"
- **Teaching.** Do we find ourselves relying heavily or even exclusively on providing only information, reading material, or "the facts?" Are we avoiding feelings and emotions?
- **Prescribing.** Do we find ourselves thinking that in a particular situation there is only one logically correct solution that should be obvious to any reasonable

person? Do we think less of a person who does not reach the conclusion we have reached? Do we find ourselves persuading too much and saying things like: "There are three alternatives, and it is obvious that the first option is best because . . . ?"

- **Saving.** Do we feel that if we take charge we can fix someone else's problem in no time? Are we feeling that we can save others from the mistakes we made? Are we saying too often, "I'll do it for you?"
- **Advising.** *Do we feel the need to have all the answers even in situations we know little about? Are we spending more time talking about another's problems than they are spending talking about their own? Do we find ourselves saying, "I know exactly what you should do?"*

When we respond to others in any of the above ways, it is possible that we have some personal conflict associated with a particular topic or crisis. One kind of conflict, for example, is a value conflict. One person may have very strong values about the importance of all individuals to live in a family and abhor the idea of any kind of out-of-home placement. Placing an elderly parent or a child with mental retardation in an institution or group home would violate that person's values. Another example is a helper whose values are such that he or she is in complete opposition to abortion under any circumstances. A third example is a helper who values independence and self-reliance and perceives individuals, couples, and families needing governmental assistance of any kind as weak. If we are faced with a value conflict, we may be tempted to interfere with another's right to self-determination and begin to judge or persuade or advise too much.

A second kind of conflict may be a personal conflict. We may have made a decision or choice in our lives that we are uncertain about or have very tender feelings about. For example, an individual who was very ambivalent about placing her child for adoption ten years ago may not be in a position to help others make this kind of decision because her personal conflict has not been resolved.

A third type of conflict may be an interpersonal conflict, a conflict between two or more individuals. The human services helper and the person in need of assistance may simply not like each other or have very little in common. Personality factors, interests, attitudes, and values all contribute to a possibility that an interpersonal conflict may emerge.

Think About It #12

1. Which kind of conflict do you most frequently experience? Value conflicts? Personal conflicts? Interpersonal conflicts?
2. What do you do when a value, personal, or interpersonal conflict arises and you find yourself behaving in ways that are not helpful?

The story following on page 26 illustrates an interpersonal conflict between two service recipients. Read the example from the point of view of Tracy, then of Doris, then of Nancy.

Illustration

Tracy, a young Caucasian woman, is a resident in a shelter for women who have been battered. Nancy works there as a peer counselor. Tracy has been supporting her cocaine habit by selling sexual activity on the streets. She has been in a very dangerous physically abusive relationship with her husband, who is also her pimp. She was accepted into the shelter from the emergency room, where she had been treated for a broken arm and jaw. Another resident, Doris, approaches Nancy, angry and in tears, demanding that Tracy be discharged from the shelter. Doris says that she cannot stay with a "woman like that." Doris is African-American, middle class, and very religious. She has been abused by her husband for several years. She says that she will leave to go home to her abusive husband if shelter staff do not discharge Tracy.

Think About It #13

1. What are your thoughts and feelings about Tracy?
2. What are your thoughts and feelings about Doris?
3. How do your thoughts and feelings influence what you would do if you were Nancy?

Conflicts of Interest

When a helper benefits in any way from working with any individual, couple, family, group, or community, it is considered unethical. It is never appropriate for a human services helper to take advantage in any way of someone who has asked for assistance. When the interest of the helper is in conflict with the best interests of the persons seeking help, this is often called "conflict of interest." These situations are problematic because helpers who take advantage of persons in crisis or trouble are exploiting vulnerable persons and betraying their trust. Examples of conflicts of interest are as follows: Having a romantic or erotic relationship with someone for whom you are a helper; requiring money (outside of salary or legitimate agency fees) or labor in exchange for services or special favors; or using your authority or power to coerce someone into doing something for you. Imagine a helper who says to a plumber, "I'll make sure your child is moved up on the waiting list for day care if you come to my house and fix the leak under my sink." Imagine a community organizer who says, "Sure, I'll help you get the garbage cleaned up in these vacant lots. But you need to contract with my brother-in-law's maintenance service to do it." Imagine a student intern in a shelter for runaway youth who says, "Yes, you can stay up past curfew tonight, but you'll have to wash my car." These are examples of highly unethical behavior.

Maintaining Boundaries

"Maintaining boundaries" refers to being aware of where one's personal life ends and where the helper life begins. These boundaries are at times difficult to recognize and maintain. Sometimes when a helper cares deeply about someone who has asked for help, it is hard to keep the appropriate distance from that person. This is not to say that it is desirable to be cold or uninvolved; on the contrary, genuineness and warmth are required for effective helping. However, it is not useful or appropriate to become so involved in the other person's situation that the helper loses all ability to be an objective problem solver and active listener. Helpers need to be very self-aware in order to catch themselves crossing the appropriate boundaries. If you are ever in doubt about how you are behaving in a helping relationship, get supervision or consultation about the situation.

Illustration

Wayne traveled to a county health department to request a blood test to detect the antibodies to Human Immunodeficiency Virus (HIV). Before his blood was drawn, he spoke briefly with Dawn, a test counselor. When he returned two weeks later for his test results, he was called in to a room to meet with Dawn. When he extended his hand to shake hers, he noticed that she was crying. Wayne asked, "Are you OK? What's wrong? What's happened?" Dawn tearfully replied, "Oh, Wayne, you've tested positive!" Stunned, Wayne was quiet for a minute while Dawn continued to sniffle and wipe her eyes. Then he reached over to pat her arm, saying "That's OK, Dawn. Don't worry. Everything will be OK."

Think About It #14

1. What do you think about Dawn's maintaining appropriate boundaries with Wayne?
2. What effect did Dawn's reaction have on Wayne?
3. Think about an alternative response that Dawn might have made to Wayne.

CHAPTER THREE

Awareness of Others

This section examines the concepts of diversity and culture and how they influence your work. Steps to becoming more culturally competent are included. Begin with the following "Think about it" exercise to help you to become more self-aware about cultural identity and diversity.

Think About It #15

1. What is your ethnic background? What is your race?
2. In what geographic location did you grow up, and what other ethnic groups resided in your community?
3. How did your family see itself as being like or different from other ethnic groups?
4. What are your feelings about your ethnic identity? How might they be influenced by the power relationships between your ethnic group and others?
5. In what ways were boys and girls treated similarly in your family growing up? Differently?
6. What were the power relationships between males and females in your family? In your community? In your ethnic group?
7. At what point when you were growing up did you realize your sexual orientation?
8. What is your religious background? What is your current religious affiliation?
9. Were other religious groups represented in the community in which you grew up? If so, which ones?
10. Were your ever discriminated against or treated differently because of your religious affiliation?
11. Do you consider yourself to have a disability? When did you become aware of your disability? Is your disability obvious to others?

12. What do you think about people who are of a different sexual orientation than you are? A different gender? A different racial group? A different ethnic group? A different age? A different religion? What do you think about people who have disabilities? Does it make a difference whether the disability is apparent or not?
13. What privileges and disadvantages did you experience as a result of a cultural characteristic (for example, your ethnicity, skin color, gender, sexual orientation, or religion)?

What Is Culture?

Culture can be thought of as a shared world view. It is a way in which a group of people organize their beliefs and make sense of life. Culture can be the "glue" that holds a community or group together. Cultural variations reflect what people hold to be worthwhile; they help to determine what is believed about what is worth knowing and doing.

A group with a shared culture has a history together. Culture is comprised of a variety of unique features of a group of people, including language, symbols, artifacts, rituals, values, customs, food, dance, music, spirituality, assumptions, myths, traditions, folklore, metaphors, accumulated knowledge, norms about behavior, and beliefs about the meaning of life and the nature of time, communication, and family. Cultures that exist for a long time are flexible and adaptable, while keeping some of the meaningful symbols and rituals that help members to feel supported and cohesive. Cultures are rich and strong, and they provide important functions for their members.

Sometimes individuals will identify their culture based on the country or religion from which their ancestors came. For example, a person may call him- or herself Irish-American, Korean-American, African-American, or Mexican-American. At other times in the United States, individuals identify themselves by a particular place or region from which they or their family came. Examples include Southerner, Northeasterner, Chicagoan, and Texan. Of course, many people consider themselves to have come from multicultural environments and to have been raised by parents of several ethnicities.

Culture can influence our expectations of ourselves, our children, our parents, and our friends. It can directly affect our roles in couples and families. Culture contributes to our beliefs about gender, religion, death, sexual orientation, and ethnicity. It can be the cause of certain beliefs about developmental phases and events. It can influence our responses to authority. It can determine our comfort with proximity, or the distance between persons. Culture can influence what we wear, what we eat, where we live, what music we like, how we dance, what we do for fun, what holidays we celebrate, the stories we know, how we define love, and how we talk.

There is wide variability between cultures, and there is diversity within cultures. Being a member of a culture means that you are in unity with your community, but you also have individual characteristics, tastes, experiences, and desires. Generalizing about persons within a culture is not useful. In any one culture, for instance, there are age differences, race differences, sexual orientation differences, gender differences, religious differences, class differences, and educational differences.

In addition, most persons inhabit several cultures simultaneously, existing in layers and collections of cultural identities. Sometimes those different cultural identities clash or conflict with each other. A poignant illustration of a clash between one person's cultural identities is the epitaph of a gay Vietnam War veteran: "They gave me a medal for killing two men and a discharge for loving one" (Thompson, 1994).

Think About It #16

1. Identify one of the cultures to which you belong. Identify one belief or behavior from that culture.
2. Make a note about the first time you ever felt culturally different. What were your feelings and thoughts?
3. Identify one cross-cultural experience that you have had that was meaningful to you or that taught you something.
4. Should cultural groups in this country try to "fit in" or "assimilate?" Why and/or why not?
5. List all the cultures of which you are a member. Where are the areas where you feel conflict? Are there any of your cultures in which you do not feel comfortable? Why? Are there particular issues or experiences about one of your own cultures that make you uncomfortable or embarrassed? What are they and why?
6. In what cultures other than your own do you feel comfortable? Why?
7. In what cultures (not your own) do you *not* feel comfortable? Why?

Mistakes to Avoid in Thinking About Culture

The deficit perspective. Sometimes we tend to label a behavior as a cultural response or label a group as a culture out of a deficit perspective rather than a strengths perspective. For example, persons who talk of the "culture of poverty," "culture of crime," or "culture of the underclass" are misusing the concept of culture. Culture is for the most part a positive, functional way of organizing the world; it is a world view that has lasted over time because it has value to its members. Of course, every culture has elements that may be undesirable, difficult, or unethical. However, if you have labeled a group as a culture and can identify no positive characteristics or behaviors of that group, then you are on the wrong track.

Stereotyping. Another unfortunate tendency is to use generalizations about groups of people as a way to perpetuate stereotypes. Generalizations about cultures should be used only as limited tools to further our understanding of human groups, not as a way to prejudge individuals. Any time that you hear of a workshop about "culture" where lists of characteristics of particular groups of people are presented, be wary. Chances are that these lists represent stereotypical ways of thinking.

Confusing culture with other concepts. "Culture" may sometimes be confused with "race," "color," or "ethnicity." Culture is a much broader concept, encompassing all of the aspects that have been discussed previously. For example, skin color varies greatly within cultural groups such as Puerto Rican or Native American.

Victim-Blaming. One of the serious traps that persons fall into when considering culture is blaming the victims of oppression for their own situation. It is a mistake to think this way about people who have been oppressed because of their cultural membership: "They must like being treated that way, or else they would put a stop to it." In order to understand why people in subordinate cultures cannot just decide to put a stop to the oppression that affects them, one must look at the consequences of taking action toward that goal. What price might they have to pay? A person or a group of people may appear to be passive or inactive to an outsider, but, in reality, had to choose survival over activism. Rather than asking "Why didn't they do something sooner?" ask, "How were they able to make progress at all?"

Think About It #17

1. Do you think that cultures should be proud of their identities?
2. Can identification with a group ever be a source of shame? Explain.
3. Who should define whether a group is a culture? Those inside or outside the group? Explain your answer.
4. Is there a hierarchy operating in this country? If so, on what is it based? Race or ethnicity? Class? Education? Income? Age? Gender? Religion? What makes you think so?

Definitions of Terms

These terms are useful to consider in a discussion about cultural issues.

- **Minority group:** a group that has less power than the dominant group. "Minority" in this context does not refer to the numbers of persons in the group as much as to whether the group is oppressed.

- **Prejudice:** the act of prejudging an individual based on an actual or imagined characteristic.
- **Stereotypes:** rigid ideas about a group; the use of preconceived ideas as a way to rationalize oppressive attitudes or behavior or the group's subordinate position in society.
- **Stigma:** a negative judgment that is placed on a person or group that is deemed to be socially undesirable.
- **Discrimination:** behavior or action that serves to limit the opportunities for an individual or group. Discrimination can be perpetuated by individuals, families, organizations, groups, or the larger society. The oppression can be obvious or hidden.
- **Racism:** attitudes and behaviors based on a belief that race, skin color, or ethnic origins are the primary determinants of human capacities, that one group is generally superior to others, and/or that certain groups do not deserve basic rights.
- **Classism:** attitudes and behaviors based on a belief that certain socioeconomic groups, particularly poor persons, do not deserve the same consideration or rights as other classes.
- **Sexism:** attitudes and behaviors based on a belief that one's gender is the primary determinant of human capacities, that one gender is generally superior to the other, or that one gender does not deserve basic rights.
- **Ageism:** prejudice or discrimination based on a particular age group, most often elderly persons.
- **Homophobia:** fear of, prejudice toward, or discrimination against gay, bisexual, lesbian, or transgendered persons, and/or a denial of the basic rights of these groups. Sometimes this is referred to as "heterosexism," a belief that heterosexual persons have more rights in society than do homosexual persons.
- **Handicapism:** attitudes and behaviors that serve to limit the opportunities and experiences of people who have disabilities. Examples of handicapist practices are facilities that are not accessible to persons using wheelchairs or programs that do not interpret for persons who are hearing-impaired.
- **Institutionalized discrimination:** oppressive and limiting practices and policies that are so ingrained in our society that most persons are not even aware of them.
- **Internalized stigma:** stigma and shame that have been accepted to some extent by the person who is stigmatized. Examples are women who believe that women are inferior to men in some way, gay men and women who believe that they are no good because of their sexual orientation, or African-Americans who believe that people in their culture will never amount to anything (Ainlay, Coleman, and Becker, 1986; Schur, 1983).

Use of Language

"Sticks and stones may break my bones but words will never hurt me." This saying, which most of us heard in childhood, is not true. Words and symbols can have a great deal of power, because they often represent respect or disrespect.

Helpers who wish to gain the trust of persons seeking help must learn the terms and labels that others find acceptable and nonoffensive. For example, it is disrespectful to call women "girls," to call men who are gay "faggots," to call people with disabilities "cripples," or to call African-Americans "niggers." Even if harm is not intended, it can occur through a careless word that is taken as an insult. It is your responsibility as a human services helper to learn what is appropriate in another culture. When in doubt, respectfully ask persons what terms and names they prefer.

As discussed in Chapter 1, this book is written in what is called "person-first" language. This means that when referring to individuals verbally or in writing, the person comes first and the characteristic or description follows. This practice helps us to think of the person first, rather than the condition. If we referred to "the wheelchair child," "the diabetic man," or "the cancer patient," we would be placing a characteristic or illness before the human being. Rather, we encourage you to say "the child who uses a wheelchair," "the man with diabetes," or "the woman who has cancer." This may seem like a small matter, or the way of speaking may seem burdensome or awkward. You may find it difficult at first to say "a boy with autism" rather than "the autistic boy" or "the autistic." However, using person-first language is respectful and will help to send the message to people who come to you for assistance that you see them as human beings first and as having a special need second.

Age and Developmental Stages

One of the important ways in which to be aware of and appreciate others is to know about stages or phases of human development. It is important to understand that individuals, couples, and families differ in many ways, but we must also consider the notion of "developmental stages." This means that people follow a life course and change over time. Needs and perceptions differ according to age. For example, a teenager who has been raised in this country has never experienced the effects of an international war; however, an older person who served in World War I, had a son who was injured in World War II, and had a grandson who was killed in the Vietnam War may have a much different perspective about peace and security.

Individual Development

Individual persons grow and develop over time in many ways: physically, emotionally, socially, psychologically (Erikson, 1963), mentally (Piaget, 1970), morally (Gilligan, 1982; Kohlberg, 1981), and spiritually (Fowler, 1981). A variety of theories use different terms to describe the developmental stages that individuals traverse over their life cycles. The commonly identified individual developmental stages are: prenatal, infancy, toddlerhood, preschool, school age, adolescence, young adulthood, middle age, and old age. Generally, each stage of individual development is characterized by typical challenges that an individual confronts and masters. These challenges are commonly called "developmental tasks."

According to developmental theory, when an individual is not fully successful at mastering a developmental task, he or she may have difficulty in mastering subsequent developmental tasks and developmental stages. Several barriers to typical development over the individual life span might include: experiencing trauma, a chronic illness, severe poverty, mental health impairment, lack of adequate nurturing or opportunities of connecting with others, or violence or war. One of the developmental tasks or challenges of adolescence, for example, is to establish one's own identity separate from one's parents or guardians. If the adolescent is successful in confronting this developmental task, he or she will be more likely to master developmental tasks of subsequent developmental stages, such as the task of maintaining intimacy in young adulthood. However, if the teenager is unable to assert him- or herself because of dependency that stems from an illness or disability or a threatening or hostile environment, forming a satisfactory intimate relationship in adulthood may be a challenge.

As a human services helper, you may find yourself helping children, adolescents, and adults with life's challenges. Challenges that block the mastery of developmental tasks may place individuals at emotional and developmental risk. For example, when a woman wants to have a biological child but is having difficulty conceiving, this may shake up her self-concept and upset her plans for young and middle adulthood. When a man wants to be in a long-term romantic relationship but has been unable to find a committed partner, he may feel depressed, disappointed in himself, and despondent about the future. If a 60-year-old grandmother finds that she must take over the raising of a grandchild because the parents have died from AIDS, this may interfere with the usual developmental tasks in older adulthood, such as reflection on one's life and planning for retirement. When a 15-year-old woman becomes pregnant and has a child, she is unable to accomplish the expected developmental tasks of adolescence, such as establishing identity and independence. In these situations, where the normal developmental time line is disrupted, individuals may need help in problem solving to find creative ways to master developmental tasks while managing their new responsibilities or coming to terms with disappointments.

The concepts of developmental stages and developmental tasks are merely guidelines to examine the similar challenges many people face at different phases in their lives. These concepts also remind us that when mastery of these challenges is thwarted, a difficult time may ensue. Although developmental stages are typically identified using chronological age as a marker, the ages for each stage are flexible. Circumstances can alter the accomplishments of a particular developmental stage or the chronological age at which the stage is experienced. For example, a long-term chronic illness such as leukemia at age 11 may extend childhood and delay the onset of adolescence.

Family Development

Couples and families are also thought to experience distinct stages of development with separate developmental tasks. Awareness of the developmental stage of a couple or family can be very useful to human services helpers.

Of course, we must take into account the wide variability of "families" in our society today. According to Friedan (1986), fewer than 10% of families fit into what has been the "traditional" family, meaning that a man is primary worker outside the home and a woman is the primary caretaker of children and house. Families consist of all variations: single people, a single parent with children, homosexual couples, those who have had multiple relationships and children from those unions, and multigenerational households. Furthermore, cultural differences influence the stages of individual, couple, and family development. Family development stages may vary widely. What is normal or typical for a family that is White Anglo-Saxon Protestant may not be typical or normal for a family that is Asian-American. In addition, the age of a person or family can effect reactions to situations. For example, a couple in their late 30s may react very differently to the experience of infertility than a couple in their early 20s, simply because of their differing ages and developmental stages. A 65-year-old grandmother may respond very differently to her role as the primary caregiver of an infant granddaughter with spina bifida than would a 28-year-old aunt.

The traditional family development stages are based primarily on the traditional, heterosexual, nuclear family. Nevertheless, it may be useful to consider that couples and families, regardless of their composition, may have time lines and goals. Six family developmental stages have been identified and will be briefly described: (1) unattached young adult, (2) new couple, (3) family with young children, (4) family with adolescents, (5) launching children, and (6) families in later life. Remember that couples and families experience these stages at different paces, and that some couples and families may be in several stages at once.

The unattached young adult. The young adult separates from his or her family of origin and forms her or his life goals and sense of self. The young adult forms intimate peer relationships and establishes financial and emotional independence. This can be a period of turmoil if the young adult and his or her parents are at odds about independence or do not recognize that the family system must change because they are now all adults.

The new couple. The major task of this stage is forming a commitment to the new family system. Other tasks include developing relationships with a partner's family of origin and friends, making the arrangements of living together, and taking on the responsibility of considering someone other than oneself.

The family with young children. In this stage, the individual parent or couple strives to be successful in meeting children's basic needs for food, shelter, clothing, and love. The division and assignment of these tasks may become problematic for families with young children, especially with the high rate of families where both parents work outside the home. It is not surprising then that this is the family life-cycle phase with the highest rate of divorce. Some couples may choose to be childless, and other couples face difficulties having birth children or with adoptions. The tasks for adults who do not have children are to find other ways to meet their needs to leave behind a legacy and to successfully manage the criticisms or pressure of others.

The family with adolescents. Here the family system goes through another marked change as the child's role as an individual and within the family changes dramatically. Not only is the adolescent changing, but the parents are also at a critical stage in their individual development, possibly making career or marital change decisions. A family system that is flexible and adaptable may more easily transverse this stage. Adolescence is not only a trying time for the maturing teenager; it is a very stressful time for parents as well.

Launching children and moving on. This stage requires that the parents and children renegotiate their relationships because all are now adults. The focus moves back to the couple's relationship, instead of the children. With people living longer and longer, many couples become caretakers for their aging or ailing parents during this stage as well.

Families in later life. There is now another shift in roles, with the parental couple moving into retirement and/or becoming grandparents. The couple may find themselves in declining health. The loss of a partner may put a person in the positions of creating a family of friends, depending more on adult children for financial and emotional support, and/or remarrying and beginning a new family life cycle later in life.

Why study family development? An understanding of family developmental tasks is part of cultural competence. This knowledge frequently helps a human services worker, student, or volunteer make sense out of seemingly confusing reactions, feelings, or behaviors. Why, for example, would a family who managed the care of a son with profound mental retardation with competence, poise, and calm, suddenly appear anxious, troubled, and incapable after the young man turns 21 years old? It could be that normally they would be expected to be launching their son into adulthood and independence and that now they must face the fact that they may be his caregivers for the rest of their lives.

Think About It #18

Review the above family stages, and think about situations where the "normal" passing through those stages might be disrupted or made more difficult.

Culture and Helping

When we stereotype based on race, ethnicity, disability, gender, religion, age, or sexual orientation, we fail to validate or recognize the individual and his or her unique stories and experiences. People often make decisions based on their cultural identities and experiences that are related to such things as ethnicity, religious beliefs, and gender. Respecting differences, therefore, means balancing appreciation and recognition of differences with an appreciation of individuality. As a helper, you should strive to approach all persons and all cultures with respect, as a learner, and in a tentative way. The world views of persons from minority groups may be different from those of the dominant culture. The world views of others may be much different from yours. Because people make sense of the world in different ways, you may not be aware that a person's culture is influencing issues such as:

- When to ask for help.
- Whether to trust helpers.
- What makes someone physically or mentally ill.
- What is appropriate if someone is sick.
- Who should care for the sick.
- What is just and fair.
- What is normal behavior.
- Whether tasks or relationships are more important.
- Whether it is important to be on time.
- Whether individual or group goals are more important.
- How social relationships are to be conducted.
- What sexuality and sexual behavior mean.
- How men and women ought to behave.
- What is appropriate at different ages.
- How parents and children should behave.
- What families are.
- How humans are to relate to nature.
- How humans are to treat strangers.
- Whether education is valuable.

As you can see, a helper must learn about and be aware of cultural issues that affect helping relationships, such as: (1) how to define the structure, definition, and function of families; (2) how one should respond to problems; (3) how one should make decisions; (4) how one should respond to a crisis; (5) how parents should discipline or nurture children; (6) how one should deal with addictions to substances; and (7) how one defines abuse or neglect.

The Road to Cultural Competence

Acquiring the skill of cultural competence is critical for effective helping. Most people do not intend to be disrespectful to members of other cultures, but they still make mistakes because they have not learned about other world views.

Cross, Bazron, Dennis, and Isaacs (1989) wrote about a variety of stages or phases toward becoming competent to work with persons of other cultures. They call this five-step journey a "continuum," meaning a line from one end to the other. Not only can individual helpers travel along this continuum, but so can agencies and organizations. The first three stages relate to being unaware and the final two involve being more appropriate and responsive.

Think About It #19

1. Do the helper and the person seeking help have to be from similar ethnic groups, backgrounds, or cultures in order for the helping relationship to be effective? What is your position about this and why?
2. What are the disadvantages or drawbacks of being from the same culture as the persons you serve?
3. What are the benefits to being from the same culture as the persons you serve?

---

1. Cultural destructiveness. When a person, organization, or culture completely and purposefully disregards another culture, it sets out to destroy that culture. Examples include genocide, or the purposeful mass killing of an entire culture. Cultural destruction occurred with many Native American tribes in this country and their subsequent forced assimilation into white European society and the Nazi extermination in World War II of Jews, homosexuals, and gypsies.

Sometimes cultural destructiveness is evident to a lesser extent when a person or a system operates out of hate and prejudice, labeling anyone different as "those people." "Ethnocentrism," meaning the view that one's culture of origin is the best possible culture, can lead to many destructive practices and attitudes.

2. Cultural incapacity. Cultural insensitivity describes the stage before a person or system is aware of cultural issues. In this stage the incompetence may be unintentional, but the individual or organization lacks the capacity to help others of different cultural membership. Sometimes people are simply unaware of differences and appear to be rude because of that (for example, wishing a Jewish or Muslim person "Merry Christmas"). A person therefore may be measured or judged based on culturally biased standards about age, gender, or behavior.

One of the symptoms of this stage is paternalistic behavior, as if one person knows what is best for everyone else. An example of this would be designing, planning, and implementing a health care program without consulting any of the members of the community for whom the services are intended.

Another effect of this accidental incompetence is to give the message or signal to people seeking help that they are not really welcome at our agency. We may behave as if these visitors are actually inconveniences, rather than the reason the organizations are open for business. An example of this is a public aid agency where persons are immediately treated rudely by a receptionist and made to wait an unreasonable amount of time before they are seen.

3. Cultural blindness. At this phase, helpers may think that they are culturally competent when they are not. This is a dangerous pitfall, because workers and agencies are then not taking steps to correct cultural biases. For example, perhaps our agency policies are not really person-driven and person-centered, but we think that they are. The result is that policies and practices that may be ageist, sexist, classist, racist, handicapist, or homophobic are never questioned. Actions are justified with statements like, "We've always done it that way." This is similar to Griffith's (1977) referring to a helper who neglects to appreciate cultural differences as suffering from "the illusion of color blindness." This is the tendency to deny a person's color and thereby treat him or her like "any other person." If we hear ourselves saying something like: "I treat everyone the same. I don't even see race," we may be guilty of cultural blindness. Griffith maintains that to deny color, for example, disregards the importance of one's cultural and family experiences and ignores the impact of racism on his or her development. As an example of being culturally blind, consider the director of one County Council on Aging who thought that he was doing all that he could to be culturally appropriate when all he had done was hire one receptionist who was African-American, the only person of color on staff. During this "blind" phase, accepting feedback from service recipients or community leaders is difficult; helpers may become defensive, hurt, or angry because "we thought we were doing it right!"

4. Cultural precompetence. Moving along the continuum toward being culturally competent, attempts to improve are made during the precompetence stage. There is a definite movement toward being more aware, sensitive, and competent, although the worker or organization may have not yet fully arrived. In this phase, the agency has realized its pitfalls, and staff are talking about what can be done to be more culturally appropriate. At this point the agency personnel may make an attempt to be more welcoming by adding reading material or posters in the waiting room that relate to a variety of cultures. For example, staff of one shelter for women who are abused realized that the program was not welcoming to lesbians experiencing relationship violence and began to correct that situation through aggressive outreach and training of staff. Another example may be a community center that has realized that its building is not accessible to people with disabilities and is trying to raise funds to install a ramp. When an organization is in the precompetence stage, staff might be discouraged by false starts, failures, and difficulties or so pleased with initial efforts that they do not go far enough. It is therefore important always to be on your guard and monitor your agency's practices and policies with cultural competence in mind.

5. Cultural competence. At this point, the worker or agency accepts and respects differences, continues to monitor behavior and attitudes, pays attention to dynamics, and strives to adapt to the demands of a diverse clientele. Organizations that are culturally competent actually welcome and celebrate diversity, holding all cultures in high esteem. An agency that is culturally appropriate tries to achieve a goodness of fit between services and service recipients. The agency structure actually supports tailoring services to diverse needs. An AIDS Service Organization that welcomes and celebrates persons of all ages, socioeconomic classes, ethnic groups, and races works to foster an atmosphere that is equally friendly to both genders, as well as to persons who are gay, lesbian, bisexual, and transgendered. Persons who are homeless, in violent relationships, or using substances also are served without judgment.

Part of being culturally competent is understanding that some universal human experiences transcend culture (such as joy or grief) and that the commonalities between cultural groups are just as important as differences. A culturally competent worker or agency validates similarities as well as celebrating differences.

Becoming a Culturally Competent Worker

What can you do as a beginning helper to facilitate your own journey toward cultural competence? We offer these three steps as guidelines:

1. Cultural awareness. The first step on your personal journey toward cultural competence is simply developing an awareness of culture and an understanding that not all persons and cultures are alike. Begin to be more open to diversity and more respectful of differences. There are probably opportunities in your community for learning about other cultures, such as festivals or religious gatherings. Strive to be aware of your own cultural values, attitudes, and knowledge. Pay attention to the limitations and difficulties in your own culture and how they affect you. Identify the strengths in your own culture, and acknowledge to yourself what nurtures you. Often when people begin to think about the concept of culture, they realize that they were not aware of having a culture or had not previously considered where their beliefs and traditions originated. At first you may not realize that your values and the values of others may not be the same. Then you grow to notice differences or conflicts between your world view and the world views of others. It is often very difficult to look objectively at your own culture, even though it is useful to do so. We can be so much more aware of other cultures when we have examined our own.

2. Cultural sensitivity. The second step is to be vigilant in monitoring your own attitudes and behaviors. Invite feedback from others about your ideas and actions. When you begin to be more sensitive to cultural differences and similarities, you can begin to make your own responses to other persons more respectful and appropriate. You begin to reflect about what part you and your culture might play in the helping relationship. You become more willing to acknowledge the validity of a variety of perspectives.

3. Cultural competence. A culturally competent human services helper demonstrates a willingness and ability to bridge differences between the agency and the person seeking help and within the helping relationship. Culturally competent workers focus primarily on what the person feels and needs, starting where the person is, rather than starting with their own agenda. An effective helper has the ability to assess and identify the strengths of a person, family, community, and culture and to use resources that already exist within his or her own environment. A culturally competent worker becomes an advocate for other's rights and for person-centered care.

What are some of the signs of a culturally competent helper?

- You are aware of your own assumptions, values, and biases.
- You change your perceptions and behaviors when you can.
- You respect the beliefs and values of others.
- You respect other's definitions of "family."
- You feel and communicate empathy (described in Chapter 5).
- You are aware of barriers that your organization presents to persons from various cultures.
- You seek information about other cultures by reading, observing, using consultants from other cultures, and by respectfully asking questions.
- You use language that is deemed to be respectful by members of the group whom you serve.
- When there are differences of opinion, you respectfully negotiate plans and approaches.
- You avoid acting on stereotypes and assumptions.
- You strive to avoid offensive or hurtful language.
- You approach each person, family, culture, community, or group tentatively; you do not jump right in without having any information.

Think About It #20

1. List three of your core beliefs about the causes and treatment of physical illness. How could these beliefs affect your work with people from different cultures or perspectives?
2. List three of your core beliefs about sexuality and/or reproductive behavior. How could these beliefs affect your work with people from different cultures or perspectives?
3. List three of your core beliefs about death and dying. How could these beliefs affect your work with people from different cultures or perspectives?
4. List three of your core beliefs about work. How could these beliefs affect your work with people from different cultures or perspectives?
5. List three of your core beliefs about spirituality or religion. How could these beliefs affect your work with people from different cultures or perspectives?
6. List three of your core beliefs about alcohol or drug use. How could these beliefs affect your work with people from different cultures or perspectives?
7. List three of your core beliefs about mate selection, romantic love, relationships, and marriage. How could these beliefs affect your work with people from different cultures or perspectives?
8. Reflect on some specific times when you realized that your cultural background caused you to feel in conflict with a group or individual with which you were associated. What was the problem? What did you do?

Culturally Competent Services

Why be concerned about cultural competence in the field of human services? Why be concerned about the cultural competence of helpers and organizations? Our values and ethics lead us to believe that all persons are entitled to social justice. Good intentions are not enough; we must make a genuine effort to meet the needs of persons asking for our help. Human services helpers are working as part of bureaucracies and organizations and should be concerned with increasing the cultural competence of those systems.

People who approach us for assistance may have different experiences than we do. These differences may be based on our ethnicity, race, social class, sexual orientation, age, gender, employment, education, disability, or religion. It is also important to acknowledge that we as helpers may enjoy certain privileges that others may not have experienced. People's experiences affect their world view and may have an impact on their motivation to participate in activities or inquire about services. The cultural context of each person must be considered each time a service plan is developed. For example, not all people have access to transportation. Therefore, asking someone to attend a meeting across town may be inadvisable.

Organizations also have certain values and ways of doing things, their own "culture." The assumptions of agencies and workers sometimes negatively influence the helping relationship. Examples of organizational culture ideas that may get in the way are:

- People who ask for help must be on time.
- Eye contact from the person seeking help is desirable.
- Technology is useful and not to be feared.
- Paperwork is essential.
- The individual is more important than family, neighborhood, or community.
- Workers should be distant and uninvolved with service recipients or applicants.
- All services are suitable for all persons.
- Everyone should be treated exactly the same.
- Persons seeking help should follow our rules.
- The causes of illness are logical and rational.
- Experts know what is best for persons who ask for help.
- Drop-in care is impossible.
- Formal settings such as hospitals and clinics are the best places in which to provide medical care.
- Visiting hours in institutions should be limited.
- Medication is good.
- Mental health problems can be dealt with by strangers.
- People should be responsible for paying for their health care.
- People should go to the doctor even when they are not sick.

Culturally competent services are not designed with the notion that "one size fits all;" rather, such programs offer a variety of alternatives and options to fit a variety of people. Persons have access to as much choice as possible in a culturally competent agency. In addition, culturally competent services and organizations realize and acknowledge that society has not always been fair to everyone and that oppression and discrimination are real. Culturally competent services have as an underlying philosophy that each and every person deserves dignity and has value. Agencies may use the following mechanisms to become more culturally competent:

- Workers, students, volunteers, and board members represent the diverse population served by the agency and genuinely have a voice in service planning and implementation.
- There is outreach to populations that may be underserved or may not feel welcome or safe in approaching the agency.
- Satellite centers are placed in neighborhoods and communities that are underserved or most greatly affected.
- There are bilingual workers and volunteers.
- The agency participates in aggressive advocacy for the rights of all persons who are affected by those social problems addressed by the agency.

Think About It #21

1. List three components of culturally competent caregiving and service provision.
2. Identify three issues in social service provision or health care provision in which cultural attitudes have an impact.
3. What personal qualities do you have that will help you establish interpersonal relationships with persons who are different than you?
4. What personal qualities do you have that may hinder you from establishing interpersonal relationships with persons who are different than you?
5. What cultural factors in your background might contribute to your having a difficult time with members of other cultures?
6. List three questions that might elicit culture-specific beliefs from someone.

A Note on Sexism and Homophobia

Women or persons who are gay, lesbian, bisexual, or transgendered are not always thought of as "cultures," yet they can feel unwelcome and misunderstood in a medical or social service system due to the insensitivity and victim-blaming that may exist. The following additional guidelines may make you a more culturally competent helper with these groups.

Guidelines for Helping Women

- Do not blindly accept traditional or societal ideas about how a woman should think, feel, or behave. For example, harshly judging a woman for working outside of the home and accessing day-care services for her children would be nonproductive. Having different standards of sexual behavior for women and men would be unfair.
- Start where the person is and do not try to convert her to your way of thinking. If you believe strongly in the equality of women, for example, but the woman seeking help from you believes that she should be subservient to her husband and father, do not try to force her to change her opinions.
- Do not assume that women can take on all of the caregiving responsibilities for children, family members with disabilities, or older relatives. Assess their needs, and help them get access to supportive services.
- Take into account the decreased opportunities that she may have had as a result of growing up and living in a sexist society.

Guidelines for Helping People
Who Are Gay, Lesbian, Bisexual, or Transgendered

- Helpers who want to be effective should be nonjudgmental and welcoming of all persons and should communicate this through the office environment and the helping relationship.
- Validate the societal discrimination and stigma that some people have felt and experienced throughout their lives.
- Help make decisions about appropriate and safe disclosure about sexual orientation and to weigh the costs and benefits of "coming out" to certain persons.
- Some persons experience the effects of internalized homophobia, evidenced by lowered self-esteem or lack of self-acceptance; developing interventions to help reduce guilt and shame may be helpful.
- Avoid heterosexism. That is, do not assume that every person you meet is heterosexual. When you ask a woman when she plans to marry, for example, you may be making it awkward for her if she is a lesbian.
- Help provide access to services that may not have been traditionally "gay affirmative," such as individual or couple counseling, as well as services to address relationship violence, if they need them.

Putting It All Together

The following illustration contains a variety of cultural issues. Use this vignette and questions that follow as a way to consider how to be a culturally competent worker.

Illustration

Selene Williams is a 10-year old African-American in New York who has symptomatic HIV disease. Her mother, Sheri Williams, has been incarcerated on drug-related charges for the past four years. Sheri has full-blown AIDS and expects to be released soon on medical furlough. When Sheri was first arrested, Selene began living with her 70-year-old grandmother, Celestine, in public housing. Celestine died suddenly last month from a heart attack. Sheri, with the help of a prison social worker, had already developed a permanent placement plan for Selene. Sheri had legally designated her cousin Jerome as Selene's guardian. Jerome is a man who is African-American and gay and lives with his lover Dale, who is white and of German descent. Jerome is an attorney, and Dale is an artist. They are in good health. They own a home in a suburban area and have been together as a monogamous couple for nine years. Jerome and Sheri were very close growing up; Sheri says that she trusts Jerome more than anyone else and knows that Jerome and Dale will make good parents for Selene. Selene has been living with Jerome and Dale since her grandmother died a month ago. Selene has always loved her Uncle Jerome and Uncle Dale, and welcomed having them as her guardians. Because Selene has just moved and changed neighborhoods, she also has changed schools. Selene's teacher referred her to the human services department at the school, because the teacher found her in the school bathroom crying bitterly and wanting to go home. School-mates had been taunting Selene about her two "fag daddies" and about the facts that her mother is in jail and has AIDS. Selene has been called "nigger" by several of the white children recently. She is terrified someone in the school knows that she herself has HIV, although there is not yet any indication this is true.

Think About It #22

1. As you read the illustration above, identify the personal, familial, and cultural strengths in this situation.
2. Identify the aspects of diversity and/or culture that have an effect on Selene.
3. Identify the different developmental stages for the individuals and families in this illustration.
4. What cultural issues might be important for you as a helper?
5. How would you help Selene to relate to you?

PART TWO

Foundations for Helping

Part 2 is an introduction to the fundamental skills necessary to become a human services helper. Chapter 4 explores the development of a helping relationship, the use of self in helping others, and communicating effectively non-judgmentally. Chapter 5 outlines methods and techniques of listening actively, especially attending to verbal and nonverbal messages and responding with empathy. Chapter 6 introduces further skills which build on the previous material: empowering others, supporting informed choice, finding strengths, setting limits, working with difficult situations, attending to your own safety, clarifying expectations. Chapter 7 illustrates the specialized skills of crisis and suicide intervention. Chapter 8 spotlights interviewing and problem solving as specific human services skills; this chapter also discusses more detailed skills, such as making referrals and home visits. The techniques covered in this section are not all that are necessary for effective helping, but they are vital for beginning helpers.

CHAPTER FOUR

The Helping Relationship and Communication

This chapter explores the importance of developing a genuine helping relationship and the principles of helping.

Relationships and Use of Self

Two fundamental and overarching concepts about helping are "relationship" and "use of self." These two ideas are intertwined; one cannot be considered without the other. Why are the issues of relationship and use of self so important? Because helpers must *connect* to other persons genuinely in a relationship in order to be effective.

The human services helping process takes place within the context of a positive, honest, and mutual relationship. The helping relationship is dynamic, meaning that it is constantly changing as the persons involved change, grow, and learn.

The formation of a helping relationship begins even before the first encounter between the helper and the individual, couple, family, group, or community seeking assistance. Thus, preparing for the first encounter is an important component of developing a helping relationship. Your anticipatory preparation as a helper involves thinking about how you feel as you ready yourself for the initial contact with another. It also involves imagining how others might feel as they anticipate you.

Sometimes helpers blame those seeking help if outcomes are negative, thinking of the persons involved as resistant, hostile, uncooperative, recalcitrant, or noncompliant. Sometimes helpers are not willing to examine their own actions when the helping relationship does not progress smoothly. Remember, however, that all helping takes place within the context of a positive and mutual relationship, and that a relationship is a two-way street. The helping process does not entirely depend on the helper or the person seeking help but rather is a product of the partnership between people. Therefore, if something goes wrong with the helping relationship, all aspects must be examined. An effective helper, therefore, realizes that he or she must be aware of the nature of the helping relationship itself at all times.

A good helping relationship is fostered by the helper's being honest, genuine, nonjudgmental, warm, accepting, interested, empathic, and respectful. An effective helping relationship facilitates the helping process and the problem-solving process.

The bond that develops between the partners in a helping relationship is often called "rapport." Having rapport in a helping relationship does not mean that the helper has only positive feelings toward someone who asks for assistance or that the helper approves of all of that person's choices or actions. Establishing rapport does *not* mean that a helper pretends to like someone. It means demonstrating genuine regard for another person's well-being, committed concern, compassion for suffering, and desire to help. Establishing a helping relationship means connecting with others in a respectful way, developing trust and fostering honest communication.

The helper's primary tool is one's self. Human services helpers depend on their relationships with others and on the ability to effectively communicate in order to affect change. Whether the system participating in services is an individual, a family, a group, a community, or an organization, helpers rely on "use of self" to make a difference. Helpers use not only their heads but also their hearts as they intersect with others. Helpers use their own personal and professional experiences and knowledge, as well as research and practice wisdom, to hone their "use of self" in competent, responsible, and careful ways.

Illustration

Lisa is a service coordinator at a local sickle cell foundation, where she connects families to needed community resources. Mr. and Mrs. Witt make an appointment to see Lisa because their 6-year-old son Arthur has just been diagnosed with sickle cell disease, after an extended hospitalization that was a result of intense joint pain and internal bleeding. They are agitated and tearful when they arrive at Lisa's office. Lisa begins by expressing regret at their recent news but assures them that they have come to the right place and that she will do all that she can to identify and obtain what they and Arthur might need. For the first several minutes of the interview Lisa simply listens to them talk about their shock and fear regarding Arthur's illness. Then she gently asks them what they think their most immediate needs are. Over the next ten minutes Lisa listens to the couple and occasionally clarifies what resources are currently available and what services they might need. The family lives in a rural farming community among many extended family members and friends. Mrs. Witt cooks at the downtown cafe; Mr. Witt farms. The small community has no local health clinic. Lisa summarizes with the couple their immediate needs: health care by a pediatrician who is familiar with the management of sickle cell disease, information about what to expect over the course of Arthur's illness, an application for Medicaid for Arthur, and the wisdom and support of other caregivers of children with sickle cell. She then gives them a list of physicians whom the sickle cell foundation recommends, a packet of information about sickle cell disease and community resources, and the schedule and location of a parent support group. From the foundation office they call the public aid office for an appointment to apply for Medicaid, and Lisa furnishes them with a Medicaid application form and instructions. Finally, Lisa offers to call the family once a week and meet them at the monthly sickle cell clinic, and they eagerly accept her offer of ongoing case management and support.

Establishing a Helping Relationship

The formation of relationships and the helper's use of self are only tools, not the *goal* of the helping process (Kadushin, 1983; Perlman, 1979). A good helping relationship is necessary but not sufficient. The point of establishing a relationship with a person seeking help is that it is a necessary step in all other steps of helping, not that it is the end result. It is possible to help someone without having a strong relationship, but, in many instances, an intervention is more effective most often if it involves two partners in problem solving.

Guidelines to establish a helping relationship are:

- The relationship is focused and has a purpose.
- The helping relationship is formed for business, not for social or personal reasons.
- The helping relationship is a professional one, even when the helper is a volunteer.
- The helping relationship is a confidential one, within the guidelines of the organization and within legal limits. As a rule, the helper cannot talk freely about another's private information or situations.
- The helping relationship is ethical, meaning that the helper does not misuse his or her authority, does not take advantage of others in any way, does not seek to control the persons seeking help, and does not pursue a personal or romantic relationship with anyone seeking assistance.
- The relationship is formed around solving problems, reaching goals, or providing support through a crisis.
- The helping relationship facilitates the formation of a plan to work together; the goals in this plan are negotiated between the partners, not imposed by the helper, unless there is a special circumstance, such as being ordered into services by a court of law.
- A helper should communicate acceptance, caring, warmth, genuineness, and empathy (these concepts will be discussed in detail later on in this chapter).
- The relationship is formed for the needs of the person seeking help, not for the needs of the helper.
- The relationship is temporary and is terminated when the mutually formed goals are met.

Illustration

Janet and Daniel Birdsong were joyously awaiting the birth of their twins. These would be their first children, and twins are considered a blessing in their Native-American heritage. During the sixth month of Janet's pregnancy, she was told by the physician that one of the twins was anencephalic, that is, his brain was not developing normally. The physician explained to the Birdsongs that one of the twins was developing typically, but that the twin with anen-

*cephaly would probably not survive more than two days after his birth. With-
out a brain, he would be unable to function in even the most basic ways. The physician also explained that, to insure the survival of the healthy twin, the pregnancy for both twins must be continued as close to full-term as possible, even
though one twin was sure to die. The physician referred the Birdsongs to an organization designed to support parents who are caring for children with special
needs and chronic illnesses. After three weeks of emotional agony, Janet called the
organization and was assigned Deborah as a volunteer helper. Deborah had experienced the birth and death of an anencephalic daughter. Deborah was available
to the Birdsongs and their extended family to listen, share her experiences when
appropriate, and make referrals when needed. After eight and a half months of
pregnancy, the twins were born. David weighed 7 pounds, was healthy, and had
no special medical needs. Mitchell weighed 4 pounds and died 40 hours after his
birth. Because of the emotional support, reassurance, conversations, and discussions that Janet and Daniel had with Deborah on a one-on-one basis, they were
able to feel more prepared for Mitchell's birth and death. They held Mitchell,
dressed him, and took pictures of him. They were somewhat prepared for the funeral and the grief, as well as the joy and celebration of David's new life.*

Think About It #23

1. Reread the illustration above, and focus on the role of the volunteer helper. Think of some of the specific ways that the helper could provide the support and reassurances the Birdsongs needed during the last two months of the pregnancy.
2. How might the support contribute to Janet's carrying the twins full-term?
3. How might the volunteer helper be available for the Birdsongs in a helpful way after the baby's birth?

Principles of Effective Helping

In this section the principles of effective helping will be explored: identification of feelings, empathy, genuineness, starting where a person is, showing respect and suspending judgment, being present, validating, communicating regard, distinguishing the person from his or her behavior, distinguishing feelings and actions, providing information, advocating, and communicating effectively.

Identification of Feelings

A cornerstone of the helping relationship is the helper's ability to assist someone with figuring out how he or she really feels about a situation. Many individuals in our culture are less comfortable in the realm of emotions than with thoughts, beliefs, and dogma. When someone approaches another for help, however, that person is likely to be experiencing intense emotions, even if the presenting problem is one of resources or finances. Because persons are often in severe crisis when they request help from an organization, helpers should be prepared to delve into the emotional worlds of those who ask for assistance.

The components to being comfortable with emotions as a helper are: recognizing that another person is experiencing emotions, being able to identify those emotions, labeling the feelings, and tentatively presenting one's perceptions of those feelings back to the person for verification or correction.

Developing a Feeling Vocabulary

In order for a helper to be skilled at identifying, labeling, and reflecting another person's feelings, the helper must develop a sufficient vocabulary of words that describe emotions. Helpers, like everyone else, are not always well versed in the language of emotions. Many persons in the United States were taught to accurately identify only a few feelings, such as anger, sorrow, happiness, fear, and tiredness. However, the English language is rich with other emotional descriptors. Helpers need to become aware of their own feelings and those of others and of the words that can be used to describe them. To increase your own feeling vocabulary, study the feeling wheel on page 54.

Think About It #24

1. To better help others clarify their feelings and emotions, you should be familiar with your own feelings and comfortable with identifying them. List ten emotions. Describe as concretely as possible what happens to you when you feel these emotions. How does your body react when you experience irritation, for example? What happens inside you? What do you feel like doing? For example, "When I feel inadequate, I look toward the ground and my eyes are heavy. My body feels as if it has no energy, and all I want to do is crawl in bed and stay under the covers. When I feel inadequate I eat too much."

2. Choose three feelings that you have difficulty expressing and write a few sentences describing the nature of your difficulty. Try to be as specific as you can, noting the situations that are most problematic for you. For example, "I have difficulty feeling appreciated. When I feel appreciated, I also feel embarrassed. I want to hide, even though it makes me feel good. When I am at my son's school volunteering and people tell me how special I am, I smile and blush. I like feeling appreciated, but it makes me uncomfortable. I feel tongue tied and at a loss as to what to say."

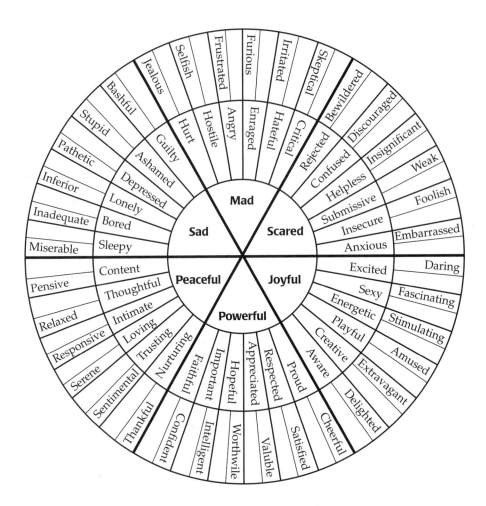

FIGURE 4-1 The feeling wheel

Empathy

After identifying the feelings of others, it is important to communicate this understanding to them. Identifying the feelings of others and then communicating this understanding is called "empathic responding." In empathic understanding, a helper "feels with" another person. The word *empathy* derives from the German word *einfuhlung*, which means "feeling into" (Brammer, 1979, p. 36). Empathy requires two equally important tasks: seeking to understand someone else more fully and seeking to communicate that understanding to that person. "Empathy involves the ability to perceive accurately and sensitively inner feelings of others and to communicate understanding of these feelings" (Hepworth and Larsen, 1990, p. 86).

Empathy is a conscious awareness and the accurate perception of another's feelings. This is more than just an intellectual understanding; it is an emotional understanding. Empathy involves an active effort by helpers to put themselves in the place of another. It consists of "walking in someone else's shoes" for a while. "Empathy may be described as the ability to understand people from their frame of reference rather than your own" (Cormier and Cormier, 1991, p. 21). Carl Rogers (1957), a pioneer psychotherapist who developed the "person-centered" model of counseling, believed that empathy is an essential characteristic of the helping process. He stated that empathy consists of entering the private perceptions of the other person and becoming thoroughly familar with them. Empathy, according to Rogers, involves being sensitive to the changing felt meanings of the other person and to the feelings that he or she is experiencing. It means listening without making judgments, sensing meanings of which the person is hardly aware. The empathic interviewer strives to feel as if he or she is in the internal frame of reference of the other person. "As if" are important words, because the interviewer should never lose sight of his or her own identity (Benjamin, 1969). Empathy therefore involves entering the world of another person without losing one's own sense of self. It is like being an emotional mirror. Empathy therefore requires a very delicate balance: It is the ability to understand how another feels without becoming that way yourself.

To help yourself begin to achieve empathy as you listen to someone else, you may want to start with imagining how you might feel if you were in a similar situation. Ask yourself: "Have I ever been in a position like this?" and "How did I feel?" or "How would I feel if this happened to me?" Then, think about how you would feel in that situation if you were someone else, perhaps a person who is a different age, gender, ethnicity, or sexual orientation or who has experienced a different set of events. This method of imagination is useful in achieving empathy, but it is no guarantee that one will be accurate in one's perception about how another person is feeling and reacting; this is why it is important to check it out through active listening techniques, which will be discussed further in the section on active listening.

Illustration

Bill, a volunteer buddy to Ethel Williams, a 60-year-old woman whose son had recently died from AIDS, used personal experience and imagination to achieve greater understanding of her feelings. As Bill listened closely to Mrs. Williams talk about how confused and upset she had been, he was thinking, "Wow, there's no way I could ever understand what it's like to lose a child." Then Bill remembered some of his own feelings when he was 14 and his 4-year-old brother drowned. He was able to use his memories of the shock, anger, and sadness he experienced to be more in tune with what Mrs. Williams was saying. He then felt more connected to her and more able to understand her grief.

Genuineness

To be genuine in a helping relationship means to be honest and free of pretension, to be transparently real. It means being ourselves but without letting our agendas and our personalities dominate the helping interview or relationship.

To be genuine is to mean what you say and to behave in a sincere way. It means to be consistent, predictable, and trustworthy. Being oneself does not mean being careless; it means using skills, training, and personalities in a serious, responsible, and earnest manner.

This quality is sometimes called "congruence," meaning that one's inside and outside match each other. To remain true to oneself, one must be self-observant, self-aware, able to tolerate differences and ambiguity, and willing to bear conflict and uncertainty (Perlman, 1979, p. 60).

A helper who is genuine does not hide behind coldly clinical jargon, mannerisms, or techniques. For example, referring to people by label or diagnosis ignores their unique experiences. Dismissing a person's situation by suggesting that "it's only a stage" may send the message that you think the person's current problem is trivial. A helper who is genuine is as human and real as possible but without letting his or her feelings or responses get out of control (Benjamin, 1969; Kennedy and Charles, 1990).

Part of being a genuine helper is to be committed to working on one's own growth. Helpers strive to be honest with themselves, to be introspective, and to know themselves. The more one achieves insight and growth, the more he or she is able to support the growth of others as separate, functioning persons.

Illustration

Gladys is a service coordinator at a community-based addictions treatment program where she coleads a support group with adult children of alcoholics. After each weekly group session she left feeling angry or sad but always tried to forget these reactions by the time she arrived home. During her sixth group session, Gladys suddenly became very agitated and upset and left the room. As she broke down in tears in the bathroom, she acknowledged that she needed to figure out what her intense reactions were about, or she would not be able to listen fully to other group members. She spoke the next morning to her supervisor, who was an addictions counselor in the agency. During this conversation, she began to cry again as she admitted that her mother had used alcohol heavily and that she had just begun to realize how enraged and betrayed she felt by her mother's neglect of her as a child. Gladys decided to seek individual counseling so that she could begin to explore her own history as an adult child of an alcoholic.

Where Do You Start?

There is a well-worn phrase in the helping professions: "Start where the client is." This means that the agenda of the person seeking help is foremost, and that unless that person determines what is of the greatest concern for him or her, the helper's involvement will not be as beneficial as it could be. To determine what is foremost on the minds and hearts of individuals, couples, and families, you must first ask them what brought them to your agency and listen carefully to what they say. Then the plan for what you will do together must be worked out mutually, based on how the person seeking help defines the problem and solution.

"Starting where the client is" also means that human beings cannot move from one point to the other unless they are *ready* to move. The best human services helpers see themselves as guides and fellow travelers, rather than experts. It is important to understand that individuals who have recently faced a crisis cannot be hurried into an attitude or a position. Much like a parent who would like to impart his or her wisdom to a teenage son or daughter so that "they do not have to make the same mistakes I did," such urgency and "advice" is rarely accepted and is frequently resented. People need helpers to allow them the time to come to their own understandings; we cannot give it to them, as much as we would like to give them this gift. Often, people gain courage and momentum from the support of others; nevertheless, persons cannot be in a particular emotional or cognitive place just because we as helpers want them to be there. Human services helpers should take into account the readiness of the individual, couple, family, group, or community to make a change.

Illustration

The following interchange between a worker and Denise, a 16-year-old woman who has run away from her foster home, has been using crack cocaine, and was living on the streets, illustrates "starting where the client is." Notice that the helper does not tell Denise what to do, does not scare her away, and does not force any action on her.

Helper: I'm glad to meet you, Denise. I know that you were sent to me by the child welfare people, and you might not want to talk to me. I'd really like to hear what's up with you, if you don't mind telling me.
Denise: I just don't want to go back to that place.
Helper: The foster home?
Denise: Yeah. I can't live with that witch.
Helper: You didn't like your foster mom?
Denise: No.
Helper: You've been on your own for the past couple of weeks, haven't you?
Denise: Yeah. I like that better.
Helper: What do you like about being on the street?
Denise: Nobody to boss me.
Helper: Nobody to tell you what to do. Anything else?
Denise: No school.
Helper: Nobody to send you to school.
Denise: Right.
Helper: Do you know why the foster care people asked me to talk with you?
Denise: Not really.
Helper: I think that people are worried about you, Denise. They're afraid that you'll get hurt somehow, either by people on the streets or by taking drugs. The world can be dangerous for a young woman on her own.
Denise: They're not worried.
Helper: You don't think they care about you?
Denise: No. Nobody has before now. Why should they?

*Helper: It makes me sad to think that you feel like no one cares about you. That
tells me that you've had a really hard time.*

Denise: (silent).

Helper: You got real quiet. I'd like to hear what you're thinking.

Denise: Why am I here?

*Helper: That's a good question. My job is to try and help you figure out what
you want to do and if there's any way to get it done. Does that make sense?*

Denise: What do you mean, what I want to do?

*Helper: For example, if you want to get off the street, I can help you find some-
place to stay.*

Denise: No, I don't want to go to no shelter.

*Helper: OK, we can talk about other places and other ideas. I was just giving
you an example. I could also talk to you about drug treatment if you
wanted to stop using crack. What do you think?*

Denise: I don't know.

*Helper: We don't have to make any decisions right now. For now, I would like
for you to just think about talking to me for another half hour or so, and
we'll figure out what to do after that. Is that OK?*

Denise: Yeah, I guess so.

Show Respect and Suspend Judgement

As was discussed in Chapter 2, being nonjudgmental and respectful are vital
components of the helping process. No one can be devoid of biases, feelings,
and beliefs, nor is that desirable. In order to be an effective helper, however, one
must learn to put aside personal opinions and judgments for the duration of the
helping interview. It is important that your agenda not get in the way of the
other person's agenda. You must respect a person's right to make his or her own
choices and to be a unique individual with a unique background and culture.

Carl Rogers (1951) states that those helpers whose personal philosophies
genuinely incorporate respect for the person and belief in the worth of the individual are more likely to be able to communicate respect and nonjudgment for
others. This is an area where helpers need to be honest with themselves about
their own biases and beliefs so that they can be prepared to suspend their own
agendas from the helping process.

Think About It #25

Sometimes suspending judgment and showing respect can be quite difficult for the helper. Imagine your feelings about working with a parent who
is suspected of beating a child; an adolescent boy in a gang who has been
violent and destructive toward other children; or a man who is seeking
help because he has physically abused his lover. How might you go about
suspending your judgment and treating the person with respect?

Be Present, Attend

In order to listen effectively, one must focus, clear the mind of other issues, and pay attention to what the person is saying and doing. Do not worry about whether you will sound clever. If you focus on the other person, the rest will follow. Although skills, values, and knowledge are vital characteristics for a good helper, none of them are worthwhile without a gentle, strong, caring presence. Sometimes the greatest gift you can give another is to be fully "there," even if you do not say anything at all.

Illustration

When Jerri received a phone call from Leslie, she knew almost immediately that all Leslie needed was a listening ear. Leslie ranted about yet another note she had received from her son's school about his disruptive behavior. What did they want her to do; sit in school with him? Jerri listened carefully and reflected empathically. After twenty minutes, Leslie was calmer and more reflective. "Well," Leslie sighed, "Now what I need to do is call the school and discuss this situation more reasonably. Thanks for letting me vent."

Other ways in which a helper can communicate that he or she is attending closely to what the other person is saying include: paraphrasing, reflecting, and summarizing. Nonverbal behaviors include maintaining eye contact, appropriate body language, and vocal cues (like "um-hum"). See Chapter 5 for a more in-depth discussion of these techniques.

Validate

An important component of good listening is to validate the feelings and thoughts of others, even though these feelings and thoughts may be painful, embarrassing, sad, or illogical. Validation means acknowledgment of the person's real situation and real feelings. Through verbal and nonverbal responses (through your breathing, your facial expressions, and your nonvocal sounds) let the other person know that you really heard what he or she said.

Illustration

The following exchange is an example of validating someone's feelings both verbally and nonverbally:

Robbie: I just found out . . . well, I just heard . . . my mother died last night.
Helper (sharp intake of breath): Oh, wow. Robbie, I'm so sorry. (Touching him on the elbow.) What a shock.
Robbie: (Silence, tears).
Helper (Pausing): It's so sad. Would you like for me to stay here with you, or do you want to be alone?

It is not helpful to try to deny the feelings of others, even though they may be difficult to observe. Neither is it helpful to try to persuade another person that his or her feelings are wrong, mistaken, or stupid. Pretending that the situation "isn't so bad" or that something "may be for the best" is inappropriate during a crisis. Do not sugar coat or falsely reassure. No matter how powerless you may feel or how difficult it may be to watch the pain and confusion of another person, doing so is important. When feelings are respected and validated, people can more easily move beyond their emotional intensity and make sense of new reality.

One mother, whose son was born with multiple disabilities and died three days after his birth, wrote 22 years after his death that she continually received the following well-meaning but invalidating message from everyone around her:

> Zach's death was for the best, be thankful, and don't grieve, he would have lived with a heartbreaking disability, and the burden of his care would have been unsurmountable. Although I could see their sadness and that they were trying to spare me grief, this was not what I wanted to hear. I needed support of my feelings, of my loss, and not reasoning. I soon learned to hide my feelings and grieve silently. I now realize that I never fully grieved my loss.

Often people ignore, deny, or discount another's feelings or try to talk them out of unpleasant or intense feelings. Most of the time this discounting is done unintentionally and out of a desire to help. However, validating someone's feelings and experiences is much more effective than discounting them. An effective helper simply listens well and stays with others as they feel things deeply.

Illustration

Notice in the three examples below that the listener does not seem to intend to be mean to the speaker, but that the replies are argumentative and therefore discount and invalidate what the speaker is feeling.

Gwen: I get so angry sometimes at my ex-husband!
Terese: Oh, you don't mean that. You couldn't possibly be mad at someone whom you loved for so long.

William: I don't seem to have any energy these days. I feel really down, like, "What's the use?"
Sydney: Man, you have so much going for you! What do you have to be depressed about? You need to count your blessings.

Cynthia: I've been feeling so stupid recently. I feel old and fat and ugly.
Eric: I really hate it when you talk like that. You're smart and wonderful, and you know it! I wish you would snap out of this mood.

The exchanges are rewritten below to illustrate a response that is accepting of what the speaker says and invites him or her to share more.

Gwen: I get so angry sometimes at my ex-husband!
Terese: You do sound mad. What has upset you?

William: I don't seem to have any energy these days. I feel really down, like, "What's the use?"
Sydney: I'm sorry you're so tired and blue. Will you tell me a little about what's discouraged you?

Cynthia: I've been feeling so stupid recently. I feel old and fat and ugly.
Eric: Wow. I didn't know you were feeling bad about yourself. What's up? Do you know what's triggered this?

Communicate Regard

Rogers (1957) identified the communication of "unconditional positive regard" by the helper as a necessary condition for effective helping. Positive regard is a non-judgmental attitude where the helper is concerned not with blame but with understanding. Communicating positive regard means conveying one's belief in the worth and dignity of each human being, an acceptance of individual and cultural differences, and concern for the well-being of others. Listening carefully and accurately, trying to understand another's view, and asking for clarification of that person's perceptions are all ways to communicate positive regard and respect.

In a good helping relationship the helper values the selfhood of the person and communicates that value to him or her. No matter what problem a person first presents us with, no matter what mistakes he or she may have made, we must remember that there is a human being before us. *This person is inherently worthy of dignity and respect simply because he or she is a fellow human being.* Helpers will be ineffective if they do not find a way to communicate acceptance and regard, gain a person's trust, and connect as individuals.

Illustration

The following exchange presents an example of communicating regard for someone.

Helper (shaking his hand): How are you today, Mr. Wright?
Mr. Wright: I'm doing OK, I guess. Could be better.
Helper: Please have a seat. What's on your mind today? What can I help you with?
Mr. Wright: I need a place to stay.
Helper: Where are you sleeping right now?
Mr. Wright: Last night, in the park.
Helper: That must have been rough. It was cold and raining last night. Where were you the night before that?
Mr. Wright: In prison. I just got out.
Helper: You were released and didn't have a place to go?
Mr. Wright: Uh-huh.

Helper: Well, I'm glad you decided to come here to talk about this. I'd hate for you to go on sleeping in the park if you're uncomfortable there. Let's start by talking about some of your ideas about where you could live.

Mr. Wright: I really don't have any. If I did, I wouldn't have been in the park.

Helper: That makes sense. Do you mind if I ask you some questions about your situation?

Mr. Wright: Go ahead.

Helper: What about any family?

Mr. Wright: No. They pretty much dumped me ten years ago when I got arrested.

Helper: You're not in contact with any relatives right now?

Mr. Wright: No.

Helper: What about friends? Buddies?

Mr. Wright: Well, there was this guy who got out a couple of months ago, and he said I could look him up.

Helper: Do you know how to find him?

Mr. Wright: Not exactly. I mean, I don't know where he is, but I know that the parole board would know where he was.

Helper: Would they release that information to you?

Mr. Wright: I don't know. Could you call and find out?

Helper: Sure can. I can give that a try in a few minutes. Let's see what else we come up with, just in case we can't find your friend. What would you think about a temporary shelter?

Mr. Wright: I hear that they're noisy and dirty.

Helper: That's probably true for some of them. Before we settle on something like that, I could take you over to look at the place.

Mr. Wright: That would be great. Thanks.

Note that in this context "acceptance" is not the same as agreement, approval, or affection. Acceptance involves acknowledging a person's decisions, behaviors, feelings, and points of view. Kadushin (1983, p. 61) states: "Acceptance implies granting the right to be different."

No one wants to be treated primarily as a problem, a disease, a label, a collection of symptoms. Most people want to be seen as valuable individuals first. Once we feel validated and accepted as worthy, we are all more likely to engage in the painful introspection and movement that is necessary for growth and healing. Sometimes the fact that someone has taken the time to communicate regard and respect to us makes us more willing to invest in ourselves and care about ourselves.

As we have said previously, one way to communicate regard for personhood is to watch our terminology. Person-first language puts the person ahead of the disease and is a way to show respect. For example, instead of saying "retarded person," say "person with retardation." Instead of saying "hemophiliac," say "person with hemophilia." Instead of saying "AIDS patient" say "person with AIDS." This may sound trivial and merely "politically correct," but the language used by helpers is a powerful way to communicate our attitudes about persons who ask for help.

Distinguish the Person from the Behavior

Separating a person's actions from the person's worth as a human being allows the identification of strengths and concerns and the development of mutual respect. Persons may *do* or *say* things that are regrettable, wrong, or illegal. An effective helper strives to assist the person with coming to terms with actions that were illegal or dysfunctional, accepting responsibility for those actions and decisions. At the same time, this is best accomplished in a relationship where the helper respects the personhood of the man, woman, or child. Saying to a young child, "You are a great kid, but what you did was not acceptable," expresses this distinction.

Illustration

The following exchange illustrates separating the worth of the person from the behavior:

Frank: I am pond scum.

Helper: What is making you feel that you are low and worthless?

Frank: The way that I ignored my family while I was drinking. I missed my kids' childhood. I'm a horrible person. I don't deserve to get them back.

Helper: Frank, you clearly regret the way you treated your wife and kids. It's obvious that you are very sorry for the way you behaved when you were drunk. But, if you don't mind my disagreeing with you, I think that shows that you are a good person. A bad person would not have any regrets and would not be trying to get sober, like you are. You did some bad things, but you are not an evil man. Can you see the difference?

Frank: Yeah. I appreciate your saying that. It's hard for me to be OK with myself, though.

Helper: I can see that. We can work on that. I'd like for you to be able to recognize your own worth as a person and to be able to take credit for your good qualities as well as take responsibility for actions that you regret.

Frank: Yeah, that makes sense. I'll try.

Distinguish Feelings from Actions

In our culture, the labels "good" and "bad" are often attached to emotions. Emotions and thoughts themselves are actually neutral; they are neither good nor bad, they just *are*. Actions and behavior, however, may be judged as good or bad, moral or immoral, functional or dysfunctional, and legal or illegal. In addition, sometimes our words are expressions of our thoughts and feelings but do not necessarily translate into behavior. For example, one may *feel* great rage during a particular moment toward one's mother and may *think*, "I hate her," and may *express* this as, "I want to kill her." The feeling, thought, and expression are value-free and may not even accurately reflect reality; however, *acting* on them would clearly be wrong. One of our important tasks as helpers is to determine these differences for ourselves and to assist service recipients with distinguishing between action and nonaction (feelings, thoughts, beliefs).

Illustration

Mildred is a middle aged woman whose mother just died from cancer. Beth is a hospice worker who does a follow-up visit to see how Mildred is doing. Mildred tells Beth, tearfully, that she feels so guilty because she had been hoping and wishing in the last few weeks of her mother's life that her mother would die. Her mother was in terrible pain, and Mildred was exhausted from caring for her mother and worrying about her. Beth listens to Mildred's sadness and guilt and then gently points out to her that her actions toward her mother were not at all neglectful. She says to Mildred that she seemed to continue to give very good care and attention to her mother, no matter how tired and frustrated she was. Beth suggested to Mildred that she try to see that her feelings of relief over her mother's death in no way change the fact that Mildred was a very good daughter to her mother while she was alive.

Provide Information

Sometimes service applicants or recipients make specific requests for educational materials, community resources, or information on a specific illness or problem. Human services helpers often respond to such requests, so it is important to have knowledge and understanding of both the formal and informal support systems and services available in the area. This type of information is very valuable to individuals, couples, and families who would otherwise be forced to spend enormous amounts of time and energy searching for resources. For example, in a survey of 892 parents contacted through hospitals, parents reported that one of their biggest problems in obtaining support services for their children was limited information about the services available (Friesan, 1989).

Advocacy

Human services helpers are frequently called upon to negotiate complex bureaucracies and social service or medical systems on behalf of individuals or groups who have less power. This activity is called "advocacy" and will be discussed in greater detail in later chapters. It is helpful if you locate and get to know individuals who are especially responsive and useful within systems that you will be accessing. For example, in one community, applying for public housing subsidies was very complicated, time consuming, and frustrating to applicants. One woman who was an experienced expert in negotiating this particular housing authority taught other individuals how to fill out forms, gave them the names of people in the bureaucracy who were helpful and the names of those people who should be avoided, and generally supported their efforts throughout the application process. This information relieved a tremendous amount of stress on people who were already undergoing the crisis of living in a temporary shelter for families without homes. Sometimes there are community-based advocates whose job it is to protect vulnerable populations and advocate on their behalf. Determine if protection and advocacy groups exist in your community, and contact them for assistance. Learning more about the services they provide and the problems they address will help you in your work as a helper and will benefit the people with whom you work.

Recognize Nonverbal or Nonvocal Communication

The communication of meaning does not just occur through the written or spoken word. Words are only partial communication. In fact, nonverbal behavior may be more revealing of our true feelings and thoughts than are words (Kadushin, 1983; Nunnally and Moy, 1989). Research suggests that as much as 65 percent of a message is conveyed in ways other than words (Birdwhistell, 1970). As will be discussed in greater detail later in the next chapter, helpers must attend to the nonverbal behaviors of those seeking help as well as of themselves.

Listen Actively

Listening is the foundation of all helping relationships. "Just listening" may seem so simple that many may not even consider it to be a skill at all. Listening may even imply passivity. Good listening, however, is a very active process and is an important ability to master. Helping requires active listening, which is very hard work. Tuning out everything but the person's concerns requires discipline and training. It differs from the passive hearing with which we usually operate, where for the most part the background noise of our lives is not fully attended to. In active listening, the helper makes a conscious decision to fully engage with a person and to focus on nothing else for the duration of the telephone call or interview. It is almost like throwing a switch in your head and your heart; you are "on" fully for another person.

The subsequent chapters on active listening and interviewing will expand these concepts and offer specific techniques that will help you communicate better with persons seeking services.

CHAPTER FIVE

Active Listening

One of the major sets of tools that helpers use to communicate empathy and positive regard, validate feelings, and support choice is called "active listening." Learning these skills will greatly increase a helper's ability to be useful to persons who ask for assistance. Without these techniques, we are in jeopardy of being well meaning but not very effective.

It is necessary, of course, that a good helper be well-intentioned and want to help, but that is not the only ingredient of helping. Helping is a combination of desire and skill. One set of skills on which helpers rely heavily is active listening.

Active listening is an essential communication skill. It is important that we communicate our interest and sensitive concern. The skill of effective listening can take two forms: verbal and nonverbal responses to listening. Without active and accurate listening, the helper will not be aware of the other person's feelings, thoughts, ideas, desires, decisions, or expectations. Listening communicates respect, interest, and regard. Brammer (1979) says that active listening includes listening not only with your ears, but also with your eyes and with a special perceptiveness.

The following poem illustrates the importance of good listening.

Listen

When I ask you to listen to me
  and you start giving advice
  you have not done what I have asked.

When I ask you to listen to me
  and you begin to tell me why I shouldn't feel that way,
  you are trampling on my feelings.

When I ask you to listen to me
  and you feel you have to do something to solve my problem,
  you have failed me, strange as that may seem.

Listen! All I asked was that you listen,
  not talk or do—just hear me.
Advice is cheap: 20 cents will get you both Dear Abby and
  Billy Graham in the same newspaper.

And I can do for myself: I'm not helpless.
    Maybe discouraged and faltering but not helpless.

When you do something for me that I can and need to do
    for myself, you contribute to my fear and weakness.

But, when you accept as a simple fact that I feel what I feel,
    no matter how irrational, then I quit trying to convince
    you and can get about the business of understanding what's
    behind these irrational feelings.
    And when that's clear, the answers are obvious and I
    don't need advice.
Irrational feelings make sense when we understand what's behind them.

Perhaps that's why prayer works, sometimes, for some people
    because God is mute, and He doesn't give advice or
    try to fix things. "They" just listen and let you
    work it out for yourself.

So, please listen and just hear me. And if you want to
    Talk, wait a minute for your turn; and I'll listen to you.

Anonymous

True communication of empathy requires that the helper has mastered active listening and is capable of accurately identifying feelings. Empathic responses include those that accurately capture the emotional and cognitive content of the other person.

Meaningful communication is quite difficult and complex. Each human being has filters through which communication must go before it is received; these filters include biases, judgments, values, beliefs, experiences, personality characteristics, expectations, and rules. As a person sends a message to us, it will be distorted through our filters, and vice versa. Only by constantly clarifying these signals can we come to an understanding of each other.

Think About It #26

1. Think about and discuss with other learners how an individual's "filters" might interfere with the helping relationship.
2. List examples of events where your behavior or the behavior of someone else was misinterpreted.

Empathy is not complete unless it is adequately demonstrated. Empathy therefore depends on effective active listening. Communication is not finished without this second necessary step where the helper mirrors back what has been presented. It is useless to understand another person if he or she does not know that you understand. Your part in the communication is to state what you think you hear, see, and feel as a result of the other person's messages to you.

Illustration

Read the following conversation, watching for evidence that the helper was listening actively.

Walter: I can't believe that my father isn't coming to my commitment ceremony!

Helper: You look and sound really upset. It seems that you're angry at your dad.

Walter: No, not really. The truth is, I'm just hurt.

Helper: He's let you down?

Walter: Yes, again. He has never seemed to accept my relationship with Jim. He has always treated us coldly and acts uncomfortable around us. Now he won't come to my wedding! If I were marrying a woman, I know he'd be right there.

Helper: You'd really like your dad's support, wouldn't you?

Walter: I sure would. He's my father, and I love him, even though we don't see eye to eye. I just wish he loved me.

Helper: This must really be painful for you.

Walter (pauses): More than you can know.

Helper: I'm really sorry that you have to feel so sad at a time that should be happy for you.

Walter: Thanks. Me too. I'm glad that you can understand what this is about for me.

Helper: I think I have an idea how you feel, Walter. I think that you would like for your parents to accept you as you are and accept your partner as part of the family, but that your father can't seem to be there for you, and that really hurts.

Walter: Exactly.

Listening Skills

The purposes of active listening skills and tools are: (1) to clarify the person's meaning, views, and perspectives; (2) to assure that both partners in the communication understand more fully the feelings and the content; (3) to check out your assumptions and perceptions as a listener; and (4) to deepen the level of communication and further the helping relationship. These techniques of active listening allow you as the listener (or interviewer or helper) to check out whether you are accurately hearing and understanding what you are being told. These tools also allow the person who is telling the story to hear back what he or she said in order to sort out how he or she really feels and thinks. It is important for people, especially when they are confused or troubled, to hear from an empathic listener what they said so that what they meant to convey is verified and so that they can examine in a more objective light what they are feeling and thinking.

Important tools of active listening are (1) verbal or vocal skills and (2) nonverbal or nonvocal behaviors.

Verbal Tools in Active Listening

Minimal prompts. Minimal prompts are responses from the helper such as "um-hum," "and?" "but?" "yeah," "I see," and "wow." As you can see, these very short responses signal that the helper is listening carefully and would like the other person to continue. Minimal prompts are the vocal version of nodding your head.

Accent responses. An accent response is a repeat of a statement, word, or phrase that the other person has just uttered. Accent responses are often made in a questioning tone of voice or with emphasis. The purpose is to clarify meaning and encourage further elaboration of the issue or situation.

Illustration

The helper uses accent responses in the following conversations.

Rose: My family was humiliated by my crying at the funeral.
Helper: Humiliated?
Rose: Yes, they were ashamed of me, because in my family we don't show our emotions in public.

George: I really don't know if I can handle my father's criticisms one more hour.
Helper: Can't handle?
George: No. I'm going to explode. I'm going to let him have it.

Paraphrasing. Paraphrasing is restating, in your own words, what the person has told you verbally.

Illustration

Examine these exchanges for evidence of paraphrasing.

Rose: I couldn't believe it when my daughter graduated with honors despite her being so sick all year.
Helper: Your daughter overcame a lot of obstacles, and you're proud of her accomplishments and resilience.

George: I feel so guilty when I get mad at my dad, since he's had the heart at-tack and hasn't been feeling well. I want him to live, but sometimes I find myself wishing he would die. I don't understand why we still aren't getting along.
Helper: You care about your father and wish your relationship were easier and that you could contain your temper around him. Sometimes it's so difficult that you can hardly stand having him around.

Note that paraphrasing is not "parroting;" that is, it is not simply a word-for-word repetition of what the other person says. As you may have experienced, having your words echoed can be annoying. Rather, paraphrasing clarifies and validates what the person has said, but in the listener's own words.

Illustration

The following dialogue occurred between Amy Washington, a young mother who has been having trouble controlling her temper with her preschool chil- dren, and Maggie, a human services helper with a child abuse prevention agency.

Amy: I just couldn't take it any more. The kids were running around the house like animals . . . hitting each other and screaming.
Maggie: Um-hum (nodding her head).
Amy: I had a splitting headache since I got up in the morning, and the kids seemed to know it, too.
Maggie: The kids were screaming, and your head hurt.
Amy: I couldn't take it. I just lost my cool and started screaming and yelling.
Maggie: You sort of lost control, then.
Amy: Yeah, I did. But I didn't hit them. I felt like spanking them, but I didn't. I screamed and yelled and sent them to their rooms.
Maggie: Wow. Even though you lost your temper, you didn't hit the kids.
Amy: Yeah. That was pretty good, wasn't it?
Maggie: Way to go! I'm proud of you!

Think About It #27

1. Read the dialogue above, and identify the minimal prompts and paraphrasing.
2. What other responses could you have made to Amy Washington that would also be helpful and encourage her talking?

Reflection. Reflection is very similar to paraphrasing. However, reflection in- volves not only paraphrasing the content of what was said but also reflecting back (like a mirror) the other person's feelings and nonverbal communication. Since empathy is communicated primarily through active listening, the helper tries to become an emotional mirror, a reflection of what the other person has sent in his or her verbal and nonverbal communication. Reflection involves ac- curately identifying feelings and communicating this perception clearly to the person. Reflection brings to the surface the emotions and attitudes that underlie the words and offers the mirrored picture back to the person in a tentative way. An example of a reflection would be, "When you tell me about being rejected for disability benefits, you look and sound very angry, and I get the sense that you feel betrayed by the Social Security system and disappointed in your case worker. Am I reading you correctly?"

These are some examples of how you might begin a reflection:

It sounds as though you are . . .
It seems as if . . .
You seem to be saying . . .

I think I understand you to be saying that . . .
It's probably pretty difficult to . . .
I'm hearing that . . .
I'm picking up that . . .
If I understand what you're saying . . .
If I'm hearing you correctly . . .
I gather from what you're saying that . . .
What you feel the strongest is . . .
So, as you see it . . .
I'm not sure I'm with you but . . .
I sense that you . . .
So you feel . . .
It appears to you that . . .
I'm wondering if you're feeling . . .
So, from where you sit . . .
What's important to you now is . . .

Illustration

Harold Baker, a 38-year-old white construction worker, lost his job three months ago. He was unable to pay his rent and was evicted from his apartment with his wife, June, and their three children (Mike, age 10; Sarah, age 9; and Miriam, age 5) three weeks ago. They lived in their 1973 Buick station wagon for three weeks until they went to a shelter for homeless families. Every week, the shelter offered a time for parents to get together and share their feelings. Several parents had been in the shelter for several months and had become very effective helpers. During the first group meeting that Harold and June attended, Harold sat in a hardbacked chair behind the rest of the group, with his legs spread and his elbows leaning on his knees. His head was down and all anyone could see was the top of the baseball cap that he wore. He was very quiet but would occasionally glance up when his wife spoke. He moved slowly when he did look up and was expressionless. June sat on a couch next to another woman who resided at the shelter. June would occasionally look at Harold but did so quickly and then looked away. She looked as if she had tears welling in her eyes. She spoke on two occasions during the group meeting. After one of the other women expressed anger at the fact that residents could only use the kitchen at certain times during the day, June stated, "My little one gets really hungry right before bedtime. I want to be able to give her something to eat so she sleeps good but I can't. I wish I could do more for my kids." The second time she spoke, June's hands went to her face and she slumped over. In a quiet voice she said, "I just don't know what we are going to do."

Think About It #28

Consider the preceding illustration, and answer the following questions.

1. Identify and list the feelings that you imagine Harold is communicating during the group meeting.

2. Identify and list the feelings that you think June is communicating during the group meeting.

3. Imagine that you are speaking with Harold, then imagine that you are speaking with June. Formulate reflections for both Harold and June. Examples are:

To Harold: "You must be feeling miserable about your family living in this shelter," or, "It really hurts to see your wife so sad."

To June: "It's a helpless feeling when you can't feed your daughter when she's hungry," or "I'm thinking that you might be feeling insecure and scared because you don't know what your future holds."

Clarification. Seeking clarification of another person's meaning periodically during a helping interview assists the helper's understanding. Obviously, it is not useful to listen quietly when one is not clear about what the problem is; likewise, it is not helpful to generate options and possible solutions if one is mistaken about what solutions would be acceptable or what the issue really is. Some confusion in communication always exists because one person can never fully understand another. However, helpers are responsible for seeking clarification for important points that may have been misunderstood. This can be done with a question (see the section below on questions) or a statement.

Illustration

The following are examples of clarifying questions or statements offered by helpers.

Helper: I'm a little confused here. Who was it who asked you not to visit your father in the hospital?

Helper: I'm sorry, I didn't hear the last part of what you said. What was the diagnosis?

Summarizing. From time to time in the interview, putting together and highlighting what has been discussed is useful. A good summary will incorporate the content and the feelings. This can demonstrate progress, summarize the important points of a problem or solution, offer the person a chance to direct the problem-solving, and give both people a chance to think through the issues. Hearing an accurate and perceptive playback of what has been said can also be very validating.

Illustration

The following summaries are offered to highlight the content, affirm the speaker, and validate the difficulty of the situation.

Helper: I'm feeling a little overwhelmed by all that you have told me about losses that you've experienced in the past year. You experienced the death of four persons who were very significant to you; you quit your job, sold your possessions, moved away from your friends and family, and came to a confusing and different place where you didn't know your way around. It's no mystery to me why you have trouble concentrating and sleeping at times. You must be missing many things and grieving deeply.

John: *Wow. You're right. I hadn't thought about it like that. Maybe I'm not going crazy after all.*

Helper: *So far you've told me that you need to make some decisions very soon about telling your parents and your lover that you are pregnant, whether to have an abortion, whether to drop out of school, whether to go into drug treatment, and whether to take an HIV test or not. You have a lot on your plate to deal with right now. Which one of those decisions would you like to talk about first?*

Helen: *You know, when you list all of that stuff like that, I realize that's why I've been feeling so panicky lately. I guess I need to figure out how I'm going to tell Paul what I'm going through first.*

Informing or educating. When it is appropriate to do so, a helper becomes a resource expert or educator on a particular topic. This is not the same as preaching or lecturing someone on how you want someone to behave. It is a way of providing needed tools or facts in a useful, nonjudgmental way. Examples of this might be: providing simple instructions about applying for food stamps or disability benefits, showing how to make sex or drug use safer in order to reduce the risk of HIV transmission, teaching how to find housing or job opportunities in the newspaper, or explaining the best way to negotiate a social service system.

Confrontation. "Confrontation" is a word that is used for inviting someone to consider another interpretation or perspective. It is a way of suggesting gently and tentatively that things could be seen in a different light. Confrontation is not an attacking, antagonistic, accusing, or angry mode of communication in the context of active listening. It is, rather, a pointing out of inconsistencies in a person's presentation of facts, or a statement of your displeasure at possibly being lied to or misused. Confrontation is not usually necessary, because most people are trying to discover the truth and trying to understand themselves and their situations honestly. Confrontation is an active listening technique that should be used with great caution. Under most circumstances it is not appropriate to be confrontive with a person just to get him or her to open up or reveal things. Persons who are allowed to disclose things at their own pace are much more likely to feel safe within the helping relationship.

If you determine that confrontation may be necessary in a particular helping interview, these guidelines should be considered:

1. Only use confrontation if a helping relationship has been established and rapport has been built.
2. Confront someone about behavior only, not about their worth as a person.
3. Be specific about the particular behavior or incident.
4. Be respectful in your language and manner.

5. Be gentle and supportive.
6. Be honest, genuine, and congruent.
7. Never confront a person to manipulate him or her.
8. Be positive and constructive.
9. Do not confront someone angrily.
10. Do not be blaming or accusing.
11. Begin your statements with "I" rather than "you."

Illustration

Examine the following helper confrontations, thinking about why some may be more effective than others.

Wrong: You never listen to anyone's opinion!
Better: It seems to me that you asked for my help but you were reluctant to accept it.
Better: I sometimes get the feeling that you are not ready to listen to someone else's viewpoint.

Wrong: You are so self-destructive!
Better: I get worried about you sometimes when I think that you are more concerned about your husband's health than your own.

Wrong: You are really irresponsible!
Better: I'm a little angry that you've missed our last three appointments, because I start thinking that this isn't so important to you. What is your take on this?

Illustration

Gloria was working as a case aide at Planned Parenthood, where she interviewed applicants about services they may wish to access at the agency. Gloria was speaking with Pam, a 14-year-old who was three months pregnant and who had received no health care previously. Gloria offered Pam a gynecological exam and Pap smear, explaining that these procedures were part of their usual prenatal care services. Pam said, "I don't want that. I don't want anybody sticking anything up there." Gloria looked at Pam for a moment and then decided to confront her. Gloria said, "Pam, I can understand that it's scary to think about having a physical exam and some tests. But it's hard for me to understand why you would not allow us to put clean, sterile medical instruments into your vagina when you allow boys to put their penises up there." Pam at first looked at Gloria with a shocked expression on her face, and then she laughed out loud. Pam said, "You're OK, Gloria. That's pretty good. I guess you're right." Gloria said, "Good for you, Pam. This will really help us take care of you and your baby. We'll be as careful as we can be to not make you uncomfortable, and we'll let you know what we're going to do before we do it."

The appropriate use of questions. Asking questions can be a part of active listening, but should probably not be your primary mode of communication. Questions are not usually the best way of getting information; rather, it is best to let people tell their stories in their own ways. When too many questions are used, the helping interview begins to resemble an interrogation rather than a mutual conversation. A question-answer pattern gives the message that the interviewer knows best what should be talked about and in what direction the interview should go. To be less directive and authoritarian, the helper should rely on reflective statements. Be aware of when you are on the verge of posing an inquiry in an interview, and evaluate whether that is the best way to phrase it.

Sometimes, specific information must be collected. When you must ask a question, making them less intrusive is more productive. Open-ended questions and probing questions are the most common techniques of asking questions in active listening. There are also forms of questions that should be avoided and reasons for asking questions that are not appropriate (Benjamin, 1969).

Open-ended versus closed-ended questions. Questions can be phrased basically in two ways: closed-ended or open-ended. Closed-ended questions can usually be answered with "yes" or "no," or they imply the answer. Examples of closed-ended questions are: "You didn't want to go, did you?" or "Were you mad?" An open-ended inquiry, which allows the interviewee to answer in his or her own way and to expand on the answer, can be framed as a question or a statement. A statement is often the most inviting form of inquiry. Examples are: "Tell me a little bit about that phone call," "I'm wondering about your experiences in college," or "I'd like to hear more about what happened today to upset you."

It is often necessary to ask direct closed-ended questions in an interview in order to get necessary information (for example, "What is your birth date?"). However, closed-ended questions do not tend to further communication. When the goal is to learn more about the person's perceptions or experiences, open-ended questions are best.

When to use questions. At times in a helping interview a question is entirely appropriate. For example:

1. When you have not heard or understood; for example, "I'm sorry, I missed the last part of your sentence. What did you say?"
2. When you must have a specific piece of demographic or financial information; for example, "How old are you?"
3. When you must complete a form and must ask direct questions to do so; for example, "For our records, what is your address?"
4. To assist in clarifying the meaning of the interviewee. Examples are: "You mentioned earlier that you had a mission in life. What is it?" "You said that you left in the middle of the family reunion. What happened?" "You said that you hung up on your lover when she called. What triggered that reaction?"
5. To probe for more information and to move the interview forward, for example, "I know you said earlier that your main goal for today is to figure out what to do about marriage counseling. Can you tell me more about what

you've been thinking about that?" (See the next section for more information about probes.)

Examples of useful probing questions. One of the most common uses of questions in active listening is to use them as probes or prompts, to deepen the level of the interview or to obtain more information. The following questions are examples of probing or prompting, or ways to phrase your questions so that they move the interview along in productive ways.

How do you feel about the fact that _____?
What seems to be the biggest obstacle?
What was your reaction?
What have you thought about doing or saying to her?
What do you think is best?
For example?
What had you thought about doing if that doesn't work?
How does this affect you?
How does this change things?
Was there anything else?
In what way?
What do you make of that?
What is your idea about why that happened?
Where will you go from here?
How does it look now?
What is an example of that?
What's happening with you right now?
What's going on with that now?

Think About It #29

Add to the above list throughout the week as you listen to others talk and as you think of ways to explore with open-ended questions.

Questions to Avoid. Benjamin (1969) lists several types of questions that are not appropriate or useful to ask in a helping relationship:

1. Do not ask double or multiple questions. Sometimes helpers mistakenly string together two or more questions in a single response or bombard interviewees with another question before they have had a chance to fully answer the previous question. These types of questions are too difficult to answer and may confuse both partners in the interview. An example is: "Did you think that you would stay in the shelter again tonight, or are you going back to your friend's house, or did you want me to try and get you a place in the SRO?" If you must ask a question, ask one, and then stop and give the other person a chance to answer.

2. Try to avoid "why" questions. "Why" has become a word in our language that denotes disapproval or displeasure. Questions that begin with "why" tend to put people on the defensive, even if the asker does not mean to be on the attack. For example, questions such as "Why did you get tested for an STD?" "Why did you take that job?" or "Why don't you get along with your mom?" are likely to cause the interviewee to think that you are questioning the wisdom of the decision or are blaming him or her for the situation. It is best to ask questions that sound less attacking, such as, "What prompted your decision?" or "What is the most difficult thing about your relationship with her?"

Illustration

Notice what happens in the following example when "why" begins the question.

Pat: I've decided to go to college.
Helper: Why did you decide that?
Pat: You don't think I'll make it? I don't have good enough grades?
Helper: No, that's not at all what I meant to say. I apologize for the way I asked that question. What I should have said is, what made you decide to go forward with your applications? I ask this because the last time we talked, you weren't sure what you wanted to do.
Pat: Oh. Well, after I talked to my mom and you and the guidance counselor last week, I realized that I really wanted to give it a shot. If I didn't try, I'd always wonder if I could have gotten in, and I might regret it for the rest of my life.

3. Never ask questions purely out of curiosity. Questions should always have a purpose that is linked to the reason that you and the other person have come together. An inquiry in an interview should be relevant, dictated by the purpose of the interview, and helpful in moving the interview forward. The interviewer should never be unnecessarily intrusive, nosy, or curious. Sometimes a helper asks an irrelevant question when feeling stuck, shocked, or uncomfortable with a silence. An example of an inappropriate and irrelevant question is:

Dave: I'm not thinking clearly right now; my daughter died in a car wreck last week.
Helper: Oh. How old was she?

4. Do not ask leading questions. Leading questions imply what you as the helper think. Examples are: "You don't really love your sister, do you?" "You wouldn't really do that, would you?" or "Don't you think that . . . ?" Even when the helper does not intend to lead a person to answer in a certain way, the phrasing of these leading questions may have that effect.

Illustration

The following examples illustrate responses that are helpful and not helpful.

A *response that is not very effective*
Dan: I just don't want to tell my mother that I have AIDS.

Helper: Don't you think she'd want to know? Don't you think she cares about you?

Dan: Well, I guess so. No, I mean, I don't know. What do you mean?

A response that is better

Dan: I just don't want to tell my mother that I have AIDS.

Helper: Tell me more about that if you can.

Dan: First of all, I don't want to worry her. Second, I think she'd really try to hover over me and tell me how to live my life. And third, I'm not really ready to deal with her reactions. I need some more time to adjust myself.

Proper use of self-disclosure. This section discusses self-disclosure, which means telling your own personal story when you are in the role of helper. Sharing personal information about yourself as a helper is sometimes useful in a helping relationship and sometimes destructive; it is important to be able to determine the difference between these two situations.

Sometimes figuring out what is appropriate to share, when to talk about your own situation, or to whom to disclose is difficult. The answer to these questions can be found when one considers the following factors: what the other person is experiencing, the relationship between you, how you are feeling, your role, the reactions of the other person(s), and feedback from consultants (such as other informal helpers or professionals or from your supervisor).

The dangers of self-disclosure. Self-disclosure should never be undertaken without a great deal of thought on the part of the helper. The major danger of your talking about yourself, even briefly, is that it removes the focus from the person who has asked for help and puts it onto you, the person responsible for offering assistance. That is not only a disservice to the person who is asking for help and a misuse of the helper's authority and role, it is unethical for a helper to use a helping relationship to work out his or her own problems. Another danger is that the other person may see you as too troubled yourself to be of much help! ("I want another case worker; this one is a mess!").

At times, people in the role of the helper disclose personal information, not to assist another, but because they are feeling very needy at that moment. This role reversal may confuse the other person, who is not prepared to help and may not reciprocate. The helpers then find themselves angry that they are not getting their needs met. In this case, the self-disclosure by the helper is inappropriate. A better solution would be for the helper to talk with a consultant or a person with whom there is a reciprocal helping relationship about his or her own situation when feeling needy.

Appropriate and helpful self-disclosure. Helper self-disclosure, when used ethically, often allows the other person to feel as though he or she is not alone. This universalism, learning that others have gone through a similar situation and that they have had similar feelings and reactions, is one healing component of this sharing experience. Self-disclosure in a helping interview, when used appropriately, may be another way to enhance rapport building, relationship development, and empathic communication. The helper's self-disclosure can facilitate another person's trust and comfort. Strangers become friends more quickly when they have something in common. In many situations, the helper will have only a brief interaction with another who is seeking support. Therefore, the helper's ability to share personal information wisely allows the short time together to be used as fully as possible.

When you share your experiences with others, they may be able to learn from them. You may model for the person how someone could handle a similar situation. Support groups and buddy systems are examples of services that provide structures for people to share their wisdom with others in a similar situation. For example, if you are a peer facilitator of a support group, you may be able to teach others through your own experiences.

One of the most valuable uses of self-disclosure for helpers is not in an interview situation at all, but in making presentations to informal helpers and to professionals. Self-disclosing takes great courage, but the educational value for others is immense. For example, one woman who had been battered throughout her ten-year marriage and then brutally raped by her husband after separating from him presented a workshop to social workers, counselors, police officers, magistrates, and ministers. Included in the audience were other women who had been battered. The audience's response to the self-disclosure was intense. One police officer said that previously he had assumed that women wanted to stay in an abusive relationship and that he really should not intervene. Now he said that he had a greater understanding of the many complicated reasons that one might stay in a violent relationship and expressed a commitment to intervening differently in the future. Others felt empowered by her courage and her willingness to risk in order to help.

Difficulties experienced with appropriate self-disclosure. Sometimes helpers have valuable personal information to share but do not do so, because they may feel embarrassed to share their own information or because they may fear that others will think less of them. In addition, helpers frequently are unwilling to re-experience the difficult feelings that may surface when they share their stories. The helper may not wish to feel so intensely at that point or may be afraid of losing control. In addition, the helper may be anxious about saying the wrong thing or not saying what needs to be said when intense emotions are involved.

Guidelines for choosing how to handle self-disclosure. The following guidelines can assist you with using self-disclosure appropriately as a helper.

1. Be brief. If you choose to share your experiences in a helping interview, do so sparingly and briefly, and immediately refocus on the person who has asked for assistance. A very brief and simple statement, such as "When my brother died, I know I felt that I was losing my mind," might be very validating to someone who is grieving. However, talking about your own grief process for a long time is not acceptable if you are the helper.

2. You have the right to privacy. You do not have to reveal anything about yourself if you do not choose to do so. If you are a helper in a particular field because you have experienced the situation yourself, then some self-disclosure may be expected. Yet, even in these cases, you still have the right to your own privacy and should not be required to talk about yourself. For example, you may be volunteering for a rape crisis center and choose not to reveal that you are yourself the survivor of a sexual assault. You may be volunteering for an AIDS service organization and choose not to reveal to everyone that you or a loved one is HIV-positive. You may not want to reveal to everyone that you are recovering from alcoholism or drug use or that you were the victim of incest. You do not have to sacrifice your privacy to be a helper.

3. Reflect. If you are asked a direct question about your personal experience, you may choose not to answer it or may choose to answer it briefly. In either case, reflect back the real concern under the question. Most often, the person is wondering if you can be trusted, if you can really understand, or what motivates you to do this work. When persons in a helping relationship with you ask personal questions of you, they are usually not asking out of idle curiosity but are signaling that they wonder what you may have to offer them as a helper. You should try to reflect the underlying concern.

Illustration

The following are examples of reflections offered when personal questions are asked of the helper.

Francine: Do you have children?

Helper: No, I don't. You're probably wondering how I can teach parenting skills if I don't have children of my own. Is that right?

Francine: Yeah. You just don't know what these kids are like. They drive me crazy.

Helper: That's a good point, Francine. I can't possibly know what it's like to be you or to be in your house. But there are some things I've learned at school and by doing this work that might be useful to you. Can we try some of them out and see what happens?

Lizzie: Have you ever been battered?

Helper: I bet you're thinking that only another battered woman can understand what you've been through.

Lizzie: Well, yes.

Helper: That's probably true. No one else can truly understand another woman's experience. But I'd really like to try to understand what it was like for you, and I'd like for you to keep telling me about your relationship with Joe.

Maureen: Do you believe in God?

Helper: What prompts you to ask that question?

Maureen: I believe in God, and I wanted to know if you do.

Helper: Your faith must be really important to you. How does it affect your decisions about whether to have this baby?

Maureen: I've been taught that abortion is a sin.

Morris: Are you gay?
Helper: Your question makes me think that you're wondering if you can trust me to understand your relationship with your lover and not judge you.
Morris: Yeah, that's what I was thinking.
Helper: Well, let's give it a try. If you ever feel judged by me, please tell me, and we'll talk about it. And if I don't understand what you're saying, I'll ask you. Is that OK?
Morris: Yeah. That sounds fair.

Delores: Do you have someone in your family with Alzheimer's disease?
Helper: Yeah, actually, my mom has dementia, and that's what got me involved with this support group. But you must have a lot on your mind right now about your dad, and I'd like to talk to you about how you've been feeling and coping.

4. Apologize if necessary. If you make a mistake and disclose too much about yourself, present personal information too early in the relationship, or accidently take the focus away from the person who has asked for help, then admit your error, apologize for it, and refocus on the situation at hand.

Illustration

The following exchange illustrates a mistake in helper self-disclosure and the subsequent refocus on the person who has asked for assistance.

Helper: As I listen to you talk about your friend, I'm reminded how frustrated I get at my own friend who has cancer, and how I wish she would take better care of herself.
Robert: You have a friend with cancer too? When did she find out? How sick is she? Have you told her how you feel? Does she listen?
Helper: You know what? I messed up. I didn't mean to shift the subject to me. You are the one who's important right now. I'm sorry. Let's get back to what you were saying about your friend being in the hospital.

5. Ask "Whose needs am I meeting?" When you find yourself about to self-disclose, stop and ask yourself this important question: "Whose needs am I meeting?" If your self-disclosure will clearly help the helping relationship, then you may consider doing so. If you are on the verge of talking about yourself because you want to get something off your chest, then do not do it. Make sure that the person has either asked you directly to share information about yourself, or that he or she is very open at this time to hearing about it. Regardless of whether the person has asked you for information, it is never acceptable in this context for you to talk simply because you need to talk.

6. Never lie. Never lie about having experienced a particular situation or problem. You may choose not to answer the question, but do not fabricate anything. For example, do not answer "yes" to a question about whether you have been raped, battered, divorced, bereaved, or anything else, unless it is true. Unfortunately, sometimes helpers will answer in the affirmative in order to try and further the trust in the relationship. Obviously, lying is not a good way to further trust. When in doubt, tell the truth.

7. Never self-disclose just to let yourself off the hook. In a helping relationship, it is not appropriate to tell a person something personal about yourself just to try to defuse that person's anger or displeasure. For example, one student intern who was helping with a support group for the adult survivors of child abuse became uncomfortable when several group members confronted her with, "Stop telling us how we feel; you haven't been through this!" Rather than accepting their feedback and looking at her own behavior, she blurted out, "Yes, I do know how you feel! I was raped by my stepfather." She then burst into tears and bolted from the room. Her self-disclosure served the purpose of getting her off the hot seat; it did not, however, further the group process or meet the needs of the group members to be heard and validated.

Illustration: Useful Sharing

Mary Catherine O'Conner, a 32-year-old, married, Irish-Catholic woman with two preschool children, has gained wisdom through her brother's struggle with AIDS and his death. She can assist another woman, Susan Allen, whose brother is ill with AIDS, by talking about her experience. Having already gone through the situation, Mary is aware of the complications of one's brother having a life-threatening disease, especially when that disease is so socially unacceptable and emotionally draining. For instance, she has seen his home after someone spray-painted obscene comments on it. Mary has experienced family and community struggles that provide Susan with clues about ways to address her current situation. Mary becomes a buddy, or peer support, to Susan. Mary and Susan meet for lunch, and Susan is feeling confused and frightened about what is happening with her brother. By talking briefly about some of Mary's own feelings when her brother was facing his death, Mary helps Susan understand that her reactions are normal. Mary may choose to talk about herself because this particular day she is feeling strong emotionally. Mary also trusts that Susan will not tell others what she said. Finally, Susan has asked for the information. The fact that Mary is willing to make herself more vulnerable by revealing her personal reactions increases the attachment between Mary and Susan and strengthens Susan.

Illustration: Inappropriate Sharing

In the above example, if Mary had called Susan to act as support, but instead talked about her brother the whole time, Susan's questions would not get addressed. In fact, Susan would find herself using her meager emotional resources caring for Mary.

Be careful of giving advice. Some people think that helping others is the same as giving them advice. Actually, the opposite is true. The role of the helper is to provide accurate information on resources and consequences, and then to support the person as he or she struggles to make his or her own decision. When the helper gives direct advice (for example, "I think you should tell him," or, "If I were you I'd leave"), the helper is not affirming the fact that the person is a unique individual who is better served by being helped to find solutions to fit his or her own culture, family, religion, expectations, values, and personality. A good helper will share resources and provide information about options but not give advice.

Of course, when a person is asking for information, resources, and advocacy, these requests ought to be handled immediately through whatever mechanism your organization provides, such as case management and referrals. The question is how to best handle a situation where it seems that the person is asking you to solve the problem or make the decision or where you as the helper feel compelled to tell a person what to do.

Even when a person asks you, "What do I do?" or, "What would you do?" he or she is not seeking to give up control or to give up the right to make choices. These words are most often a cry of confusion and bewilderment, not an invitation for the helper to take over. When someone asks for help, this is *not* really delivering one's life into the hands of another. This is an expression of helplessness, a frightened attempt to avoid the problem, a desperate cry for help, or an expression of pain.

Giving advice, rather than giving information, is very dangerous in a helping relationship. If the person does not take the helper's advice, the helper may be hurt, feel rejected, or even get angry at the person and find it more difficult to work with him or her. If the person does take the helper's advice, the outcome is not much better. If the accepted advice does not work or gets the person in trouble, then the relationship has been damaged, sometimes beyond repair. (For example, "You told me to go home and tell my mom and dad that I'm pregnant. I did what you said, and they kicked me out of the house, and now I don't have anywhere to live."). The best possible scenario is that the other person does as you ask and your advice works; but this can lead to dependence on you as the helper to solve problems. This has not supported the person in learning how to make decisions or to solve problems, nor has it increased their pleasure at mastering the situation.

What If Someone Asks Me Directly for Advice? If a person says, "What do I do?" reflect the concern, like you reflect everything else in active listening. "It sounds as though you're feeling confused, overwhelmed, lost, frustrated, and you'd love for someone else to take this on and solve it for you. It's a pretty difficult situation." Offer to be with the person in developing options and facing consequences; problem solve with them. But don't tell someone what to do! That is disrespectful because it disregards the person's right and ability to solve his or her own issues.

Illustration

Consider the following conversation, and think about how this helper handles being asked for advice.

Ronnie: What am I going to do? What do you think I should do?

Helper: I can't tell whether you are serious about wanting my opinion, or whether you're expressing your frustration and confusion.

Ronnie: I'm asking you. What am I supposed to do now?

Helper: Well, I'm going to try and help you figure that out, but I'm not actually going to tell you what I think about what you should do.

Ronnie: Why not?

Helper: Because, although I really care about what happens to you, it's not my life, so it's not my decision. Only you can figure out what will fit for you. What I want to do is help you list all of the possible solutions and see what might work best for you.

Ronnie (laughing): I actually knew that's what you'd say. I believe that if I asked you what time it was, you would ask me what time I thought it was.

Helper (also laughing): That's very funny. You really have me pegged. But does what I say about this make sense to you?

Ronnie: Yeah. I know that I have to make up my own mind. It's just hard, that's all.

Helper: I know it's tough. We won't give up until we have some idea about what you should do. I'm thinking that we should start by talking about what you've already tried. Is that OK?

Ronnie: OK.

Recognizing and Interpreting Nonverbal Behavior

As was mentioned briefly in the earlier section on principles of helping, non-verbal behavior is a vital ingredient. An effective helper will be observant of all nonverbal signs, because that provides more information for the assessment of the current situation. Common channels of nonverbal communication are:

1. **Smell.** Has the person been drinking? Does the person smoke? Does the person have adequate personal hygiene? Is the person incontinent?
2. **Sight.** Observe the person's posture, gestures, facial expressions, clothing, and appearance.
3. **Sound.** This includes the tone, clarity, and pace of the words, the volume of speech, breathing, tears, laughter, and sounds that are not words.
4. **Proximity.** How does the person use space? Does he or she appear to want to be close to or far from you?

In addition, the helper should reflect back the nonverbal behaviors that seem to be relevant. For example, "You look and seem really sad to me today. Is that how you're feeling?" "You seem a little jumpy and nervous when we talk about your boss; am I right in thinking that he frightens you a little?"

It's important that we not assume anything about anyone's nonverbal behavior; we need to check it out instead. Many times personal or cultural factors create particular appearances and behavior, and it is not useful to interpret nonverbal behaviors purely from our own perspective. For example, Joe, a middle class white male, observed that Mim, an elderly Japanese woman, was not looking at his face or into his eyes during the interview. He thought that perhaps she had something to hide or was lying about something. When he discussed his suspicions with his supervisor, she explained that in many Asian cultures it is considered unacceptable for a woman to look directly into the face of a man who has authority and is not a member of the family.

Think About It #30

Unobtrusively observing other people's behavior can be enlightening. Find a location where you can sit and watch others without disturbing them, such as the zoo, a mall, an airport, or a hotel lobby. Do people's touching behaviors (i.e., parts of body that are touched and/or frequency) vary by their age? Their ethnicity or race? Their gender? Their disability?

The Helper's Nonverbal Communication. It is not only important that the helper attend to and reflect the nonverbal behaviors of those who have asked for services, but that the helper send nonverbal signals as well. The communication of empathy involves nonverbal as well as verbal responses from the helper. Several nonverbal or nonvocal responses can facilitate rapport-building and help to communicate the helper's attention and regard.

Eye contact. "Eye contact" simply means looking someone in the eye and having your eyes on the same level as his or hers. For example, sitting on the same level as a child or a person in a wheelchair communicates collaboration, acceptance, and respect. Looking in the face of persons with visual impairments allows the helper to recognize nonverbal facial cues. It is also crucial for a helper to look directly at people who have appearance differences, such as burns, cerebral palsy, or disfiguring characteristics.

Many beginning helpers make the mistake of thinking that the persons who have asked for help should maintain eye contact. It is the helper's responsibility to keep his or her eyes on the person's face and eyes; it is not the responsibility of the person seeking assistance. Appropriate eye contact does not mean staring. Rather, when the person is ready to glance up, he or she needs to find the interviewer focused and attentive.

Posture and proximity. An open and relaxed posture (rather than crossed legs and arms) is thought to be more inviting, as is leaning forward toward the person. Try to maintain a comfortable distance between people, without artificial barriers such as a desk. Sit in a way that communicates attention and respect, that is neither too close to nor too distant from the other person, and that mirrors the demonstrated comfort zone.

Facial expression.   A neutral and concerned expression is most useful for most situations. Of course, the appropriate expression of feelings like sadness or humor is part of a genuine interaction.

Voice and language.   The helper's voice should be soothing and modulated, not too loud and not too soft. You should not talk down to a person or be patronizing. Neither should you use jargon, obscenities, or slang.

Use of silences.   In the helping relationship, silence can be a very important and powerful form of communication. Silence is often not the absence of communication; frequently it is a specific type of communication that can be understood in the context of the relationship and the topic under discussion. Understanding and appreciating silence is a vital and meaningful part of active listening, effective interviewing, and understanding another person's experience.

Many of us have difficulty understanding the meaning of a silence during a helping interview, interpreting it as disinterest, boredom, hostility, or resistance. This is, of course, occasionally the case. It is valuable to learn, however, when silence is communicating other thoughts and feelings.

Many people in our culture feel uncomfortable with silence, feeling tension if words are not filling up what seems like empty space. Many helpers are afraid of silences in interviews and hurry to fill them. The prevailing perception is that talking is better than silence. Yet silence is actually an important component of communication and should be honored and utilized.

If the silence has followed something that you as the helper have said, then it is probably significant for the person and should not be interrupted. Of course, if the silence has occurred after the other person has spoken to you, then it may simply mean that it is your turn to talk! If you have no response at the time, then encourage the person to continue with "Go on," "I'm listening," or, "And?"

Pauses in the conversation can be a signal that the other person is:

- Thinking or feeling about something that has come up in the interview.
- Carefully contemplating a response to a question or reflection.
- Trying to make a decision.
- Struggling with how to express something.
- Uncomfortable with the topic.
- Organizing his or her thoughts.
- Trying to decide whether it is safe to share with you a particular thing that is sensitive, painful, or private.
- Thinking of the validity of what was just said.
- Needing time out.
- Trying not to cry.
- Disturbed by what was just said.

It is important for the helper to allow persons the quiet time and the space in which to think and feel. Interrupting someone else's meaningful silence is as rude as interrupting someone's words. A respectful helper will allow an individual the time and consideration that he or she needs. It is important that helpers learn to tolerate silence without experiencing extreme discomfort. Manage short periods of silence with calm and patience.

Pauses can be handled in two ways. One, the helper can reflect the silence in the same way that he or she would reflect anything in an interview, tentatively and with a calm and soothing voice. Examples of commenting on the silence might be:

- "You got quiet. What are you thinking right now?"
- "You seem really sad. Is this a sensitive topic for you?"
- "I'm wondering if your silence means that you aren't sure how to answer my question."
- "You seem to be finding it difficult to go on."
- "It's OK to cry if you feel like it."
- "Whenever you feel ready, I'd like to hear what's on your mind."
- "I'm interested in what you are thinking. Can you share it with me?"
- "It's fine if we just sit here and think for a while."
- "This is all very overwhelming. Can you try to put some of your feelings into words?"

Two, the worker can simply wait out the silence. Since most people are uncomfortable with silences, if the helper is able to sit quietly and wait, the person is likely to say something, rather than tolerate the quiet. This should not take the form of a game between the two interview partners, but silence can frequently be effectively and appropriately used to move the interview along. Be sure to pay attention to whether the person is becoming too uncomfortable with the silence, however, and check that out. For example: "I'm not sure whether you'd rather me give you some more information, or whether you're needing to think about what we've already talked about. What would be most helpful to you right now?"

Illustration

The following conversation illustrates one way to deal with silence.

Helper: You've told me a lot about how sad you've been recently about finding out that your child has autism. You've also told me that you have not told your parents or sisters about this news. Have you thought about whether you will tell them?

Sally (looking down, is silent for a minute): I've been thinking about it, but I don't know what I'm going to do.

Helper (observes that Sally is struggling to speak and has tears in her eyes): Take your time.

Sally (another minute of silence): Well, I don't really know how to break it to my mother (cries silently).

Helper says nothing, but hands her a tissue and waits quietly.

Sally (after a time of silence and crying): She was really looking forward to her first grandchild.

Helper: You think she'll be disappointed.

Sally (silent for another minute): Not exactly disappointed, but she will really feel this loss. (More silence.)

Helper: Sally, I don't want to intrude on your thoughts or on your feelings.
     Would you rather I left you alone right now?
Sally: No, I'd like you to stay. If you don't mind my not being real talkative,
     I'd like to have your company.
Helper: Of course. I'd be glad to stay. Please let me know, though, if there's
     anything that I could do that would be more helpful to you.

Touching. Physical contact is very significant and symbolic in our culture. It is not appropriate to move into a person's personal space without first clarifying whether that person wants to be touched. Touching someone who is in a helping relationship with us should never be intrusive, coercive, abusive, erotic, sexual, or uninvited. It is equally important to honor someone's refusal to touch you. In some religions a man would not want to shake the hand of a female helper, for example. Similarly, if someone is struggling with moving a wheelchair forward, always ask if you can help, instead of assuming that your assistance is wanted.

Being touched without warning or permission is very threatening. When touching is used appropriately, however, it can be validating and comforting. Always check out with the other person how he or she feels about your touching him or her. For example, after a conversation that has been very emotional, you may want to say "I'd like to hug you. Is that OK?" It is important that you allow the person to comfortably refuse this offer. There are many less intrusive gestures that you can make, such as briefly touching someone on the elbow, shoulder, or knee, which communicate concern. If you get the sense that your touch has been unwelcome or has made the other person uncomfortable, then take care not to do it again.

Other Tips for Active Listening

Focus on others. Keep the person to whom you are listening as the focus. Do not go off on a tangent about third parties and persons who are not in the room.

Illustration

Notice the difference in the following two responses. Who is the focus of each one?

Wrong: So your son is leaving his wife. How does your daughter-in-law feel
     about that?
Better: You're concerned about your son and his family because it looks as
     though there might be a divorce.

Remain neutral. Support the person, and strive to understand his or her perspective without taking sides.

Illustration

Notice the difference in the following two responses. One is choosing sides and one is more objective and neutral.

Wrong: Your grandmother must be a real bitch.

*Better: It seems as though you and your grandmother are having a lot of con-
flict between you right now, and you're feeling punished and criticized
by her.*

*Wrong: Obviously you're completely in the right here. Your lover had no
right to demand that of you.*

*Better: You seem to be really hurt by this argument with your lover, and to
feel that he was being unfair.*

Respond to defensiveness. When a person sounds or acts defensive (defending himself or herself against an attack), reflect and paraphrase it as you would anything else. Do not try to argue with the person about his or her feelings or perceptions.

Illustration

*Notice the difference in the following two responses. How do they differ in de-
fensiveness?*

Wrong: You're taking what I said all wrong. Don't be so touchy!

*Better: I sense that you're feeling offended by something I've said, and I'd like
to clear that up. I didn't mean to be hurtful, and I'm sorry. I'd like to
talk about it if that's OK with you.*

Respond to anger. When a person seems to be angry with you, it is not useful to become defensive, hurt, angry, or tearful in return. A person who is angry is a person who is expressing real needs and genuine feelings. There are two important concepts to remember when dealing with an angry person:

1. Do not allow yourself to be abused emotionally or physically. You have the right to set limits and to enforce them. Your safety is important. If someone seems to be on the verge of becoming violent, then leave and seek help. If a person is calling you names or shouting at you in anger, then calmly state your position that you will not be treated in that way. Say something like, "I can't talk with you if you shout at me." If the person continues to yell or threaten you, then terminate the interview, and state clearly that you will be glad to talk later when he or she is calmer. (See Chapter 6 for more information on saying no and on your own safety).

2. In the vast majority of angry situations that helpers encounter, the other person is not abusive or violent but simply angry. In that case, use active listening. You may reflect and paraphrase the anger in the same way as you would any other emotion.

Illustration

*The following examples demonstrate ways to reflect anger toward you as a
helper.*

*Helper: You seem to be ticked off at me because I haven't located a shelter
that will take you in tonight. I understand that you're scared and mad.*

I'm frustrated about this too, and I'd like to keep working with you on this problem.

Helper: You're really mad at me. I'd like to hear from you what that's about.

Stay in the here and now. Keep the interview focused on the present, the here and now. In most helping situations in which you will be involved, it is not useful to delve too deeply or for too long in the past. The current problem and feelings are the concern for you and the person who has asked for help. Gently bring the person back into the here and now.

Illustration

These statements from helpers attempt to bring the conversation into the present.

Helper: What happened recently to bring up these feelings of guilt?

Helper: What prompted your call to me today?

Helper: Something seems to have upset you about this quite recently. I'm wondering if you have a sense of what it was.

Be aware of the environment. Try to maximize privacy and quiet when you undertake an in-person active listening conversation. This is not always possible, of course, and you may find yourself speaking to someone about very personal and distressing matters in a hospital hallway or a bus station. However, be aware of how the environment plays an important part in the person's sense of safety and confidentiality and in your ability to listen well and to focus. If possible, minimize distractions such as television or radio playing; interruptions (have your door closed, and have your telephone calls held if you are in an office); or other persons in close proximity. You may want to go for a walk or talk on a porch if you cannot get privacy any other way. Give the other person as much control over this as possible. For example: "I'd like to speak with you in private. Where would you suggest that we go?"

Learn how to say "I don't know." It is not possible or necessary for a helper to have all of the answers at his or her fingertips. Typically a helper will find it necessary to search for resources or information for the persons. It is vital that you become comfortable with the words, "I don't know, but I'll find out," and that you follow through with finding out what you have promised.

Do not interrupt. Do not interrupt, and when you do so, apologize. If you must interrupt in order to redirect the interview, then say that you are sorry to have to interrupt, and explain the reason. For example, "I'm really sorry to have to stop you from finishing the story about your last doctor's visit, but I'm aware that we're running out of time today, and I needed to get some more information from you so I can finish this application for you to get the medication at a reduced cost. Do you mind if I shift focus?"

Do not talk too much. Be aware of how much you as the interviewer are talking. The person who has sought help should be doing the vast majority of the talking. If you find yourself talking frequently, ask yourself silently if it is necessary.

Be clear. Avoid jargon, euphemisms, slang, cliches, false reassurances, empty phrases, and words that are not readily understood outside of your own professional and personal culture. Communication should be clear and real. Label what is really going on without being harsh. Say "suicide" rather than "harm yourself." Say "die" rather than "pass." Say "cancer" rather than "your illness." This gives the person the message that you are not afraid of hearing and talking about the truth and that you are strong enough to bear his or her pain.

Do not fall into these communication traps. Avoid these roadblocks to helpful communication: ignoring, put-downs, threats, warnings, commanding, moralizing, persuading, criticizing, minimalizing, blaming, ridiculing, shaming, diagnosing, and humoring. In addition, do not use "should," "would," and "ought" statements. Examples of how *not* to respond are: "You should . . . ," "You ought to . . . ," "You ought to have . . . ," "You shouldn't have . . . ," "If I were you, I would . . . "

Summary of Active Listening

Effective helping incorporates genuine concern with needed techniques and skills. Care and concern must be demonstrated to be of use. Both sincere presence and effective responding are vital. An insincere listener who uses good techniques will be experienced as phony and unable to be trusted. Conversely, a well-meaning helper with no skills is just as ineffective. Your task as a helper is to practice using these active listening techniques so that you can incorporate them into your style and integrate them with your personality, so that you can use them in a genuine, natural way. Just as a highly trained dancer appears to move effortlessly, an effective listener is so comfortable with active listening techniques that he or she appears to be reacting spontaneously. You will know that you are using these techniques well when neither you nor the persons with whom you are working notice that you are using them.

CHAPTER SIX

Special Helping Skills

As a human services helper, you will benefit from the acquisition and practice of special helping skills such as: (1) helping others become more empowered; (2) supporting informed choice; (3) identifying strengths; (4) knowing how and when to say no; (5) meeting the challenge of difficult or reluctant persons; (6) paying attention to your own safety; and (7) clarifying expectations of the helping relationship. This chapter is designed to teach you these new helping skills, enhance skills you currently have, and reinforce for you that you have the ability to be an effective helper.

Empowerment

Not long ago, "helping" was seen as "doing for" another person or a group, without involving those who were most affected in the planning or decision making. This had the side effect of making those who had the least power in our society feel even less powerful. Those who were suffering from stigma or discrimination were further robbed of their rights to guide their own lives when outside helpers made decisions for them. For example, until the 1966 Freedom of Information Act, medical, psychological, and public school records were not available to parents or consumers. One can hardly make a good decision without being fully informed.

Fortunately, the philosophy of "empowerment" is more common now (Saleebey, 1996). The philosophy of empowerment states that we are all capable of making decisions for our own lives and futures. The empowerment approach views individuals, couples, families, groups, and communities as being essential participants in services and in problem solving, not just as passive recipients of care. Individuals, families, groups, or communities can become empowered; that is, at any level helpers can assist other persons with the process of increasing personal and political power to improve one's situation (Gutierrez, 1990).

Empowerment also implies a belief that people are competent, capable, and valuable. A helper who operates from a place of empowerment offers persons seeking help, as well as their loved ones, the right to be partners in the service care delivery system. In the empowerment model, helpers accept the description and definition of the problem as presented by the person seeking assistance. Effective helpers lend a hand so that someone in temporary need of support can identify and use his or her own strengths. Helpers also teach problem-solving skills and advocate on behalf of those with less power (Gutierrez, 1990).

Sometimes the social service and medical systems for which we work and volunteer are guilty of taking power away from people. Service recipients are often required to adhere to certain rules or practices in order to gain access to necessary financial, emotional, and social support; some of these requirements or restrictions may make you uncomfortable. This is part of the reality of service delivery (Dunst et al., 1988). Helpers sometimes become busy and overworked and therefore tend to reward passivity and dependency, rather than inviting the input and opinions of those who have asked for our assistance. To whatever extent possible, be aware of this possibility, and strive to help the consumers of systems and services have a voice in the design and delivery of those services. The empowering worker can help others exercise their rights and assume responsibilities of decision-making.

Examples of practices that *do not* promote empowerment include: unnecessary use of professional jargon; failure to describe the purpose of treatment or services; eagerness to do something for someone rather than explain how he or she can obtain it; and proceeding with a developed plan of action rather than involving the consumer in developing the most appropriate plan.

The importance of greater participation by consumers of health, social, and mental health services has increased over the past 20 years. More people are being cared for in their own homes and communities due to such factors as the community mental health movement and deinstitutionalization. More patients in health care facilities are demanding to be fully informed about their medical conditions. Furthermore, consumer participation is now mandated by some laws (for examples, parent involvement required in P.L. 99-457 and the involvement of persons with HIV on Title I planning councils funded by the Ryan White Comprehensive AIDS Resources Emergency Act).

People in the United States are generally taking a more active and responsible role in the care of themselves and their families. Responsibilities that were once the sole responsibility of professionals are now being taken on by the lay public. Helpers cannot be expected to make the best or most appropriate decisions. Most people are not willing to entrust their futures entirely to a doctor, a lawyer, a social worker, or a psychologist. Dependence on service providers is diminishing; therefore, one role of a helper is to support efforts toward empowering the consumers of services.

Illustration

Thomas Caldaron, a 33-year-old Cuban-American newspaper reporter, was involuntarily hospitalized in a state hospital for persons with severe mental illness. His behavior had become so erratic and unusual that he was picked up by the police while wandering in a park in a medium-sized western town. During his hospitalization, he was diagnosed as having paranoid schizophrenia and was prescribed three types of medication. Thomas refused to take this medication because he was aware of the potentially dangerous and permanent side effects. He demanded a hearing by a panel of three psychiatrists to determine the advisability of this medication. He was entitled to this hearing by law. The review panel determined that the medication was not advisable.

Helping that Leads to Empowerment

How can we help persons participate effectively in their own care or in the care of family members? Dunst et al. (1988) defines "effective helping" as:

> [The] act of enabling individuals or groups to become better able to solve problems, meet needs, or achieve aspirations by promoting acquisition of competencies that support and strengthen functioning in a way that permits a greater sense of individual or group control over its developmental course (p. 1).

Specifically, Dunst et al. (1988) identified helping behaviors that are associated with empowering and competence:

- Helping to clarify and prioritize the needs of the person asking for services.
- Offering rather than waiting for help to be requested.
- Offering help that is not demeaning.
- Offering help that matches the person's appraisal of his or her problem or need.
- Using partnerships for meeting needs and enhancing competence.
- Allowing decision making to rest entirely with the help seeker.

The importance of empowerment cannot be overemphasized. It puts each of us in charge of the course of our own lives with the help of others of our choosing. The following story is typical of experiences parents of children with special needs have.

Illustration

Donna and Barry looked forward to the birth of their second child, Cheryl. Their oldest child, 3-year-old Michael, was a happy and healthy child and brought them great joy and pride. After Cheryl's birth, the family settled into a busy but content routine. Barry taught high school math, and Donna cared for the children at home. When Cheryl was about 3 months old, Donna suspected that something was "not quite right" with Cheryl. She shared her concern with Barry, and they both took their questions to the pediatrician at Cheryl's next checkup. The pediatrician dismissed Donna's concerns as being those of an overprotective mother. He told her to relax and to stop worrying. By the time Cheryl was 6 months of age, both Donna and Barry were concerned. Cheryl was not even close to following the developmental stages that Mike had. She wasn't holding her head up or smiling, and she seemed listless and unresponsive. Once again, Donna and Barry shared these very specific concerns with the pediatrician. Once again, the physician reassured them that absolutely nothing was wrong. The pediatrician told them in an admonishing tone of voice that they were comparing their daughter to their precocious son and they should love Cheryl for who she was. Of course, Barry and Donna were relieved to hear that nothing was wrong with their beautiful little girl. They went home and wanted to believe that everything was just fine, but the nagging doubts continued, and they worried when Cheryl did not babble or roll over or sit up. When Cheryl was 9 months old, a friend of Barry and Donna who worked at a developmental clinic strongly recommended that they take Cheryl for a developmental evaluation. In 15 minutes, the diagnostic team suspected seizures and immediately referred Cheryl to a pediatric neurologist for further testing. Cheryl had a severe and serious seizure disorder that had gone undiagnosed and untreated for over 6 months.

Think About It #31

1. What are your reactions to the approach of the pediatrician?
2. What other options did the pediatrician have for responding to Barry and Donna?
3. What did Barry and Donna need?
4. What might have been done by the physician to help Barry and Donna feel more empowered?

Fostering Empowerment

The following worker activities can help a person find ways to exercise more control over his or her life:

1. Encourage others to acquire information and become as informed as possible about their concern; for example, maintain a file or have access to information that can be shared with others.
2. Help others translate professional jargon into everyday, understandable language; for example, keep a "dictionary" of terms and their translation on your desk.
3. Inform others of their legal rights to participate in meetings, have access to services and information, and acquire resources; for example, keep a file of laws and policies that can be passed on to others.
4. Encourage others to assertively express their opinions and concerns; for example, support efforts of persons with mental retardation to sit on committees or in their own Individual Education Plan (IEP) meetings.
5. Encourage active decision making based on the unique needs of the individual and family; for example, treatment for leukemia may differ based on individual or family values and circumstances.
6. Seek out professionals who encourage and promote consumer participation; for example, professionals who are supporting an empowerment philosophy

can be asked to participate in policy-making committee meetings and/or address consumer groups.

7. Continually send messages to individuals and families that they are capable, competent, and able to make sound decisions; for example, assure individuals and family members through praise and encouraging statements such as, "You are doing an excellent job of caring for your mom. Your decision to get a housekeeper to help you was just what your family needed."

8. Affirm the perceptions that others have about their experiences, hunches, and conclusions; for example, believe what others tell you. If a mother is talking to you about how distressed she is that a social worker told her to institutionalize her handicapped daughter immediately after birth, be outraged, and be willing to explore their options for addressing this behavior with the social worker and/or agency.

Think About It #32

1. What are some other ways that service providers promote dependence rather than encourage full participation?
2. What have your experiences been in this area?
3. Have you been encouraged to acquire information and participate in the decision-making process?

Supporting Informed Choice

A good helping relationship supports the person's autonomy and right to make choices. A good helping relationship does not foster dependency; it is not a way for the helper to take over someone else's decision making. A helper should therefore support a person's informed choice. This means that, after providing the person with the necessary information and listening to the person express his or her own views of the problem, the helper must support the person's right to choose a course of action if it does not harm another person. This concept is often called "self-determination," meaning that each adult must be allowed to make decisions for him- or herself.

Goals, plans, and programs should be developed based on what the service recipient desires for his or her life. This does not mean that helpers and administrators are passive or apathetic; rather, helpers should provide education on the issues and discuss a wide range of options. Helpers must assist persons obtain what they need and want, not what the helper desires for them.

Sometimes supporting a person's informed choice is very difficult. When a person whom you care about decides not to continue cancer treatment, for example, it may hurt you very much. When a person continues to use alcohol or drugs, continues to practice unsafe sex, or stays in an abusive relationship, you may feel scared and disappointed. At these times, make sure that you talk with your supervisor about your feelings, and make sure that you are able to remove your judgments from the helping process. Recognizing how another person's choices and behaviors affect you is critical, so that the helping process is not derailed by your own feelings and opinions.

Supporting another's right to choose does not mean that you cannot express your own views in a nonjudgmental, nonaggressive way. It is sometimes useful to state clearly and nonjudgmentally how you view a situation. This is safest to do when the helping relationship is a strong and long-lasting one. For example, a helper may say, "I support your decision to discontinue your chemotherapy, and I understand how important it is to you to live your last few months comfortably. It's hard for me, though, because I care about you, and it's difficult for me to think of your dying." You may say something like, "It's hard for me to see you drinking so much, because I care about your health and safety. How do you feel about my telling you this?" When your views are expressed in a concerned, loving way, service recipients are most likely to be able to accept them without being unduly influenced by them and will be able to better understand how their behavior may be affecting the people around them.

Finding Strengths

Using a strengths perspective is a way of looking at the world that points out more readily the inner and outer resources that individuals, couples, families, and communities possess. We are all accustomed to having people notice and comment on our deficits, failings, and weaknesses. It is more rare for most persons, unfortunately, to experience being in a relationship with someone who seeks out and validates strengths, coping mechanisms, successes, and positive characteristics. This is like the old adage about the water glass being half empty or half full; it is both, of course, and people choose to focus either on the empty part or the full part rather than seeing the whole picture.

An effective helper will take the perspective of catching persons doing something right, rather than catching them doing something wrong, and then commenting on that strength or success. You will find, not only as a helper but as a person, that when you begin to focus on the strong points of persons and systems and to comment on them, your perspective may change as well.

The strengths perspective is not the same as being falsely reassuring or an unrealistically optimistic Pollyanna about problems and people. A Pollyanna approach is not at all useful to persons who need help. Taking the strengths perspective is simply another way of viewing reality, a way of putting on another set of lenses that allows you to identify the functional characteristics and components in someone's life (Saleebey, 1996).

Illustration

Allen is a junior high school student who has difficulty with reading and writing and tends to require more time than his classmates do to produce assignments and papers. His teachers have found him to be bright and motivated and have usually allowed him to turn in assignments late when he shows them that he has been working on them steadily. He attends the after-school program where Millie works, and she finds him one day with his head in his hands, very discouraged about his progress in school. As Millie listens to him talk, she realizes that Allen is worrying about the struggles he has with writing and on the amount of time it takes him to research a topic and write a paper. However, he does not seem to readily realize his own courage, motivation, and integrity in sticking to his educational goals. Millie gently says, "Allen, I can hear how frustrated you are with your pace, and I realize that you sometimes get tired. But I'm thinking that you don't see what others see when they look at you. What I see is a determined, energetic, bright young man who always finds a way to open a window when he runs into a closed door. You are really a remarkable person."

Think About It #33

1. Do you focus on strengths as much as you could?

2. How can you improve your ability and willingness to identify strengths and to communicate them to those seeking help?

Assertiveness and Setting Limits

Being assertive, which includes being able to say no, increases a helper's ability to care for himself or herself. Assertive behavior does two very important things: expresses feelings in an honest and direct way while being sensitive to the needs of others. Assertiveness is the ability to make known your thoughts, needs, and emotions with self-confidence, skill, and caring.

Many people confuse "assertiveness" with "aggression." These methods of relating to people are far from identical. People who are exhibiting aggressive behavior are not being sensitive to the feelings or the rights of others and are most likely to be focusing primarily on themselves. Passive behavior is not assertiveness either. Persons who are behaving passively are ignoring their own rights and feelings, not taking responsibility for themselves. Assertiveness is a method of communicating that takes into account the feelings and the rights of everyone involved, including your own.

TABLE 6-1 Continuum of Passive-Aggressive Behavior

Passive	Assertive	Aggressive
Others are more important than I am. What I have to say and what I feel are not valuable.	The needs and feelings of others and my own needs and feelings are equally important.	I am more important than others are. What I want and feel should be the priority.

Illustration

The following example from a family may serve to point out the different types of responses to a request. A retired couple (Sam and Jill) are preparing for and looking forward to an out-of-town vacation together when they receive a frantic telephone call from their son and daughter-in-law. Their son asks Sam and Jill to babysit with their three small children for the weekend so that the younger couple can go on a last-minute ski trip. The son and daughter-in-law are very excited about the chance to go on this trip with friends and assume that Sam and Jill will agree to their request. The son says, "I know you have plans, but since your schedules are so flexible, I know you won't mind rescheduling to stay with the kids."

The Aggressive Response: *Sam angrily shouts, "You've got a lot of nerve! You are so inconsiderate! I can't believe the way you use your mother and me. You never even thanked us for all the other times we've helped you out. We're not your servants, you know. Forget it!" Sam hangs up, angry and hurt, not thinking about the effect that his words have had on his son.*

The Passive Response: *Sam says, "Sure, no problem. It's OK, I guess. Bring them over, and I'll tell Jill." Sam hangs up and faces his disappointed wife. They complain to each other about their son's insensitivity and selfishness, but when their son arrives with the grandchildren, they act as if everything is fine.*

The Assertive Response: *Sam says "Son, we love the kids, and we enjoy being with them whenever we can. We also like to help you out whenever we can. But now is a bad time. Your mother and I have planned this trip, and we would be disappointed if we had to cancel it. I realize that you were counting on us, and I'm sorry about that. Next time, if we don't have plans, we'll be happy to babysit those terrific kids."*

There are specific situations for most people in which it is difficult to be assertive, such as: saying no to a request, setting limits, expressing strong feelings, or defending oneself. Some people have difficulty saying no to requests because they are accustomed to responding to the needs of others and denying their own. Sometimes, because we want others to think well of us, we respond affirmatively to all requests. In addition, requests are usually made sincerely and because a real need exists, and we do not feel able to turn down a request in these situations.

How can you protect yourself from being a helper 24 hours a day, seven days a week, which would quickly wear out anyone? Your task as a helper is to assist someone else in a similar situation with their ability to problem solve, not to actually solve all their problems. It is important that we learn appropriate communication techniques, not only to protect ourselves, but to model to others that one can be responsible for one's own needs without hurting others.

Steps in Deciding Whether to Say No

1. Determine if saying no is appropriate on a case-by-case basis, based on your own needs, the guidelines of the organization, and the situation of the person requesting assistance.

2. If you are uncertain about whether or not to say no to a person's request, then ask for time to think about a request before responding. This is a good time to discuss the request with someone else, if you wish assistance in your own problem solving. Always get back to the person with your answer as soon as possible.

3. After you have considered the situation and perhaps asked for assistance from your supervisor, if at this point you wish to say yes, go ahead. If you have decided that you cannot grant the request at this time, say so.

Refusing a Request

If you are faced with a situation as a helper where you must refuse a request, the following guidelines may ease your way:

• Language and tone are important. Begin your statements with "I" rather than "you" to express your concerns. Instead of beginning with, "you never," or, "you didn't," say, "I can't do that right now," or, "I need to say no to that right now." When you are saying no to a request, use a firm but gentle voice and manner; do not scream or whine.

• If you say no and you wish to explain why you have refused, do so. It is probably best not to get into lengthy explanations. For example, "I'm going to have to say no to going with you to the funeral, although I didn't want to let you down, because I really care about you. But I just feel too raw still about my own friend's death, and I need not to go to another funeral right now."

• Be aware of and sensitive to the feelings of others, as well as your own. Never assume that you know the other person's feelings, situation, or motivation. Ask about their perceptions in a warm, interested way. For example: "I really need to take care of some personal business this afternoon, and I'm sorry that I can't go with you. I hope that you're not disappointed in me. What are you thinking about this?"

• Try not to sound blaming or angry at the other person for making the request. We all have the right to make requests, just as we have the right to refuse requests. When you are assertive you do not accuse or blame the other person for making the request; rather, you communicate with that person about that request in a manner that is honest and respectful. For example, it would be damaging to your relationship with the person seeking help if you responded angrily with, "I can't believe you would ask that of me. Can't you see how busy I am? Can't you see how many people I have to take care of?" It is more productive and more respectful to explain, "I wish I could help you with that today, but I'm booked solid."

• You might also offer an alternative solution to the request. Some people will not be interested in your helping them generate other possible solutions; if not, then let it go. But if the person is open to your suggestions, give it some thought, and present other options. One way to do that is by offering yourself in another way or another time. For instance, if someone calls at supper time needing to problem solve, you might say, "I can't talk right now; may I call you back at 9:00 after the children are in bed?" Another example might be, "I won't be able to visit you in the hospital today, but I would really like to come see you tomorrow night. How would that be?" At other times, suggesting that the person contact someone else, either in your organization or another organization, may be more appropriate than your fulfilling the request yourself. It is even more helpful for you to offer to make the connection, if that is feasible. Examples are:

> My agency does not have a program where someone can accompany you to court, so I won't be able to do that. But the Rape Crisis Center has a good court advocacy program for people in your situation, and I will help you get in touch with them if you wish."
>
> "I can't take you to the doctor today, but perhaps one of the other volunteers can. May I call one of them for you?"

Illustration

Alice, who was grieving the death of her brother and anticipating the end of her marriage, felt overwhelmed with grief. She asked her friend Tammy, whose sister had also recently died, to attend a grief workshop with her as her grief partner. At the workshop, partners were to assist the individual identified as the "griever" in exercises to work through their many emotions and thoughts. Tammy very much wanted to help her friend Alice, but was just beginning to stabilize her own life following her sister's death. She said no to the exercise in order to protect herself from feeling her grief so intensely in a public place.

Think About It #34

1. How do you feel about Tammy's response to Alice?
2. What would you have done in a similar situation?
3. Under what circumstances would you say no to a request by the person you are supporting? What words would you use?

Working with Difficult Situations

Not every person who crosses our paths will want our help or be easy to work with. One of the difficulties that helpers may face is a person seeking help who appears to be hesitant or reluctant to ask for help from strangers or to accept the services or suggestions offered.

When service recipients do not behave as we expect them to, there are a variety of labels that are used, such as angry, hostile, in denial, stupid, lazy, resistant, difficult, recalcitrant, aggressive, noncompliant, uncooperative, involuntary, demanding, and difficult. Think about the ways in which people are sometimes labeled and sometimes even dismissed, and remember that there may be more to the situation than meets the eye initially. It is probably more useful to examine what is behind their behavior, rather than getting frustrated.

Think About It #35

Think about all the teachers you have had in your life—in elementary, middle, junior high, or high school. Think of the one teacher whom you consider to have been the worst teacher you ever had. Recall the details of being with that teacher in that class.

1. How did you behave in this class and with this teacher? Did you get angry? Act like a clown? Challenge the teacher? Be disruptive? Perform poorly on assignments or exams? Skip class? Drop out?
2. Imagine that the teacher or another member of the school staff referred to your behavior as "resistant," "noncompliant," "lazy," or "difficult." How would you feel if you overheard these words?
3. How would you explain the meaning of your behavior in this class or around this teacher?

What does the word "difficult" actually mean in helping situations? Are there difficult service applicants and recipients? Difficult helpers? Difficult situations? Could all three situations occur at once? Could the difficulty be an interaction between the three?

People, as well as problems, issues, and situations can be difficult. However, given the fact that helping happens in the context of a mutual relationship, difficulties in that relationship cannot always be the fault of the person who asked for help. Examining our own roles in the conflict or resistance can be fruitful. Think about these questions: Have I failed to hear the person's concerns and worries? Have I failed to build rapport and safety well enough? Have I been ignoring or invalidating what a person has been trying to say? Have I not explained adequately to the person a particular process or course of action?

In addition, many circumstances may complicate matters. What characteristics and behaviors might make a helping situation more difficult? Examples are:

• The person is using alcohol or drugs.
• The person is very angry.
• The person won't do what you ask.
• The person continues behavior that we consider dangerous or destructive.
• The person seems to not recognize the extent of the problem.
• The person is reluctant to tell other persons about the situation.
• The person is having sex or having babies when we think that is unwise.
• The person does not want to take medications or seek medical treatment.
• The person might be mentally ill.

Think About It #36

List other difficult situations, behaviors, or characteristics in addition to the ones listed above that you can imagine or have experienced.

People seeking help naturally want to protect and defend themselves from pain, shame, and blame. The helper's job is to create a climate where the persons' sense of safety grows so that he or she can gradually be more vulnerable and open to change. Trying to change or give up a pattern of behavior or a perception that has felt comfortable and normal is scary. Human beings normally want to maintain their autonomy and dignity and therefore may be unhappy about having to accept help from a stranger.

Things may not go well with a helping relationship for many reasons. Using empathy to imagine what might be going on with someone who appears to be reluctant to accept our help is important. Why might a person be reluctant to approach a stranger or a group of strangers for help? Possible reasons from the perspective of the person seeking help are:

- I believe that needing help is a sign of weakness.
- It's difficult to admit that something is wrong with me or my family.
- I'm afraid of being judged or ridiculed.
- I don't want to face your curiosity about my behavior or circumstances.
- I want to solve my own problems; I prefer to make my own decisions; I value independence.
- Facing change is scary.
- I don't know if I can trust you to keep my private information private.
- I'm scared to give you power over my life.
- I do not want help; I resent needing help.
- I feel afraid (of illness, change, death, uncertainty, or loss of privacy, freedom, and dignity).
- I do not feel listened to or validated by you.
- I do not want to openly share my private feelings or vulnerabilities with a stranger.
- I'm feeling confused, ashamed, helpless, out of control, anxious, inadequate, or embarrassed.
- I need to reject you before you can reject me.
- I feel like a failure.
- I feel silly.
- I do not trust you or your agency.
- I'm sad, hopeless, or depressed.
- I'm feeling dehumanized by medical and social systems.
- I'm tired, frustrated, and fed up.

Think About It #37

1. How many of the above situations seem realistic to you?
2. List other reasons that someone might be reluctant to request and accept help.
3. Think about a time when you were forced to ask for assistance from someone when you were in crisis. What was it like for you to ask for help? Have you ever felt any of the above as you decided whether to ask for help?

Checking It Out

It is often useful for a helper to try to figure out what is going on when a person appears to be hesitant or difficult. The following are guidelines for the assessment and clarification of a person's hostility:

• Try to determine whether this behavior is part of the current situation or is more likely to be part of the person's personality.
• Explore what the hostility is about. Examples are:
• You seem real upset with me; what's wrong?
• I'm feeling really helpless here; I don't know what to do. Can you help me out?
• It seems like whatever I do or say, you get angry. I'm not trying to frustrate you, and I'm getting frustrated too. Can you help me to figure out what to do next?
• I really care about what happens to you, but I have a hard time with what seems to be hostility toward me or my agency. How can we work together better?

Helpers' Reactions

Think not only in terms of why other people may not change easily, but also why we workers may get stuck. Sometimes helpers are one of the most resistant groups to be found; sometimes using the job duties and titles, good intentions, experience, or degrees as justification for demanding compliance from persons seeking help. It is difficult to let oneself feel vulnerable, to admit that it is not possible to have all the answers, and to ultimately let people choose for themselves how they will live. In difficult situations, the goals as helpers should be:

1. To determine and assess possible sources of hesitancy in service applicants and recipients.
2. To be able to recognize our own responses to others' reluctance.
3. To assess our part in what may go wrong in a helping relationship.
4. To be able to be more honest with ourselves about what makes us hesitant to change as helpers.
5. To develop greater empathy and understanding about the views, perspectives, and experiences of others.

What are possible reactions to service recipients who seem to resist help? Helpers might become hurt, sad, angry, defensive, frustrated, or bewildered. Helpers might feel devalued, thinking that service recipients are not showing proper gratitude. Helpers might pout, sabotage services, avoid or blame the person being helped, or delay returning telephone calls or making home visits; at worst, helpers might refer them inappropriately to another resource, refuse care, or breach confidentiality by airing frustrations to persons other than supervisors. When helpers are feeling frustrated with someone who is depending on them for a professional helping relationship, helpers need to be very careful to behave appropriately.

Principles for Considering Difficult Situations

Here are some reminders of principles that have been covered elsewhere but that can help to put so-called "resistance" into perspective.

1. **Self-determination, support of choice.** Each adult has the right to make decisions about his or her life. Helpers should therefore strive to discover what persons seeking help really need, how they view their situations, and what options might fit for them.
2. **The dynamics of the helping relationship.** The worker is as much a part of what happens as is the person seeking help; communication or miscommunication takes two.
3. **The need to clarify biases.** Workers need to assess their own responses and reactions and deal with them in constructive ways, perhaps through therapy or supervision.
4. **A strengths perspective.** Helpers must acknowledge that often a person seeking help is doing the best that he or she can at the time, that he or she does have something to offer, and that his or her survival skills are strong and valuable. Remember that it is often the fighters, the persons who try to maintain some control, who tend to maintain better physical and mental health.
5. **Access to services and social justice.** Not just pleasant or blameless people deserve help, but anyone who has a need for our services. It is therefore unethical for helpers to refuse or delay services just because of difficulties in a helping relationship, or because they find a person to be reluctant or hostile.

Given the above principles of helping behavior, it is not appropriate for helpers to use the labels of "difficult" or "resistant" to avoid working on a helping relationship. Rather, the helper can strive to develop empathy for another's feelings and behavior and to offer services in a fair and unbiased manner.

Here are three additional ideas to keep in mind as you encounter difficult situations or with persons who are difficult for you to work with.

1. You should not let go or give up on someone until you have done your best.
2. After you have done your best and it still has not worked out, then realize that you cannot reach everyone all the time, you can't fix everything, and you can learn when to let go.

3. If someone is truly unable to cooperate, not ready for you, then accept that and withdraw, and explain your withdrawal to that person in a nonjudgmental way. Do not take action of this sort without the knowledge and backing of your supervisor. This may be a good time to explore making a referral, offering the person an alternative to working with you or your agency.

Paying Attention to Your Physical Safety

Not every person or situation that you will encounter will necessarily be safe for you. It is important for you and your supervisors to pay attention to safety issues in service delivery, to assess danger, and to have plans for making helpers more safe. The following guidelines may be useful:

1. If someone is consistently or repeatedly or intensely abusive, inappropriate, drunk, high, threatening, or dangerous, set limits and stick to them. Call the police if necessary. Inform your supervisor and agency as soon as possible if these measures are necessary.
2. Have safety plans in place at your agency. You may want to have predetermined codes, safety alarms, and plans of action if someone feels or is threatened.
3. Be careful with home visits and work in the field; leave the telephone number and address with your office, don't go at night, and leave if you are feeling unsafe. Some agencies are providing cellular telephones for field staff for safety reasons.
4. When doing work outside the office, especially if service applicants or neighborhoods are unknown to you or your agency, meet in safe and neutral places (such as shelters, police stations, clinics, other agencies).

Clarifying Expectations

It is important to determine what the person's expectations are of you and your organization. Did he or she misunderstand the eligibility requirements for your agency's programs? Did he or she have the wrong information about what the agency does? Did he or she come to your organization expecting that you would provide housing although you cannot? Do some of the service recipients expect that counseling will make them feel better and find that it makes them feel worse? Did the person expect that you would fix or prescribe the problem away?

The process of clarifying expectations is applicable to persons asking for assistance and to providers. When confusion is avoided about what can be realistically expected from both parties, then the helping relationship will develop more smoothly.

Helpers should explain as soon as possible what the agency can and cannot do, what the eligibility requirements are, and what the application process might entail. Otherwise, the longer service applicants or recipients remain in the dark, the worse the disappointment will be when it finally becomes clear to them that they have been wasting their time in the wrong place or doing the wrong thing. There are simple assessment questions that you may ask, such as "What have you heard about us?" "What were you thinking would happen once you came here?" or, "What have people said to you about this program?"

Just as specific service guidelines must be clarified, it is also useful to consider what can be expected in general. Think about your own place in this. What are your expectations as a helper? Do you expect all persons to be pleasant? Do you expect gratitude from every service applicant or recipient?

What can a helper reasonably expect from a relationship with a person seeking help? You have the right to expect respect and appropriate behavior. This means, as was discussed previously, that you must be prepared to set limits on behavior or to say no if necessary without being mistreated or abused. However, helpers do *not* have the right to expect gratitude from service recipients. Helpers also cannot realistically expect to like all service recipients or to be liked by all persons seeking help.

What can service recipients reasonably expect from helpers? They should be guaranteed confidentiality, fairness, respect, active listening, a nonjudgmental stance, equal access to services, advocacy for rights and access, constructive and mutual problem solving, the separation of your biases and beliefs from your practice, the separation of the person's behavior from his or her inherent worth as a human being, your partnership in his or her care plan, and professional expertise, conduct, ethics, and demeanor. However, a person seeking help does not have the right to expect you to take abuse or to break the rules.

CHAPTER SEVEN

Crisis Intervention

Often human services workers, students, and volunteers are called upon to respond to crises (including suicide threats). This chapter introduces you to skills necessary to respond to someone in crisis or who is suicidal.

What Is a Crisis?

A crisis is defined as a brief transition period during which the person in crisis has the potential for heightened growth or deterioration. "Crisis" means a turning point, a fork in the road, that is both a danger and an opportunity. Crises are triggered by specific, identifiable incidents; these can be single traumatic events or an accumulation of events. The way a person perceives a situation or event will partially determine whether he or she experiences it as a crisis. For example, some events are so traumatic (such as sexual assault, witnessing a violent death, or being kidnapped) that the vast majority of people would have a crisis response. Other events or experiences may or may not cause a crisis response, depending on the interpretation of the situation by the individual. Examples of these events include divorce, moving to a new location, or an automobile accident.

Entering a state of crisis is a normal, anticipated response to a real or imagined trauma. Crises are considered to be temporary. The acute crisis state with high emotional turmoil usually lasts a month or two, and then the person tends to level out as the situation gets resolved (Caplan, 1964; Golan, 1978; Puryear, 1979; Rapoport, 1970; Roberts, 1990).

Persons who are experiencing a crisis usually try to ease their discomfort and solve their problems through the methods they are accustomed to using. Often, these customary coping mechanisms do not work, so anxiety grows.

Persons in the middle of a crisis are more open to help, because the emotional pain is so great that they are more willing to turn to someone else. In this sense, offering intervention during a crisis is highly useful. People tend to react and change quickly during a traumatic time in their lives. Old ways of thinking, feeling, and responding have been shaken up, and the person is searching for new ways to act. The disequilibrium created by the crisis often allows two people to attach to each other and rapidly develop a close helping relationship. Helpers will probably find themselves in the midst of personal information and intense feelings very quickly when someone comes to them in a crisis.

Many times a person who is experiencing crisis feels confused and disorganized. This disorganization may affect the individual's functioning at work, at home, and in social activities and relationships. Sometimes crises create physical symptoms, such as headaches, stomachaches, backaches, tiredness, nausea, or diarrhea. People experiencing crisis also sometimes feel inadequate, anxious, panicky, tearful, irritable, or helpless. Agitation and trembling may be apparent. Changes in eating and sleeping patterns are also common.

As the acute crisis phase passes, adaptive or maladaptive responses become part of the person's behavioral repertoire. Some people emerge from a crisis at a lower level of functioning than before; others become more capable of coping. Good crisis intervention can facilitate the development of better coping skills and deeper personal insight.

Crises are classified as developmental, situational, social, or compound. Developmental crises involve the normal tasks of maturity and life phases (such as graduation, marriage, getting a first job, or becoming a grandparent). Situational crises are caused by specific traumatic events (such as losing a job, experiencing the death of a friend, getting divorced, or being the victim of an assault or a natural disaster). Social crises involve a person's conflict with his or her society or culture (such as experiencing discrimination, stigma, or injustice). Compound crises are caused by a current trauma that is intensified by memories of previous issues and losses that have been reactivated by the present difficulty (such as the experience of receiving a diagnosis of a terminal illness causing someone to finally try to come to terms with incest in his or her background).

Because many persons contact helping organizations while they are in a crisis state, it is important to understand the nature of this distress and the principles of crisis intervention.

What Is Crisis Intervention?

While persons are in crisis and emotional turbulence is high, they benefit from immediate, intense, and frequent contact with supportive helpers who validate their feelings and help them problem solve. Crisis intervention as a model is *not* the same as longer-term services, such as case management or counseling, where weekly, monthly, or quarterly contact is sufficient. The first principle of crisis intervention is to make supportive systems (such as volunteers, support groups, social workers, crisis counselors, therapists, or hotlines) available immediately and often.

What Should the Crisis Intervention Helper Do?

- **Listen.** Active listening techniques (see Chapter 4) are very important throughout crisis intervention, most especially the identification, labeling, and validation of emotions.
- **Assess the intensity of the crisis.** Involve the person seeking help in discussions about how stable he or she feels, and decide together whether the situation calls for a higher level of intervention. Refer the troubled person to a professional counselor, social worker, psychologist, psychiatrist, or mental health worker if the crisis is intense and feels unmanageable.
- **Identify strengths.** Throughout the steps of crisis intervention, your demeanor should communicate to the person seeking help that you believe him or her to be a capable, functioning person who is in a temporary difficult situation and who is naturally overwhelmed with extreme stresses.
- **Normalize.** "Normalizing" means letting the person seeking help know that his or her crisis responses are to be expected and are not a sign of pathology or mental illness.
- **Be directive.** Helpers often must be quite directive in the beginning of the crisis intervention relationship, simply because the person in crisis feels immobilized and out of control. However, as the person regains some stability and confidence, the worker should withdraw, encouraging the person seeking help to do more of the problem solving him- or herself.
- **Problem solve.** Teaching problem solving is also vital in crisis intervention. Teaching someone to divide overwhelming problems into manageable tasks is a valuable intervention during a crisis. (See Chapter 8 for more information on problem-solving techniques).
- **Decide whether it is appropriate for you to help.** Sometimes crises arise that are unrelated to the original reason the helper and the other person met each other. The helper must evaluate whether it is appropriate to address a crisis that is unrelated to the initial problem; at this time, consulting supervisors and agency guidelines can be helpful. Remember, if you decide that you cannot help a person with a particular crisis, appropriate and timely referrals are vital.
- **Follow up.** Regardless of whether you and your agency provide the crisis intervention services or refer the person seeking help to a therapist or another organization, try to contact the person frequently over the next few weeks to see how things are. If you know who the person in crisis is and how to find him or her again, offer follow-up support where appropriate.

Illustration

Catherine, a woman in her fifties, participates in a support group for people whose loved ones have cancer. Her son has been diagnosed with leukemia. Each time he sees the doctor, gets a cold, or feels weak, Catherine experiences a jolt and looks to others in the support group for modeling about how they have survived similar situations. She is also present for consoling and hugging when another group member has a rough time. Catherine feels especially close to Pat, the human services helper who facilitates the group. Although Catherine joined the group to receive support for dealing with cancer, life's ups and downs unre-

lated to her son's illness also continue unabated. Her brother, who is a self-employed taxi driver with no health insurance, needs a hearing aid. Where can he find financial assistance to obtain an aid? Her father-in-law, who has become an alcoholic since his wife's death five years ago, is dying from alcohol-related illness and has been through local treatment centers so many times that they refuse to accept him any more. Catherine is at a loss about how to help him. She feels responsible and worried. She turns to Pat for a listening ear when these family crises seem to be too overwhelming and too difficult for Catherine to carry alone.

Think About It #38

1. What feelings do you have when thinking about Catherine's situation? Her son's? Her brother's? Her father-in-law's?
2. If you were Pat, what would your first response be to Catherine when she talked about her "troubles?"

Dealing with Suicide

Persons who are experiencing severe trauma or depression may question the meaning of life, as a part of their making sense out of what has happened. This can be a very intimate questioning, including consideration of why they are alive and whether they want to be. Suicide is a topic that raises intense emotional and philosophical issues for everyone. Perhaps no other problem frightens helpers as much.

It is vital that your personal beliefs about suicide are separated from your responsibility as a volunteer or paid helper. Clarify for yourself your own values, fears, and beliefs about suicide. As in all helping work, you must learn to recognize the buttons that get pushed for you, so that you can work more effectively with a person seeking help.

It is also very important that you involve a supervisor immediately if you suspect that the person is considering suicide. Even if you have been well trained as a crisis intervention helper, do *not* attempt to make all the decisions about a suicide situation yourself.

What Is the Helper's Stance on Suicide?

Many people feel strongly that each individual has the right to decide for himself or herself whether to live or die and that suicide is an individual choice, an alternative that all possess as human beings. Others feel that it is not an acceptable option no matter what the situation. Regardless of how you view this topic as a person, you have the ethical responsibility *as a helper* to accurately assess each potential suicide situation and to intervene if at all possible. If you are in the role of helper, part of your responsibility is to save lives if you can.

You must be very clear about the policies of your organization or agency about suicide intervention. If you adhere to these rules and act in good faith within the scope of your training, you are much less likely to have legal or ethical problems.

The Nature of Suicide

Remember that everyone has problems at some time or another. The suicidal person sees suicide as a way of coping with, or an answer to, his or her problems, because at this particular moment there is no other option that looks feasible. The suicidal person is not sick or bad but in despair. Some persons who make suicidal attempts have mental illnesses, but this is not true in the majority of cases. For the most part, suicidal thoughts and feelings are experienced by persons who are normally well functioning and stable but who are in an enormous crisis where they feel completely out of control.

A person who is feeling intensely suicidal often sees no alternatives, no real hope for the future, and no other way to escape the tremendous and unbearable emotional or physical pain that he or she is experiencing. These feelings can be triggered by any situation or feeling that is currently overwhelming and intense: such as the loss of a relationship, the death of a loved one, shame, or the pain of an illness. Persons who have attempted suicide previously, who have a history of suicide in their family, who use drugs or alcohol, or who are generally impulsive or aggressive are at greater risk of making a suicide attempt as well.

Some helpers and agencies make a distinction between persons who are not immediately terminally ill and persons who are. There are passive techniques that a person or family may choose to use when someone's death is imminent, such as not aggressively treating an infection or cancer or choosing to end a treatment.

Signals of Suicidal Ideas

All threats of suicide must be taken seriously. As will be discussed in greater detail later, factors that enhance the chance that the threat is serious include having a plan of action, lethality of the planned method, the lowered probability of rescue, and the frequency and duration of suicidal thoughts. Other possible warning signs include (Maltsberger, 1986; Victoroff, 1983):

- Sadness, depression, or nervousness.
- An unkempt appearance that is a change from the person's usual appearance.
- Weeping for no apparent reason.
- Expressing a pessimistic attitude; feeling hopeless, particularly about the future.
- A loss of interest in one's usual activities.
- Feeling bad about one's self-image.

- Stating that he or she would be better off dead or that the world would be better off without him or her.
- Giving away prized possessions.
- Being unable to reason, argue, tell or laugh at a joke.
- Feeling unduly victimized.

Certain personality characteristics such as impulsiveness, risk-taking, and a tendency toward aggression correlate with an increased risk of suicide (Goodwin and Brown, 1990). People who are at increased risk for suicide are those who: (1) have experienced recent, important losses (for example, the death of a loved one, the loss of a job, or deterioration in health); (2) have made previous suicide attempts; (3) have a history of suicide in their families; and/or (4) are using drugs or alcohol.

If you suspect that a person is thinking of suicide, consider the risk factors. However, do not put so much weight on risk factors that you do not listen well to what the person is telling you; sometimes persons who are in despair do not fit the above profile.

Suicide Intervention

Two important rules about suicide prevention to remember are: (1) You are required to do your best to save someone's life, and (2) you are not responsible for another person's decisions or actions. This means that if you have assessed the situation and intervened appropriately, and the person still makes a suicide attempt or dies, then you must strive to accept this, even though it is likely to be painful and sad for you.

The short-term goal of suicide intervention is to get the person through the acute crisis stage, connect them with support and therapy, and restore a measure of hope and control. The most important goal for a helper is to determine whether someone who is feeling suicidal is in danger right now. To do that, helpers should go through a process called "assessing lethality." Lethality means the likelihood of death, the possibility that a life is in danger soon.

Assessing Suicide Lethality

The following guidelines for assessing lethality apply to any situation in which you think that a person may be contemplating suicide.

Step 1: Check it out. If you hear a hint of suicidal despair, then check it out with the person. Ask, "Are you thinking about suicide?" If you are wrong, then the person will tell you so. You will not cause someone to commit suicide just by raising the issue. Do not be afraid to confront someone on this if you genuinely get a feeling that it would be appropriate to do so.

Here is an example of a telephone confrontation about someone's suicidal ideation:

Jay: It's no use. There's no hope.
Helper: You're sounding really depressed. How do you see your life right now?
Jay: I'd be better off dead.
Helper: You're thinking of suicide?

The following two examples illustrate how a helper may proceed.

Example: I have heard you say twice in this conversation that you don't want to go on like this. So I have to ask if you have been thinking about killing yourself.

Scenario one:
Person seeking help: "Well, the thought does cross my mind sometimes, but I don't really want to die. I just want the craziness to stop."
Helper: "So you don't have a plan for taking your life?"
Person seeking help (crying): "No. I really don't. But I'm so tired of suffering."
Helper: "Well, I needed to know what you were planning. I was hearing so much emotional pain from you. I appreciate your talking to me about this. Will you promise me that you'll tell me if you ever do find yourself making a plan?"
Person seeking help: "Yeah. OK"
Helper: "Will you talk with me some more now about how tired and sad you are?"

Scenario two:
Person seeking help: "Well, this is hard for me to say, but yes. I think about it a lot."
Helper: "How did you think that you might kill yourself?"
Person seeking help: "I was thinking about pills."
Helper: "What kind of pills?"
Helper: "I hadn't gotten that far; I don't know exactly."

Sometimes the hints that someone is very depressed, overwhelmed, and possibly suicidal come not from verbal but from nonverbal cues. For example, if someone who has been very ill or depressed or has experienced a trauma begins to show a marked change in mood or behavior, there may be a cause for concern that should be checked out. These changes can include giving away prized possessions, filing a will, or withdrawing completely from other people. Often people with severe depression are not even consciously aware of the depth of their own despair, so that significant others may be the first to note a problem. For example, one woman, after her marital separation, hid all of the kitchen knives from herself. Obviously, she knew where she had put them. But the symbolic gesture of protecting herself from suicidal impulses was meaningful.

Step 2: Assess the lethality. Three levels of suicidal ideas or actions exist along a continuum. These levels are:

1. **Suicidal ideas, feelings, or thoughts.** This is sometimes called "suicidal ideation." This means that the thought of killing oneself has entered someone's mind.
2. **Suicidal plans, gestures, or threats.** This means that someone is beginning to talk to someone about the suicidal thoughts or feelings, that someone has be-

gun to develop specific plans about how to kill himself or herself, or that someone has begun to put the plan into place.

3. Suicidal attempts. These are planned actions that are possibly life threatening. Whether persons intend to die at this time or not, this is a very dangerous stage.

In determining the level of lethality, remember that ideas, feelings, and thoughts are not the same as actions. Although they can lead to life-threatening behavior, ideas in and of themselves are not dangerous.

To think about lethality, consider a person's intentions to die, along with whether the chosen method of suicide is dangerous. The following table illustrates one way to think of the levels of danger.

TABLE 7-1 Classifying Suicidal Lethality

	High lethality	*Low lethality*
High intent	1	3
Low intent	2	4

Level 1. Person has high intent and a dangerous method; this is the most dangerous; for example, shooting or hanging when alone and not expecting anyone to arrive.

Level 2. Person does not intend to die but miscalculates lethality of method or chances of rescue; for example, dying from drugs or wounds by accident.

Level 3. Person intends to die, but method is not lethal; for example, sublethal dose of drugs or an unsuccessful attempt to die.

Level 4. Person has a low intent and uses method with low lethality; for example, making a gesture in order to make a cry for help.

To assess suicidal lethality, determine the following:

• **Danger.** How serious do you assess this threat to be? How lethal is the intended weapon? How deep and current is the stress and depression?

• **Plan.** If the person is thinking about suicide, is there a plan? When and how will this plan be put into place? How concrete and specific is that plan? Will the plan work? Does the person have access to the method of choice? Does the person know how to make the plan work? Examples of questions to ask about the plan are:

- Have you thought about how you might kill yourself?"
- What are you planning to do?"
- Do you have a gun? Do you have bullets? Do you know how to load it? Is it loaded?"
- "What kind of pills do you plan to take? How many? Do you already have the pills? Do you have enough? Do you have a prescription? Do you know of a drug store that's open this time of night?"

- **Emergency or not.** In a suicide situation your immediate assessment must include the vital question, "Is this an emergency?" If a life is in imminent danger, then you will need to act quickly and decisively. Has the person already swallowed the pills, loaded the gun, or slit his or her wrists? Do you need to call the police and ambulance, or do you have time to talk to the person for a while? Examples of questions at this point might be, "Have you swallowed the pills? What kind? How many? How long ago? Are you drinking alcohol? Is there anyone home now? Are you expecting anyone?"
- **When do they plan to commit suicide?** If someone is talking about killing himself or herself in the distant future (for example, "I plan to take pills if I become incapacitated," or "I don't know what I will do to myself if Joe ever dies") then you do not have an emergency, or even a current suicide situation. In this case, active listening is the appropriate path to take. For example, one man, after being diagnosed with AIDS and quitting his job, told his volunteer buddy that he was planning to commit suicide whenever he was no longer able to care for himself. The buddy, after making sure that there was no immediate plan for suicide, spent several hours listening to this man talk about his fears of being bedbound, incontinent, and helpless. This person did not commit suicide right away but rather went on to have seven relatively productive and healthy years. One day, seven years after the suicide conversation with his buddy and after a very discouraging visit to his physician, he fatally shot himself. He did so without mentioning to anyone what he was planning to do.

Step 3: Take the emergency path or the active listening path.

The emergency route. An emergency situation means that you are with a person or on the telephone with a person who has specific and lethal plans to kill himself or herself very soon or has already taken some action toward that end.

Emergency Procedures

- In the case of an emergency situation, then you may need to get aggressive, directive, and loud.
- If you are talking to someone you don't know, try to obtain some form of identification. Try to get a name, an address, a telephone number, a location. Also try to find out if the person is alone. Push for all of this information, but not to the extent that a caller hangs up.
- If you are talking to someone you already know (and you already have an address and telephone number), find out where he or she is at this time and who is nearby.
- Stay on the telephone or with the person. Never leave someone who is feeling actively and immediately suicidal.
- Push the person for permission to call the police and ambulance. If the person passes out or hangs up, call the police even if you do not have the person's permission. If you are on the telephone, you may need to pass a note to a companion or coworker to call the police from another telephone.
- In the cases of high lethality suicide situations, the customary confidentiality concerns do not apply. Saving a life becomes more important. If you do call

the police or ambulance, the only facts you need to disclose have to do with the suicide attempt or threat. Say only as much as you need to in order to obtain help. You do not need to break confidentiality about physical or mental diagnoses or the reason that the person is involved with your organization.

The nonemergency route. If you are not dealing with an emergency, then a conversation over the telephone or in person about suicide is for the most part no different from other active listening situations. You should assess the situation and respond to feelings. Try to engage the person in talking about the intense emotions of sadness or hopelessness that have led to thoughts or plans of suicide. Allow the person to freely express his or her thoughts and emotions, an element that will already probably be an integral part of your helping relationship. Everyone wants to be understood and listened to. Many times the suicidal person feels that no one understands him or her. Sometimes, listening with undivided attention, even if you don't completely understand, agree, or have any advice, can help someone else feel understood. The suicidal person does not necessarily want or need advice.

Illustration

Faye calls the crisis intervention hotline one night, very tearful and feeling out of control. Her husband of 16 years has just informed her that he wants a divorce and then left the house. She had been unaware until that evening that he was dissatisfied with the marriage. She is feeling distraught, depressed, and very frightened. She is unwilling to tell any of her friends, family members, or in-laws about the situation at this time, because she is feeling ashamed and as though she is a failure. She confesses to the hotline volunteer that she is feeling desperate and out of control and can't stop thinking about driving her car off a bridge.

Think About It #39

1. What would you do during this telephone call with Faye?
2. What are some of the questions that you might ask Faye to determine the extent of her crisis and suicidal thoughts?
3. What feelings do you have when you think about listening to Faye?

Principles for Responding to a Suicide Situation

1. Take every suicide threat seriously. On rare occasions, someone will be making a suicide gesture or threat as a cry for help and not as a genuine decision to die. If you are lied to or manipulated by a person seeking help, you will get over it. It is always best to err on the side of safety.

Several points are important to remember. People do sometimes die by accident, sometimes from a suicide gesture from which they wished to be rescued. Another issue is that suicidal actions tend to get more and more lethal (dangerous) as they are repeated, so that someone who has made previous suicide attempts and may appear to be "crying wolf" may actually commit suicide in the next attempt.

2. Use the person's ambivalence. The reason that persons talk to helpers about suicide, instead of just doing it, is that they feel ambivalent or undecided. It is that ambivalence, that inner wish to live, that helpers are trying to reach and use. Even if someone is 98 percent sure that he or she wants to die, rather than 100 percent, you must work hard to reach that 2 percent of the person that may want to live.

3. Most of the time, suicidal thoughts are temporary. Helpers need to remember this, not only so that they can help people make it through those temporary impulses, but so we can let people know that the deep and acute pain will not last forever. However, suicidal thoughts and feelings often resurface during a crisis, and helpers should prepare people for that. One of the most useful approaches is to help the person put together a plan to get support when the suicidal ideas come back.

Specific Techniques of Suicide Intervention

- Label what you are really talking about as soon as you suspect that the person may be feeling suicidal. Say "suicide" or "kill yourself." Do not use euphemisms like "do away with yourself" or "harm yourself." Suicide is extremely serious and unpleasant, and the situation should not be sugar coated.

- As was stated earlier, if you determine that the suicide is not imminent and that this is not a high lethality situation, then listen, validate, and talk about how difficult things feel. Externalizing these concerns may be enough for the person to feel more in control, and the suicidal feelings may be abated just by talking about them. Let the person explore all emotions, no matter how negative they seem. Journey with the person into the depths. Do not shy away from hate, fear, anger, revenge, jealousy, shame, guilt, or pain.

- Do not use such cliches or meaningless phrases as, "Suicide is a permanent solution to a temporary problem," or "That's no answer." People have heard these things and know them already.

- Likewise, do not use false reassurance, such as, "Things will get better," or "It can't be that bad." To the suicidal or depressed person, these do not sound like reflections of his or her reality.

- Do not try to change the subject, to be cheerful, or to try and talk them out of the mood. This will probably convince the person that you do not really care about or understand his or her pain.

- Do not try to invalidate someone's intense feelings. It is not helpful to say such things as, "I know plenty of people who have it worse than you do."

- Do not preach or moralize. Do not interject your own religious, moral, or philosophical beliefs. What counts at this time is the views of the troubled person. When persons are feeling intensely suicidal, the fact that suicide is sinful or illegal tends to mean nothing to them.

- Although it may be difficult to do so, remain calm and clearheaded. Do not overreact or underreact. Do not panic, and do not be too nonchalant.

- Focus on strengths and on hope when you can. Reflect, when you can do so genuinely, the positive or optimistic aspects of what the person says or what you observe. You need to grasp any glimmer of hope in order to find ways to engage the person in fighting for his or her own survival. Examples of things you might say are:

  - "So you're saying that there are some things that you really love about your life. I'd like to hear more about those things."
  - "You seem to love your daughter very much. I'd like to hear more about your children and how much they mean to you."
  - "You're telling me that you're not entirely sure that you want to die right now. Can you tell me more about that?"
  - "Is that a dog I hear in the background? Tell me about your pet" or, "Have you thought about who would take care of him if you are dead?"

- Explore with the suicidal person what support systems or loved ones may be available. For example: "Who in your life do you trust right now?" "Who do you usually turn to for support?" "Who can I call for you right now?" Help persons to access this support. You may be able to call or notify sources of support for them if they are feeling immobilized.

- Never dare anyone to commit suicide, and never make fun of or mock the suicidal person.

- Try hard to refer the person to further counseling or therapy or even inpatient hospitalization (see the section below on "Making Referrals").

- Help the person to search for ways that he or she can regain some sense of control. Help persons to divide overwhelming tasks into small steps. For example, negotiate with the person that he or she will call one person that night to ask for support or call a crisis hotline the next day. It is too overwhelming at times to plan too far in advance or take on too many tasks.

- Some helpers find it useful to negotiate a verbal agreement with the person, often called a "contract," which serves to get someone through an hour or a night. Contracts may include:

  - "Call me back in an hour."
  - "May I call you in an hour?"
  - "You seem to have a suicide plan in place, and it will work, and you can use it any time. Will you agree to wait 24 hours before you make a decision? And you will agree to call your therapist tomorrow?"
  - "I want you to make a deal with me that you will go to the emergency room when you leave here."
  - "Let me call your therapist right now."
  - "Promise me that you will call this crisis line or your therapist if these suicidal feelings return."
  - "Let me take you to the mental health center now."
  - "Let me call someone to spend the night with you."

Making Referrals

Human services workers, students, and volunteers should refer the suicidal person to others who are more prepared to address the complexities of someone who is thinking of ending his or her life. If the person at risk for suicide is already connected with a mental health professional (such as a psychiatrist, psychiatric social worker, psychologist, or therapist), encourage him or her to speak with the professional immediately. The professional is in the best position to make a decision about the seriousness of the situation and to take responsibility.

Most cities have hotlines or helplines to assist people in crisis. A counselor is available in an imminent suicide situation and can provide resources to the person at risk and others who are concerned. In addition to hotlines and already existing relationships with mental health professionals, emergency rooms of hospitals, personal physicians, counseling centers such as Family Service Centers, local mental health clinics, newspapers, and friends can provide services or make appropriate referrals.

Often the suicidal person is unable to make decisions, and you may need to make the phone call to the helpline, therapist, or local mental health center. It is important to stress that help is available, and that people do care.

Taking Care of Yourself

More than at any other time or with any other issue, a helper in a suicide situation must take care of himself or herself and get support. If you have had an upsetting experience with dealing with suicide, get in touch immediately with your work or volunteer team, your supervisor, or a supportive loved one. As previously suggested, it is very important that your supervisor be involved as early as possible in a suicide situation.

Survivors of Suicide

You may be asked to provide emotional support to someone who cared about someone who killed himself or herself. The survivors of suicide are likely to feel guilty, angry, bewildered, and/or ashamed, in addition to feeling grief over the death. As always, your role is to listen intently and actively, validating the painful feelings while dealing with the reality.

If someone you know commits suicide, you will probably feel intense regret, rage, and/or guilt as well. Get help from coworkers and supervisors, as well as from friends, family, or persons who advise you spiritually. It is important that you begin to divide the issues of responsibility from sadness. You can feel these grief reactions without feeling that you are responsible for another person's actions.

Illustration

Joan, a human services helper who works as a parent support person for a private nonprofit organization that connects the parents of children with special needs, makes her first contact with Ruby, mother of a child with special needs. Ruby is currently experiencing stress: a recent divorce and the temporary placement of her children in foster care. In addition, she feels depressed. Ruby says that she appreciates Joan's concern and warmth, and starts calling Joan frequently, at unusual hours. Then she tells Joan that she has no reason to continue living.

Think About It #40

1. How do you feel when you think about yourself in Joan's place?
2. What is your opinion of the mother in the story?
3. What verbal responses would you make to the following statements?

Ruby: I'm really lonely. I miss my children.
Helper:
Ruby: You don't understand. Your child still lives with you!
Helper:
Ruby: I'm sorry I woke you up. I know it's late. But I need to talk with someone, and you are the only person I have to talk to.
Helper:
Ruby: I don't have any reason to live.
Helper:

4. Locate the crisis number in your community. Call them, explain that you are a human services student, and ask them how they handle crisis calls.

CHAPTER EIGHT

The Helping Interview
and the Problem-Solving Process

A common mechanism to demonstrate the components of a helping relationship, such as empathy and active listening, is a helping interview. Another common activity in helping is problem solving. This chapter presents basic information about interviewing and problem solving.

A Helping Interview

A helping interview is different from an ordinary informal conversation; an interview has a purpose, and the persons participating have specific roles. The following chart points out some of the differences between a helping interview and an informal conversation with a friend, coworker, or family member.

Characteristics of Conversations versus Helping Interviews

Conversation	Helping interview
unstructured	somewhat structured
informal	formal
no special roles	roles of person seeking help and helper
no specified purpose or goal	purpose is to benefit the person seeking help or to accomplish that person's goals
equal exchange or dialogue	the person seeking help talks more
spontaneous, unplanned	most often scheduled, planned
relationship is ongoing, social	the relationship is temporary and businesslike
no rules govern the communication	there are ethical guidelines and communication techniques

A helping interview can have several purposes: gathering information, assessing the nature of a problem, delivering information or education, providing emotional support, assisting with generating alternatives and solving problems, and counseling (Kadushin, 1983). Often an interview serves more than one purpose, and there are beneficial side effects that are not necessarily related to the main purpose of the interview. For instance, when the helper's main goal is to provide emotional support, he or she is also likely to receive much information from the person seeking help during the interview because that person feels safe to open up. Likewise, an interview structured around problem solving can feel very supportive to the person seeking help.

What Is an Effective Helping Interview?

Echterling and Hartsough (1980, 1989) offer a useful model to explain what a helping interview (in person or by telephone) might look like. They divided a typical helping interview into four separate steps. They believe that going through these phases in this order is vital in an effective interview. Study the four phases or steps in the accompanying table. Note that Steps 1 and 3 (the left side of the table) involve mostly feelings ("affect"), and Steps 2 and 4 (the right side) involve mostly thinking ("cognition").

These four interview phases should most often occur in this order, because certain tasks must be accomplished before others are possible, and different stages require different helper behaviors. The timing of a helper's responses is often as important as the responses themselves. For instance: (1) rapport building should happen initially and quickly, but a helper should pay attention to this throughout the interview; (2) assessment is much more useful in the beginning of a call or interview than at the end; (3) listening actively and responding reflectively to feelings is a continuous process but is most important in the

Interactive Dimensions (Echterling and Hartsough, 1980, 1989)

Step 1: Affective Dimension: Rapport building. In this phase the helper establishes the initial relationship, creating a safe environment and a helpful climate.	**Step 2: Cognitive Dimension: Assessment.** In this phase the helper explores the basic crisis or problem and how the person seeking help views the situation. This is a time to gather information and facilitate the person's telling you what prompted him or her to seek help.
Step 3: Affective Dimension: Feelings. This step, the core of helping, concerns the identification of, accepting of, validation of, and reflection of the person's feelings.	**Step 4: Cognitive Dimension: Problem solving.** This last step, which is not always reached, involves generating alternative solutions with the person seeking help, discussing possible resources, and making a plan for action.

middle of the interview; and (4) problem solving is most useful if the bulk of it occurs toward the end of the encounter rather than at the beginning (Echterling and Hartsough, 1989).

These four phases do not always proceed in a simple linear way; in other words, an effective helper may move back and forth between the feeling and thinking dimensions depending on what the situation calls for. The helping interview may move quickly through these phases, or one lengthy interview may focus primarily on one portion (especially Step 3, which explores feelings).

The advantage of thinking of the helping process in the above framework is that during the interview you can ask yourself questions such as:

- Have I created a safe climate?
- Do I understand the person's definition of and perception of the problem, issue, or situation?
- Do I understand how the person feels about the problem, issue, or situation?
- Am I problem solving too early? Is the person seeking help ready to talk about options?

More on Teaching Problem-Solving Skills

The first three phases of the above interactive dimension are handled through the active listening which we have already explored. Building rapport and a helping relationship, listening to the person's view of the problem, and reflecting feelings are concepts that are familiar to you. The last phase of this model, problem solving, is a slightly different skill and will be discussed in this chapter.

Sometimes persons come to helpers with a specific task to accomplish or a specific problem on which to work. At times, because the person is experiencing a crisis, has not yet learned to problem solve well, or does not trust his or her own powers of problem solving, helpers may need to provide support and education on how to go about solving problems. Teaching problem-solving skills is a specific method of helping someone.

Remember what was said previously regarding problem solving in active listening: The helper is not giving advice or rescuing persons but is helping them to clarify the issues and develop alternatives for themselves. When you model this process and help people work through their problems, you are assisting them to be more confident and independent. Furthermore, when a person has had some successes with managing problems, then he or she may be more likely to feel competent to manage future crises.

When in the midst of a crisis, one frequently develops a narrow view ("tunnel vision") and sees only one alternative, often an undesirable alternative, and the crisis feels worse because of the perceived scarcity of viable choices. Increasing problem-solving skills enhances a person's ability to consider more than one alternative and puts him or her back in the driver's seat. Problem solving with another person or in a family or group also increases the ability to think about more than one alternative. Problem solving achieves two goals: (1) it increases the likelihood that a satisfactory conclusion will be reached; and (2) the person can gain a sense of control.

The helper who is teaching these skills should not act all-knowing, superior, or patronizing. Rather, the helper should take the stance of a *partner* in problem solving. The helper should view the other person as a capable human being who is temporarily stuck and overwhelmed. The helper often serves as a resource person; that is, the person with access to information, knowledge, expertise, referrals, and services.

The following steps in problem solving can be taught to others.

Step 1. Defining the problem. The following are among the issues that should be discussed or considered:

- What the problem or worry is.
- How serious the problem or worry is.
- What happens at the time of the problem or worry.
- Who is involved in the problem or worry.
- Why these persons are involved.
- Where the problematic situation does or does not happen.
- When the problematic situation does or does not happen.
- What has been tried before to relieve or remedy the situation.
- What has worked and what has not worked in the past.
- What is wished for.
- What can be done, by whom, and when.

Step 2. Generating alternative solutions. Possible plans are listed; this is accomplished through brainstorming, without deciding the pros and cons of any of the alternatives. As the next step illustrates, after one has listed all possible options, no matter how far-fetched the ideas seem, one can go through the list and decide what is and is not feasible.

Step 3. Evaluating the alternatives. The options that were generated are now judged for possible consequences, barriers, and feasibility. The alternatives can be ranked in priority order, from most to least desirable. In this way one can begin to see some ways in which the problem or worry can be realistically addressed.

Step 4. Trying the option chosen. Once an alternative has been chosen as likely to work, then it can be tried or tested.

Step 5. Evaluating the alternative. Once the option has been rehearsed or tried, figure out whether this alternative is satisfactory and workable.

Step 6. Trying another option. If the first option was not ideal, try one of the other viable plans.

Problem solving is an automatic process in most cases; people act without examining the thinking that got them to that point. When faced with new problems to solve, though, consciously moving through the steps in the problem-solving process can stimulate thinking about creative solutions and often makes people feel more in control and empowered.

Illustration

Lucy is a college student who has become quite concerned about her sexual health recently and asks Noel, a counselor at a peer sexuality center, for guidance. Lucy tells Noel that she does not consistently practice safer sex, although she has had vaginal and anal intercourse with several men whom she has dated recently. Lucy says that she takes birth control pills. She is aware of her risk for contracting HIV and other sexually transmitted diseases through unprotected intercourse but has not felt able to make the necessary behavioral changes on her own. Noel first congratulates Lucy on her courage in asking for help and offers to go through some problem-solving steps with her. Lucy agrees. Noel then asks Lucy to list all the possible alternatives that she can imagine. She lists three solutions that seem feasible to her: (1) abstaining from sexual intercourse, (2) using male or female condoms each time that she has vaginal or anal intercourse, or (3) mutual monogamy with a man who has tested negative for HIV and other sexually transmitted illnesses. Lucy then, with Noel's help, evaluates each option. After discussing all options with Noel, Lucy decides that monogamy is not realistic for her at this time of her life, because she does not have a steady lover. She also realizes that she is unwilling to abstain from intercourse at this time. She settles on the alternative of using male or female condoms as barriers each time that she has intercourse. Noel and Lucy then rehearse how Lucy might ask a sexual partner to use a condom. Noel then talks to Lucy about the proper way to use the male and female condoms and gives her a few of each to take with her. Lucy agrees to practice using condoms and to approach her lovers about using them. Noel and Lucy schedule a follow-up appointment to see how her plan is going. When Lucy returns for this appointment, she has shopped for condoms at the drug store, purchased male and female condoms, practiced putting a male condom on a banana, and practiced inserting a female condom. She says that she is disappointed, however, that she "chickened out" when she was spending the night with a date and did not request that they use barrier protection. Noel then offers to rehearse with her having such a conversation. After Lucy role plays asking a sexual partner to use a condom, she says that she feels more prepared to negotiate safer sex with a partner.

Problem-Solving Exercise

Choose a problem you are currently thinking about that you think has potential for being solved. Work through to a solution using the worksheet provided. This exercise may take quite a bit of time. You may need to work on it two or three times before it is complete.

1. State the problem.
2. Describe what is happening at the time of the problem.

    a. List who is involved or concerned.
    b. Why are they involved or concerned?
    c. Where does the problem occur or not occur?
    d. When does the problem occur or not occur?
3. List what you and others have already tried.
    a. What was tried?
    b. Did it work?
4. How serious is the problem?
5. What do you wish would happen?
6. Restate the problem, based on what you are thinking right now.
7. Brainstorm alternatives. Do not think about the pros and cons of any of the alternatives. They do not need to be rational; creative and silly options are encouraged. List at least ten alternatives.
8. Rank the alternatives from the most desirable to the least desirable.
9. What must be done in order to implement the most desirable alternative, by whom, and when?
    a. Alternative chosen:
        i. What is to be done?
        ii. By whom?
        iii. When?
10. Try it!
    a. Congratulations!
11. Are you satisfied with the outcome?
12. Would you like to try another option? If yes, go back to your list, reorder the choices with the first option omitted, and go through the process again.

Think About It #41

1. What is your reaction to having completed this problem-solving process?
2. Under what circumstances might you use this problem-solving process as a helper?

Sometimes when helpers get to the problem-solving phase of an interview, they forget the principles of "support of choice" and "starting where the client is" and begin telling the person what to do. Problem solving must be a mutual process, embedded in the relationship that has developed between the helper and the person seeking help. The role of the helper is simply to assist the person seeking help with developing alternatives, options, ideas, choices, and resources that may fit for that person's life.

Study the following steps to mutual problem solving, paying attention to the examples of how to approach each phase:

1. Identify what the specific problem is that the person wishes to work on today.
For example, "You've told me that you're worried about losing your job because

you are getting sicker, that you'd like to change doctors, that you'd like to talk about how to apply for disability, and that you're almost ready to talk to your family about the fact that you have cancer. Which one of these issues seems most important to talk about today?"

2. Determine the person's view of the problem and feelings about the situation. For example, "I know that you are pregnant, but I'm not sure how you feel about that and what you want to do next. Can you tell me a little bit about how you see your pregnancy?"

3. Establish the person's priorities about this particular issue. For example, "You've been telling me about how you dread the time when your children are all in college and out of the house, and how depressed you'll feel about not having them around you. What has been important for you about being a full-time mom?"

4. Break the task down into smaller, more manageable pieces. If the person seeking help is feeling overwhelmed by a big task like applying for a job for the first time, it might be useful to break up the whole process into steps for which you can prepare and practice together; for example (1) getting the newspaper and marking possible positions, (2) making an appointment with job service, (3) enrolling in a typing class at the vocational school, (4) working on a resume, and (5) planning what to wear to interviews.

5. Develop and list possible options, along with pros and cons of each one. Include in this step what the person has tried before, in this situation or similar situations, and how those attempts have worked. For example, when a person is trying to decide on a course of action to take, it might be useful to help the person write down all of the possible ways of proceeding and then weigh the costs and benefits of each alternative. This process can make the possible results of one's decisions and actions clearer and thus make the choice itself easier to make.

6. Explore which solutions would be comfortable or possible for the person to try. Even when a course of action seems feasible, the person seeking help might not feel that it fits for his or her particular situation, personality, family, or culture. For example, it may seem logical and practical for a woman to use condoms to guard against pregnancy, HIV, and other sexually transmitted diseases, but if she is devoutly Catholic, she may not embrace that solution. Check out with the person whether a possible solution is realistic for him or her.

7. Develop resources. Share with the person seeking help what community resources may be available, accessible, and appropriate, and then let the person seeking help make the choice that is best for him or her. For example, if the person seeking assistance, after agonizing over all possible alternatives, has decided to have an abortion, it is your job to furnish her with information about how an abortion can be safely obtained, regardless of how you feel personally about her choice.

8. Develop an action plan, agreeing on what each of you will do. There are often tasks that the person seeking help agrees to do and task which you as the worker will agree to do. Make sure that these are clear, and make sure that you do what you promise to do. For example, "Let's see, you have said that you will make a doctor's appointment as soon as possible, and I have promised to send the disability forms to your doctor. We have also said that we will talk again on Thursday."

Components of an Interview

The usual way of looking at a helping interview is to examine the process in three distinct parts (beginning, middle, and end) and to think about what must happen in each of these (Benjamin, 1969; Kadushin, 1983). Remember that helping interviews are flexible and fluid, and that the helper should, as much as is possible, gently guide the process without taking control away from the person seeking help. Given those caveats, the three parts of an interview follow.

Beginning/Introduction

Greeting and meeting the person seeking help is actually the end of a process of preparation on the part of the helper, the agency, and the person seeking help. "The interview begins before it starts" (Kadushin, 1983, p. 106). First, the person seeking help must recognize that assistance or support is needed and must gather the courage to call or see a stranger who will be asking for personal and private information. Second, the way in which the first telephone call is answered, the way in which the agency receptionist greets the applicant, and the helper's handshake and first words all serve as important components of the development of rapport and trust.

The first thing to do as you meet a person seeking help is to thank him or her for coming to your organization and for trusting you. You may choose to assure the person seeking help that he or she has taken an important and courageous step in seeking support. This is unfortunately the opposite of the greeting that many persons encounter when they take the plunge and come forward to request help from an organization. Often persons are treated rudely or with indifference by the volunteers or employees of the organizations that are designed to welcome and serve them.

When someone asks for assistance, let him or her state the purpose for the call or visit as well as his or her perception of the problem. You can invite this explanation with phrases such as "Please tell me what prompted your call," "Please tell me what is on your mind," or "How may I help you today?" If the interview is initiated by the interviewer or agency, then you should state the purpose and the circumstances immediately.

The helper should try to explore and clarify the person's expectations about what the organization can and cannot do and explain what services and referrals are possible. There may be some confusion about this, especially if the person has been sent to you by someone else. If the person has a mistaken impression about what you can do, it is best to provide the correct information early in your relationship.

While we are discussing the beginning of an interview, a reminder about referrals is in order. If your agency does not provide the services the person needs or requests, always make an appropriate referral (see the section that follows on referrals); for example, "It sounds as though you thought that we would be able to provide housing for you. We are not able to do that through this office, but I can put you in touch with other agencies that may be able to help you with that. In the meantime, are you still interested in participating in our support group?"

Clarifying expectations is also important at this stage. After both the helper and the person seeking help understand the general nature of the problem, the helper should suggest or explain what will happen next; for example, "Perhaps we could talk for a few minutes about your situation and then begin to brainstorm together about some options which you might have," or "I will need to complete a brief intake sheet, getting some information from you, and then we can talk more about what you had thought about doing." (See Chapter 5 on clarifying expectations).

The Middle: The Core of the Interview

The bulk of the interview should be spent exploring the problem, talking about how the person seeking help feels and experiences the current situation, and generating options for solutions (the last three steps of the interactive dimension). During this time, the helper should be assisting the interviewee with identifying feelings and thoughts, clarifying views and perceptions about the situation, and listening intently to statements about the nature of the problem. Throughout the interview the helper must maintain a delicate balance among (1) directing the interview somewhat and allowing the person seeking help to tell his or her story, (2) eliciting content and reflecting feelings, and (3) formulating a possible plan of action and inviting the person seeking help to develop his or her own options.

The End/Closing/Termination

Both the helper and the person seeking help should be aware that the interview is drawing to a close; the ending should not be abrupt or a surprise. As the helper is terminating the interview, it is most productive to briefly summarize what has been done or decided in this interview, schedule the follow-up interview or call, and restate the action that both of you have agreed to; for example, "We've talked quite a bit about the pros and cons of your quitting your job and applying for disability at this time, and you have said that you need to talk it over with your lover and with your boss before you reach a conclusion. So, we agreed that you will read over the material I gave you and we will talk by phone next Wednesday."

Other Interviewing Issues

There are several issues related to the helping interview: documentation (using forms, taking notes, and keeping records), being on time, follow-up, making referrals, and handling interruptions.

The Use of Forms

Depending on the setting and the purpose of the interview, it may be necessary to complete certain forms or make notes about the information obtained. Remember that filling out a form is hardly ever the primary goal of a helping interview. The form is simply a tool for data collection or accessing services. Therefore, state as soon as possible that some information must be gathered as part of your time together, approximately how long it will take to get this information, and what the form or information is used for. If you are able to do so, conduct the interview without relying on the form. This means that you must be familiar enough with the intake form or assessment outline to guide most of the interview without referring to the form. In some cases, the organization for which you work may require that the forms be read to the person seeking help and that their answers be recorded on the form. If that is the case, try to make this process as unobtrusive and noninvasive as possible. If you are able to conduct the interview focusing primarily on the person seeking help and not the form, then do so.

Note Taking and Record Keeping

The most effective interviewing technique to develop is the ability to listen intently and remember what the speaker tells you. However, certain types of information not as easily remembered, such as dates and medication names, are often jotted down during the interview. Key words and phrases may occasionally be written down as well. Explain that you might be taking a few notes and why. For example, "I don't have a good memory for dates and details, so I might need to jot something down on my pad as we talk. Is that OK?" Or, "The agency requires that I get birth dates and addresses on everybody, so I'm going to ask you for some information, which I will then write down." You may also choose to offer to let the person seeking help see what you have written down, especially if someone seems uncomfortable with your taking notes. This might also be a good time to explain once again your agency's confidentiality policies.

It is vital that the interviewer write down the gist of the interview as soon as possible *after* the interview ends, especially plans for action or follow-up. Schedule time for yourself after each interview to make notes. If you are doing a home visit, pull your car over a block or two from the house, and jot down the important points of the interview or record the highlights into a audiotape recorder.

Being on Time

Whether the interviews are scheduled for the office, on the telephone, or the person's home, be as prompt as possible. Call if you are running late or cannot make it. Unfortunately, sometimes helpers inadvertently send the message that their time is more valuable than that of anyone who might seek assistance. Be as respectful of time as you can.

Making Plans for Follow-up

An interview often ends with a plan of action of some kind. It can be as simple as, "I will call you tomorrow," or as complex as a service plan that covers several weeks or months. If you promise to do something, *do it*. There are no exceptions to this rule. Do not promise something you are not sure you can deliver. If you say you will find out about something, do it. If you say that you will come by or call at a certain time, then do it.

Making Referrals

Dumping a person by referring him or her to another resource because we cannot or will not deal with the problem is inappropriate. It is, however, appropriate and often necessary to provide access to other resources when someone needs more than you or your agency can provide. As a matter of fact, handling a situation for which you are not trained or prepared is unethical. Making appropriate referrals is often the way in which helpers facilitate persons getting what they need.

Always follow up on a referral you have made, either by asking the referred person if the connection was made and whether the resource seemed able to meet his or her needs or by calling the agency to see how the referral worked out. (This should only be done with the person's written consent.)

Handling Interruptions

If possible, schedule interviews so that they are not interrupted by telephone calls and coworkers, with your attention focused on the person seeking help as much as possible. If you are speaking to someone in your office, ask that your telephone calls be held, and close your door. This is not always possible in human service agencies, which are often crowded and under-staffed. If it is necessary for you to be interrupted, apologize and resume the interview where you left off. Sometimes interruptions are problematic during home visits. If you can, explain that it would be better for your concentration if the radio and television were not playing and the children were not in the room. As will be discussed in the section on home visits, however, you are a guest in someone else's home and must always respect their wishes about where and how to conduct the interview.

Interview Types and Settings

A helping interview can happen in a variety of places. One way in which a human services helper can increase the likelihood that a positive relationship will develop is by being aware of external or environmental factors that might affect the development of a helping relationship. For example, the first contact should be made at a time when the helper is relaxed, has some extra time, is not in a hurry, and is not distracted by other people or things competing for his or her attention. Whether the first contact is a telephone call, a home visit, or an interview in the office, you should be prepared to spend at least an hour. Set aside this uninterrupted time. This section will look briefly at the three most common environments for an interview: an agency office, the telephone, or the person's home. Each has its own advantages and disadvantages.

The Agency Office

The advantages of seeing a person in your organization's office include (1) having a safe, clean, private, convenient, and professional environment for the interview, (2) having supplies, forms, and resource lists handy; and (3) being able to introduce the person to other helpers who may have contact with him or her in the future.

The major disadvantages of an office visit for helpers is that they are not able to assess people and their family members in their own environments, so the helper will not have a clear picture of how living arrangements have an impact. Disadvantages of an office visit for the person seeking help can include the inconvenience of getting there (especially if transportation or child care are problems), the increased nervousness of coming into a business environment and facing many persons, the stigma of being a mental health or social services consumer, and the possible difficulty of leaving sick or dependent family members at home.

Telephone

Interviews and interventions by telephone are common in all agencies. Some organizations, such as crisis hotlines, use this method of interviewing and intervention exclusively. Other agencies conduct all of their follow-up contacts by telephone. When you have already met in person at least once, it can be useful and convenient to make subsequent contacts by telephone.

The major advantage of the telephone interview is its efficiency. Business can be conducted quickly, and time can be saved, because neither of you is required to travel. The primary disadvantage of the telephone call is that there are fewer nonverbal cues. Because you cannot see your caller's face or body, it is sometimes more difficult to interpret what is happening. Note, however, that you can still attend to a person's breathing, tone of voice, and pace in order to identify their emotions, such as joy, strain, sadness, or anger. In addition, the caller cannot see your nonverbal behavior, so you must work harder verbally at communicating your interest and attention. Your telephone partner cannot hear your head nod, for example; he or she must rely on your nonverbal cues and tone of voice. Remember also that not all service applicants and recipients have access to telephones.

Home Visits

Depending on the policies of the sponsoring organization, you may be visiting the home of the person or family to whom you are referred. It is vital that you remember that you are the guest in someone else's home. This is a good thing to remember when you visit people in hospitals, nursing homes, prisons, and other institutions as well. When an individual or a family invites you into the place where they live, treat this invitation as a generous gift.

Home visiting is a "reaching out" technique and has several advantages. For example, the person or family you visit will not have the inconvenience of leaving their own home. Leaving home may mean additional stresses or costs such as arranging for child care or transportation. In addition, the individual or family you visit may feel more comfortable in their own environment and may therefore be more open to discuss their situation and their concerns. Frequently the informal nature of the home visit intensifies the relationship between the helper and the person seeking help. A home visit is really the only way for a helper to get an accurate sense of a person's living environment and neighborhood or to meet other members of the household or informal network. Bloom (1973) states that, "seeing first-hand the concrete situations that families faced increased empathy for that family and its struggle" (p. 67).

There are also some disadvantages of home visits, particularly for the residents. Even though it is inconvenient for some people to leave their homes, it is also inconvenient to have guests. Cleaning and straightening for a new acquaintance is time consuming and energy consuming. Some families may feel embarrassed by their homes and feel uncomfortable with strangers visiting their houses, rooms, apartments, or neighborhoods. In addition to individuals and families feeling uncomfortable or embarrassed, individuals, couples, or families may feel that inviting strangers into their homes is too intimate and private. Insisting on a home visit when another person is hesitant is often inappropriate and can jeopardize the helping relationship.

An additional disadvantage of home visits is the potential for multiple distractions, such as children, neighbors, the television or radio, and telephone calls. These disturbances can detract from the ability to focus, concentrate, or speak about private matters. If a home visit is full of chaos and commotion, the helper may have gained valuable insight into the household routine but little else may be accomplished. In this case, the helper may want to suggest alternative places or times for future meetings (for example, "What would you think about meeting in the morning next time when the kids are in school?").

There may be times during a home visit that you, the helper, feel uncomfortable. For example, you may visit an environment that does not match your expectation for cleanliness or neatness. There may be garbage, insects, or mice in evidence, or there may be unpleasant odors. It is crucial that the helper in these circumstances maintains a nonjudgmental and supportive stance.

Sometimes you will find that the people you are visiting feel frustrated or helpless with their environment and would like your assistance with problem solving. You may want to respectfully and sensitively check out with the residents if they desire some help with safety or sanitation (for example, "Ms. Wilson, I'm noticing some roaches around here. Is there anything I can do to help you with this?"). If people do request or accept your assistance, you may find yourself helping them rid their homes of roaches, mice, or fleas; treat children for lice infestations; or find ways to make their living quarters more safe and healthy. Helpers may offer to support the residents in contacting landlords or managers who have been unresponsive to requests for repairs, sanitation measures, or pest control. Helpers should only offer these interventions if two conditions are met: (1) the residents identify these situations as concerns that should be addressed, and (2) your agency has the funds, resources, or referral sources to assist with these problems.

If you are in the midst of a home visit and find the sanitation situation (sights or smells) to be so distracting that you cannot concentrate on the interview, you may want to gently suggest an alternative to the resident, still maintaining a nonjudgmental stance. Examples of this approach are: "Mrs. Jones, I have a headache and could really use some air. Do you mind if we move out into the yard to talk?" or "Mr. Smith, I'm feeling uncomfortably warm. Could we go out on your porch and sit?" Of course, there are times when these alternatives are not possible, such as when you are with an older person who is bedbound and incontinent and is not being adequately cared for at home. In such a situation, when the state of the house is relevant to your agency's response, observe and document the environmental concerns carefully.

A home visit may sometimes (but rarely) be dangerous for the helper. If you feel uncomfortable or unsafe going into a neighborhood, apartment complex, or home by yourself, or if the persons you intend to visit feel that the area might not be safe for you, the safest decision is to discuss alternative meeting places. Your safety does not need to be placed in jeopardy. (See Chapter 5 for safety concerns.) For example, if you know that a member of the household has been violent in the past or that gang activity is common in a particular neighborhood, you should not plan to make a home visit by yourself.

Illustration

Rita and Larry Madison live in public housing on the south side of Chicago. They have three children: Maria is 5 years old, Lloyd is 3, and their baby, Elena is 6 months old. About a year ago, Larry was diagnosed as having AIDS. Rita was diagnosed as HIV positive after the birth of Elena, who also is HIV posi-

tive. Larry had been a heroin addict and poly-drug user for most of the couple's marriage. However, he has been participating in a Narcotics Anonymous program offered by the neighborhood association and has successfully remained drug-free for the past two years. His diagnosis came as a severe shock to the family, and Rita's and Elena's subsequent diagnosis of positive HIV status has been devastating. Larry had been working full-time as an assistant manager of a dry cleaning store, and the family had hoped to be able to move out of the projects in the next several months. Larry and Rita were concerned about violence and drug sales in their neighborhood and specifically in their 12-story high-rise building. One of their neighbors had several bullets enter his window during a gun fight, and two women have been sexually assaulted in the elevator during the last two months. Even the police are reluctant to enter this public housing complex, so the Madisons feel little protection against a dangerous living environment. This violence has been regularly reported by the newspapers and television news. Rita and the children spend most of their time in the apartment until Larry gets home from work when they all go grocery shopping and run other errands. The family is currently involved with two helpers. One helper is Larry's partner with Narcotics Anonymous; they have been helping each other stay drug free for most of the past two years. Rita recently contacted an AIDS Support Group that is sponsored by the local hospital. Rita was informed of the regular meeting times and was also told that she would be contacted by one of the group members, Beverly Hogan, who had been participating as a support person for the past three years. Beverly and her 4-year-old son are also HIV positive. When Beverly called Rita to arrange for a meeting, Rita suggested that Beverly come to her home since she couldn't afford a baby-sitter, she didn't want to drag three children around, and Elena wasn't feeling well. Beverly was aware of the problems that the neighborhood was having.

Think About It #42

1. Jot down what you think may be the feelings of Rita and Larry.
2. Imagine the feelings of the helper, Beverly Hogan, as she anticipates meeting the Madisons. How do you think Beverly might feel about making a home visit? What do you think Beverly's alternatives are? What do you think are her best choices?
3. How would you feel about making this home visit? What might you do to make yourself safer?

PART THREE

Models for Helping

Two common models for human services helping are introduced in this section: (a) working with individuals, couples, and families through case management or service coordination and (b) supporting communities through advocacy and organizing. Human services workers, students, and volunteers must be aware of issues at both of these levels in order to effectively address the needs of persons who may need assistance.

CHAPTER NINE

Case Management

One of the important tools of the human service worker is case management. This chapter will define case management and discuss its history, principles, and activities.

"Case Management" in Human Services

Currently, *case management* is the generally accepted term for helping an individual, couple, or family negotiate the complicated systems that are designed to deliver medical, legal, and social services and to help them receive appropriate and timely assistance (Rubin, 1992). The concepts and tasks usually labeled "case management" are sometimes called "case coordination," "care management," "care coordination," "service coordination," "service integration," or "service management" (Kane, 1990; Weil and Karls, 1985). Whatever label your human services organization uses to designate the activity, case management is one of the most important helping functions we can offer to those persons who have approached us for assistance.

Case management is an ongoing process, not a single event. Case managers are committed to provide services to recipients over the long run. The intensity of the contact and activity may vary over time as well, depending on the situation and on the person's needs at the time (Intagliata, 1992). Case management can be very intense in the beginning of the relationship with the person seeking help or when that person is in a social or medical crisis. Contact can be less frequent when things are going well.

As is true of many terms in human services work, "case management" is both inadequate and controversial. The activities associated with case management could perhaps be more appropriately called "person-centered service coordination." "Case managers" (the common term for human services workers who perform the activities of case management) do not really "manage" a "case." Case management is not an attempt to control the person seeking guidance or assistance. Rather, case managers arrange for, coordinate, develop, provide, and/or oversee *services* for the person or persons who might benefit from accessing resources. The way human services workers arrange and coordinate resources is based on the requests of the persons who have requested case management services.

"Case management" and "case manager" are examples of human services jargon; these concepts may not mean anything to persons who have asked for assistance, and the phrases may even be offensive to them. For example, the parent of a child with developmental disabilities said, "I'm not a case, and I don't want to be managed." The use of these terms continue among human services agencies, however, because funding sources still designate certain activities as "case management" and because the term has been in use for so long that human services workers and managers immediately recognize what it means.

The use of jargon is rarely useful when speaking with persons who have applied to receive services. Earlier we talked about the fact that someone who asks our organizations for assistance may be offended by the use of the terms "case manager" and "case management." Other words and phrases serve to depersonalize the helping process. Imagine what it feels like to hear statements such as the following from a human services worker: "I have to do an intake on you"; "First I have to assess you"; "I'd like to enter you into case management"; or "I'll be monitoring you." Rather than sound distant and clinical, try to speak plainly and clearly whenever possible. Compare the following statements with the ones above: "I'd like to talk to you for a while about what brought you here"; "I'm wondering if you can give me an idea about what you think would help you right now"; "What would you think about my trying to get some services in place for you and your family?"; or "Is it OK with you if I call you from time to time to see how things are going?" Which type of language would you rather hear from a human services helper?

One purpose of case management is to assure quality of care to persons, couples, and families who are temporarily vulnerable. A second is to facilitate fair access to appropriate services and to link persons with the organizations that could serve them. A third is to provide services in the most cost-effective way possible (Kane, 1990; Weil and Karls, 1985).

Case management can be a confusing concept because the above goals may seem to be contradictory. It may sound like a conflict to talk of quality of care in the same breath as being cost conscious, but case management is a tool to balance these two concerns. The overall goal is to meet the needs of the applicants for programs while struggling with the realities of scarce resources. The "advocacy" role and the "gatekeeper" role must therefore be blended, and that may be difficult at times for a case manager to accomplish (Kane, 1988).

It is important to distinguish case management in the realm of human services from other common uses of the phrase. Sometimes "case management" is the term used by insurance companies and health maintenance organizations to describe "cost containment"; that is, "case managers" in these settings monitor a person's medical care and limit spending in order to decrease the costs to the insurance company or to minimize excessive or unnecessary expenditures on health care. "Case management" or "care management" in the human services field in no way means "managed care" (American Hospital Association, 1992).

It is important to understand that case management as provided by a helping professional is *tailored to the needs of the person who has asked for help, not the needs of the sponsoring organization or agency.* True case management in the human services field is person-driven, not system-driven.

Similarly, sometimes the term "case management" is used by medical personnel to describe the medical management of the care concerning an individual patient's particular disease or symptoms. In human services, the case management is not purely oriented to medical treatment; rather, the management of the care takes into account all aspects of the person's life and environment, not only the biological problems.

Explaining to service applicants and service providers what is meant by "case management" in the human services field can take the following form: "Case management doesn't mean that I'm going to try to manage you in any way or to treat you as just another 'case'; it's a term meaning that I will do all I can to help you find the services you may want or need." You may also find yourself having to define case management to people in other professions. For instance, the chief executive officer of a large hospital said that he did not understand why case managers would ever be necessary for any patient who was leaving his hospital. He said, "Our physicians do the case management around here." When it was explained to him that case management in the human services field involves not just medical care but also locating community resources and referring persons to those resources, he replied "I could do that myself with a telephone book!" He was then told that true human services case management was offered by persons who were trained to take the side of the patient at all times, rather than being focused on the needs or desires of hospitals, insurance companies, physicians, agencies, or even human services workers. At that point he seemed to understand that this was indeed a different perspective and value orientation than he had previously imagined.

Since "case management" is a term with several possible meanings, one cannot assume that everyone who is called a "case manager" is actually doing human services case management. Conversely, some persons who are performing human services case management are not called "case managers" (Rubin, 1992). Not every service recipient has access to true case management, regardless of what the hospitals, clinics, institutions, or agencies may claim. When you are dealing with someone who has requested assistance with accessing services, it is important to explore whether that person is really receiving comprehensive service management and advocacy.

The range of activities the case manager offers and performs depends on the current needs of the person asking for help and the resources in the community. Effective service managers link persons to existing resources, advocate for them when the service systems are not adequate or responsive, and sometimes develop or provide a service or resource if it cannot be found elsewhere (Rubin, 1992).

Case management should be provided confidentially and with the signed consent of the service recipients. Case management is an especially important service to offer persons or families who are vulnerable and are having difficulty advocating for themselves or understanding the complexities of service delivery (Rothman, 1994; Rubin, 1992). Examples of vulnerable populations are persons with mental retardation, developmental disabilities, or autism; persons who have a mental illness or physical impairment; persons who are frail or terminally ill; or persons who have little social and political power, such as minorities, elderly persons, or children (Rothman, 1994).

Illustration

Jeff, a case manager with the local Association of Retarded Citizens (ARC), is contacted by an attorney who is representing Gloria, a 50-year-old woman with moderate retardation whose 80-year-old mother recently died. Gloria's mother had been her only caregiver all of her life, and now Gloria is in need of an alternative plan for how she is to live. Gloria has inherited her mother's house and does not want to leave it. However, she does not yet know how to cook or manage a budget and has never been employed.

Think About It #43

1. If you were Jeff, what would be your first steps?
2. What are the initial questions Jeff could ask Gloria that could make her situation clearer to him?
3. How could Jeff involve Gloria in the development of her care plan?
4. Who else might Jeff want to involve in Gloria's care plan?
5. What services or agencies might become involved in Gloria's care plan?

History of Case Management

The term "case management" is fairly new in the realm of human services but is actually a traditional activity in this country. Offering support to vulnerable persons has a long history in the United States, emerging in the mid- to late 1800s with friendly visitors, case workers, and the settlement house movement (Weil, 1985). With the boom in federally created and supported services in the 1960s and 1970s, case management as a term, concept, and human services activity became prominent. The attempt to coordinate and integrate care came as a reaction to the growth of specialized social service programs during that time. It had become increasingly difficult for persons who were vulnerable or in crisis to know what to do in a fragmented and uncoordinated maze of social service agencies (Rubin, 1992). Case management is now required in the field of developmental disabilities because of the 1975 Developmental Disability Act (PL94-103). The Omnibus Budget Reconciliation Act of 1981 created Medicaid waivers so that long-term care for older persons or persons with disabilities could be provided in their homes. After that piece of legislation passed, most of the home-based health care programs included case managers (Kane, 1990).

Case management as a function is now commonplace in the human services, mental health, and health care fields.

Principles of Case Management

Several essential principles underlie case management:

1. Participation, empowerment, and mutual planning. The objective is to involve the person, couple, or family in setting goals and developing a care plan that best fits their needs and preferences, in order to restore control and choice to persons whenever possible. A case manager is a partner in care, not the sole controller of care. In addition, throughout the helping process the case manager should be teaching the service recipients how to negotiate systems and essentially manage their own care.

2. Advocacy. The case manager's goal is to represent the needs and choices of the service recipients above all else. As will be discussed later in detail, human services workers advocate for individuals, couples, and families (called "case advocacy") and for groups of people who have less power or representation (called "class advocacy").

3. Appropriate and high-quality care. Facilitating the provision of social service and health care delivery accommodates the current and future needs of the individuals, couples, and families who request assistance and ensures that services are timely, appropriate, effective, and responsive to the real needs of vulnerable persons.

4. Continuity of care. The goal is to offer comprehensive services along a continuum of needs so that persons, couples, and families do not fall between the cracks as they move from agency to agency or as their needs change. This is sometimes called "seamless" care, meaning that even if there is a shift to another service organization, the person does not notice the "bump" or "seam" of that transition.

5. Access to services. Access to services should be as easy and as fair as possible. It is important to provide whenever possible a single point of entry into the service system so that persons are not forced to make multiple contacts with agencies and work with multiple workers. Persons requesting services should also be treated equally, without favoritism or discrimination. Sometimes case managers are instrumental in developing or revising service programs so that the needs of their communities are better met.

6. Integration of services. Since many persons have multiple needs that must be addressed through multiple sources, the goal is to ensure that the care is coordinated, nonduplicative, appropriate, and of high quality.

7. Comprehensive view. One objective of case management is to take into account all of the aspects of a person's life, including the person's social network, economic situation, ethnicity and culture, vocation, spirituality, emotions, and mental and physical health. One of the important tasks is to support and augment the informal support network (family, friends, and volunteers) rather than replace or discount it.

8. Evaluation and monitoring. Case managers or service coordinators must provide ongoing follow-up and counseling throughout the helping relationship, periodically reassessing and evaluating the service recipients' progress, strengths, and needs; the appropriateness of the plan; and the responsiveness of the service delivery system. Through ongoing contact with other agencies and with the persons who have asked for help, case managers monitor the quality and appropriateness of care (Woodside and McClam, 1998).

Illustration

A hospital discharge planner telephoned Georgia, a case manager at an AIDS service organization, to refer Lydia Youmans, who had been admitted several weeks previously. Lydia was preparing for discharge from the hospital but had no clear-cut plans for how to manage her illness and her care once she left the hospital. Georgia immediately arranged to go to the hospital to visit Lydia, who over the last month had become progressively incapacitated by an HIV-related infection of unknown origin. Georgia found that Lydia was very tearful and agitated. Lydia had recently lost the ability to move her arms and legs and had lost the sight in one eye. Lydia said that she was sad, frightened, and confused. Lydia had been living alone and wished to return to her own home, even though she knew that she was quite ill. After spending an hour with her, listening to her talk about her increasing fear about the future, Georgia developed a list of concerns she thought she had heard from Lydia. Georgia outlined these issues with Lydia, who agreed that these were her priorities:

1. *Arrange for visiting nurses to monitor her blood pressure and reactions to medications.*
2. *Arrange for home health aides in her home each day to feed her and bathe her and assist her with toileting.*
3. *Develop and coordinate a schedule of friends and relatives who were willing to be with her each day and night.*
4. *Help her to tell her employers that she could not return to work and assist her with applying for Social Security disability.*
5. *Arrange for home-delivered meals once a day.*
6. *Contact a medical provider about administering medications intravenously each day at home.*

Georgia promised Lydia that she would begin right away, with the help of the hospital discharge planner, to put this care plan into place. Georgia also asked Lydia if she would be interested in talking to a counselor or pastor about what her rapidly declining health would mean to her; Lydia said that she had grown to like and trust a priest on the hospital staff and would be interested in continuing her contact with him after she went home. Georgia agreed to ask the priest if he would consider home visits. After two days, Georgia had arranged the following services:

1. *Visiting nurses from the county health department would come twice a week to monitor Lydia's blood pressure and reactions to medications.*

2. *A home health aide from the local HIV clinic would come to her home each day to take care of her personal care needs.*

3. *With Lydia's input, they had developed a list of friends and relatives who were willing to be with her each day and night and had located a friend of Lydia's who was willing to coordinate the scheduling of these visitors.*

4. *Georgia sat with Lydia as she called her employer to disclose her illness and resign from her position. Georgia had also completed and submitted applications for Social Security disability, with Lydia's input.*

5. *The local Council on Aging had agreed to deliver a hot meal in the middle of each weekday.*

6. *A private health care provider would administer medications intravenously each day at Lydia's home; Lydia's insurance from work would cover the cost.*

7. *The hospital chaplain whom Lydia felt close to agreed to visit Lydia in her home in the evenings whenever she wished.*

Think About It #44

1. How would you have felt in Georgia's position? Does Lydia's situation seem overwhelming to you?
2. In what ways was Georgia able to "partialize" the crisis so that she and Lydia could manage it better?
3. What principles of case management did you notice Georgia using? How did she make these principles a reality in Lydia's care?

Elements of Case Management

Regardless of the dispute over terminology, there is wide agreement on the components of case or service management as a human services activity. Following are the elements commonly considered to be part of service management (Intagliata, 1992; Kane, 1990; Rothman, 1994; Weil, 1985; Weil and Karls, 1985).

"Case" or "Client" Identification or Finding

Persons reach service organizations through a variety of methods. Sometimes agencies advertise their services through public service announcements and brochures. Volunteers, students, or staff may go out into the community or other organizations to invite eligible persons to apply for services (this activity is called "outreach"). Other human service or medical organizations make referrals. People also locate programs through the telephone book, an information line or hotline, or through word of mouth. The case manager's role in case finding is to make sure that the agency's services are as accessible as possible to those who may be in need of them and eligible to receive them.

Screening, Intake, and Enrollment

A person who approaches a service provision system and requests assistance must be screened to determine eligibility. If eligible the person must sign a consent to receive services and be enrolled in the program. Case managers determine eligibility and link people with appropriate services if they are found to be ineligible for the agency's programs.

Comprehensive Assessment of Individual Needs

With input from the person, couple, or family, the case manager determines the current or urgent unmet needs and priorities. As was explained earlier, "comprehensive" means that all aspects of the person's life should be considered when making these decisions: social, biological, spiritual, psychological, emotional, mental, and financial. It is important to understand the person's definition of the problem or problems and to solicit his or her ideas about what solutions might be feasible.

Crisis Intervention

If someone is currently feeling traumatized and off balance, crisis intervention (and possibly suicide assessment and intervention) accompanies case management activities. Once the current crisis has begun to resolve, the case management contact is not as frequent or intense. (See Chapter 7 for an in-depth discussion of crisis and suicide intervention.)

Development of Goals and Service Plan

With the input of the person, couple, or family seeking assistance, determine the steps to take in order to gain access to the services and care the person or family needs. Talk with them about what services or resources they would consider using; offer them all of the available options, and help them choose the most appropriate path for them. Work with the individual, couple, or family to assure that the goals are reachable and doable; it does no good to set one's sights so high that one is doomed to failure.

Delivery of Services

The case manager arranges for services to be delivered, identifies resources, makes referrals, implements the care plan, and coordinates services. If programs outside your organization are necessary, guide the person, couple, or family through the referral process, explain what might seem to be a confusing maze of programs, and facilitate the linkage with the other service systems. Case managers may also help persons apply for benefits and services. Coordinating services with other agencies avoids duplication and ensures that the service recipients get the best possible care. The advocacy role may be important at this stage, as case managers work to change systems that are not responsive to the needs of vulnerable populations. If, over time, several applicants ask for a particular type of program that does not exist in the community, advocacy may be needed to form and fund that service.

Provision of Advocacy

As will be discussed in Chapter 10, there are two types of advocacy in the human services field: case advocacy and class advocacy. Case advocacy refers to taking the side of an individual, couple, or family who needs a representative to help them negotiate and arrange for services. Advocating for someone who is frustrated, sick, or scared is an important component of service coordination. Often the human services case manager is an ally to a person who is in crisis or in need. Examples are: helping a battered wife to prepare a legal request for a restraining order for a family court appearance, helping a man with cancer to appeal a disability determination decision, helping an older person fight an involuntary commitment to a nursing home, or helping a woman recovering from drug use to ask the child welfare agency for probational custody of her children.

Illustration

Helen is a middle-class, middle-aged Caucasian woman who approached Rob, a case manager in the public health department, for help with getting approved for Social Security disability and Medicaid. She tells Rob in the first meeting that she has been widowed for four years and has two adult children. Helen's husband died from AIDS, and Helen, who tested positive for HIV when her husband did, now is experiencing symptoms as well. Her physician has told her that she is now eligible for disability, based on her extreme fatigue, low T-4 cell count, and rapid weight loss. She applied for Supplemental Social Security income (SSI), with her doctor's cooperation and documentation, and was rejected by the Social Security Administration. Her application reconsideration was also turned down. She has learned from the Social Security Administration that she has one more chance at appeal. Helen tells Rob that this is the first time in her life that she has ever approached a human services agency for help. She asks Rob to guide her through her appeal and help her secure disability income and insurance.

Follow-up

The case manager follows up on the appropriateness of services and referrals, monitoring service delivery. "Monitoring" refers to periodically checking with the service recipients and/or family caregivers to determine whether the service plan is working, the care is of high quality, and connections were made with other agencies as planned. Ask or survey service recipients about their satisfaction with how they have been treated. If you have a signed consent granting specific permission to contact another agency, call and ask if the person you referred was helped there. The specific time frame for monitoring in case management depends on the dictates of the agency or the needs of the service recipient.

Reassessment

A case manager must reassess periodically to determine changing needs. People's situations are not static; expect change. Make sure that every month or so you conduct a brief reassessment to determine whether service needs have changed. Be proactive; try to anticipate what crises or challenges could arise and plan for them. You may want to ask the service recipient how often he or she wants you to call to check in; some people do not want their case manager calling them and will agree to call the agency when they have a question or need.

Termination

Helping relationships terminate for a number of reasons. This final stage is not always reached in case management. Case records can be closed when the individual or family requests termination, when the service recipient dies or moves away, or when the person has improved and no longer needs services. Sometimes a case is closed because the payment source has ordered that the case be closed. In that case, do your best to help the person connect with services elsewhere if they are still needed. Never terminate case management or other services without the knowledge and consent of the person or family who requested services.

Illustration

Nan is an Asian-American woman in her late 60s who called the county Council on Aging requesting help with her 85-year-old mother, SuLin, who came to live with Nan recently after having a stroke that left SuLin paralyzed on her left side. Nan is close to tears as she requests an appointment with Brian, one of the case managers. Brian asks Nan if she can come in to see him that afternoon. She replies that she will ask a friend from church to stay with her mother so she can do so. When Nan arrives at Brian's office, she is still tearful and states that she is quite tired, after having been up all night with her mother. Brian discovers as he listens to Nan describe her current situation that she is feeling overwhelmed and unable to continue to provide care for her mother because it is more demanding than she ever anticipated. She wishes to care for her mother at her home rather than a nursing home, but she says that she is "at the end of my rope." Brian and Nan settle on a plan involving home-delivered meals from the Council on Aging, a visiting nurse from the county health department, a home health aide from the organization that manages the Medicaid waiver program, occasional respite care in the adult day care center operated by the county nursing home, and transportation to doctors' appointments from the regional transit authority's program for the handicapped. Nan also decides to attend a support group for family caregivers sponsored by the Family Services Center. Brian offers Nan ongoing periodic contact with him for moral support and to see how things are going, and she accepts with relief. In three days Brian has arranged for all parts of the service plan to be implemented and calls Nan with the details. She tells Brian that she never would have had the time, knowledge, or patience to find all of those resources herself and that she now feels much more optimistic and able to cope.

Think About It #45

1. In the illustration above, what were some of the elements, functions, or components of case management which Brian utilized?

2. How do you think that Brian was able to utilize resources from such a wide variety of places?

3. What resources do you know of in your community that might have been helpful to Nan and SuLin?

Functions performed by a case manager depend on the environment in which he or she works. If there are many services in a community available for a certain population, then a lot of time will be spent in arranging for those services to be delivered and coordinating with other agencies. On the other hand, if there are few services available, then more time is spent in planning and creating new services (Intagliata, 1992). For example, Sylvia, a human services worker, had different experiences with case management in her two different internships while she was in school. During her first field placement, she was a case manager in a county Council on Aging. Although not all needs of all service applicants could be met, for the most part the services designed for the elderly population were plentiful, available, and accessible. Her second placement, however, was with a small AIDS-service organization in the same community. The majority of her time there was spent in creating and implementing support groups, transportation services, volunteer buddy programs, and a telephone hotline, because no other organizations or programs offered these services.

The intensity of case management also varies according to the circumstances. Sometimes peoples' situations require that they have frequent contact with a case manager. Examples are persons with severe retardation or autism, persons who have disabilities, persons who are chronically or terminally ill, or families who are at high risk for substance use or physical abuse or neglect. If an agency has adequate funding in these cases, the case managers will have smaller case loads (fewer service recipients assigned to them) and will be able to give more attention to the needs of these vulnerable populations. Persons and families can be considered to be vulnerable and in need of intense case management when their "emotional, physical, or psychological well-being is endangered" (Weil and Karls, 1985, p. 12).

Within a case management program, there are often varying degrees of contact with a person or family. Variables that may determine intensity of case management include the person's existing social support, physical and medical status, involvement with law enforcement agencies, ability to advocate for him- or herself, complexity of needs, knowledge about resources, and existence of other complicating problems.

Case Coordination

Case managers often coordinate their activities with other workers within their own organizations and in other agencies. Many case managers work in teams. Sometimes the care teams involve several types of professional helpers, such as physicians, nurses, psychologists, social workers, and clergy persons.

Case managers do not just arrange for services within the organizations that employ them. An effective service manager will work closely with other agencies to arrange for and coordinate care across a wide range of service systems (Kane, 1990). It is important to develop a network of allies throughout the health care, social services, and mental health fields in your community so that you can facilitate the arrangement of services for individuals and families regardless of where those services are found. Case management is "interorganizational"; that is, communications, informal sharing, cooperation, and transactions between agencies must be present and viable (Norman, 1985).

Whether the case coordination occurs with your own coworkers or with service providers in other agencies, the following are important guidelines for relating to other helpers.

External Case Coordination

When you meet with helpers who are not employed by your agency, whether individually or in a team, it is vital that confidentiality and consent rules are in place, to protect the persons' rights and privacy. It is unethical for a case manager to speak to another human services helper about the person without a signed consent from him or her. Even when you have consent, it is still advisable to share information in the interagency group using only first names or pseudonyms. If case studies are distributed, these papers should be shredded after the meeting. If notes are made on the blackboard, these should be erased when the meeting has closed.

The reason to participate in interagency case meetings is to coordinate care when several agencies are working with one person or family. Sometimes the purpose is to work out solutions to difficult situations, using a wider range of resources. The interagency group members should probably establish guidelines that insure that they treat each other with respect; it is not useful, for example, to participate in meetings where people simply blame each other for letting persons fall through the agency cracks. The goal of the meeting should be constructive problem solving and resource sharing.

Internal Case Coordination

When you have a team meeting or staff meeting within your own organization, the confidentiality concerns are different. Your commitment to service recipients is that information about them will not be shared *outside* your agency without their permission. However, it is a good idea to let people know how information is used within your organization as well. Explain to the applicant at intake that information will be shared with your supervisor and/or your team, but that no one who does not need the information will have access to it. For example, service recipients and applicants need to feel secure that the agency receptionist and janitor do not have access to their case records.

"Case staffings," as these internal meetings are often called, are meant to update and inform other case managers about concerns that may arise for individuals and families, so that everyone on the team can be prepared to help them. These meetings are also a mechanism to ask for advice from your coworkers about problematic situations and to get support when you need it. Reports on situations should be organized and brief, hitting the high points and clearly outlining the potential concerns. You may want to request a brainstorming or problem-solving session with your team or supervisor if you are having difficulty with a particular person or service plan.

Illustration

Nina is a service coordinator in a small nonprofit agency that serves persons with sickle cell anemia. She receives a call from James, a 25-year-old African-American man, who says that he is calling from a pay phone in one of the state mental hospitals. He has been diagnosed with schizophrenia and symptomatic sickle cell anemia and has been hospitalized in the mental health facility so that his psychotropic medications can be monitored and regulated. He tells Nina that he has just learned that he is to be discharged that day, but he has nowhere to go. The reason he has not been taking his medications is that he is homeless and has no income. James wants to have a place to stay or at least be sent to a shelter but says that the discharge planner (Linda) has not seemed receptive or cooperative. James reports that Linda has told him that she intends to discharge him back to the street, and that what happens to him after he leaves the mental hospital is no concern of hers. Nina asks James to call her back in an hour. Nina wonders if she should first try to get James into a shelter for the night, but then she decides that the wisest course of action is to try to lengthen his stay at the hospital until there is time to help him locate permanent housing. Nina calls Linda, the discharge planner, and explains that James has contacted her because he is worried about being back on the street with no way to take care of his mental and physical health. Linda tells Nina that she cannot do anything about this, that James's time is up. Nina tells Linda as clearly and as professionally as she can that this plan is not acceptable and that she will have to talk to Linda's supervisor if Linda cannot help to find a workable solution. Linda angrily tells Nina that rules are rules and nothing can be done. Nina then calls the director of the state mental hospital and explains the situation. The director promises to call Linda himself and get James's stay extended for a week. When James calls back in an hour, he says to Nina with a chuckle, "I hear you put your boxing gloves on!"

Think About It #46

1. Consider the above example. What do you think about Nina's decision to work on changing the situation rather than accepting the situation?

2. Would you have been comfortable doing something similar to what Nina did?

3. What do you think about Linda's response to Nina's efforts at coordinating James's care?

The Role of Documentation in Case Management

The "management" of a "case" involves keeping an accurate written record of the person's work with the agency. Most human service agencies have minimum standards for maintaining a case record, such as having a consent form signed by the applicant for services, filing a progress note for each personal or telephone contact, updating the file when a person's situation changes, and having a consistent system for the order in which forms are filed. Documentation is important because of the following reasons:

1. Consistency in communication and service provision. Other case managers on staff can pick up the case record and understand the service plan, thus being able to assist the person when he or she calls or drops by.

2. Data for evaluation reports and funding requests. Agency administrators must satisfy contract and grant requirements by providing accurate data on services and must have accurate information to develop grant proposals.

3. Monitoring quality of care. Supervisors can use case records as one tool for determining whether workers, students, and volunteers are responding to persons, couples, and families most appropriately, and whether community resources are adequate. In this way, gaps and errors can be caught, and unmet needs can be discovered.

4. Tracking progress and changes. Reading a case record can give case managers, coworkers, and supervisors a good sense of how a person has been doing over time and whether the goals of the service plan have been met.

5. A "tickler system." By keeping a schedule of when visits or calls should occur in the future, case managers are reminded to maintain regular contact.

These systems of reminders, often called "tickler files," can be kept by computer or in an index card file.

It is important to balance paperwork with service provision. Some human services workers make the mistake of acting as if the case record and the forms are more important than the individual, couple, or family receiving services. The case record and the forms are merely *tools* in case management, designed to document activity and track needs and progress. On the other hand, it is a mistake to be careless, nonchalant, or negligent about record keeping. In human service agency work, if it is not written down, it did not really happen! A good case manager is accurate, thorough, and succinct with record keeping, so that the work is documented but the time with the individual or family does not suffer.

Application of Helping Skills to Case Management

Several helping skills are vital to the provision of case management: (1) active listening, (2) empowerment and self-determination, and (3) cultural competence. Each of them is discussed briefly below.

Active Listening

As was examined in detail in Chapter 5, through nonjudgmental, attentive listening, helpers learn the needs and desires of those who ask for assistance. It would be impossible to be an effective case manager if you did not concentrate on what the service applicants and recipients were saying. All of the active listening and interviewing skills that have been discussed in previous chapters are applicable in the provision of case management services.

Empowerment and Self-Determination

The concepts of empowerment and self-determination are key to the provision of effective case management (Rothman, 1994). The case manager is not a mechanic or a physician who has all the answers and can fix things; rather, the service manager is a partner in the planning and the care. The message at all times must be, "We are in this together" (Kisthardt and Rapp, 1992, p. 118). As was discussed in Chapter 2, the informed choice of the person, couple, or family must be supported when developing the service plan. The service plan must take into account the uniqueness of each person, each bringing a different array of strengths and needs (Intagliata, 1992). Persons who ask for assistance from human service agencies must be free to refuse services if they are not a threat to themselves or society by doing so (Kane, 1985).

When people request assistance with obtaining services and with negotiating the confusing maze of agencies, this does not mean that those individuals, couples, or families are ready to give up control of making decisions about their lives. People who are feeling vulnerable and possibly ashamed about asking for help need to have a sense of hope, dignity, and self-worth. Case management should be delivered in a way that maintains, encourages, and augments a person's independence to the greatest extent possible (Intagliata, 1992). Not everyone will be able to achieve total independence; some persons may need ongoing support because of being mentally or physically impaired. People in these situations still need to maintain as much control of their lives as possible (Weil and Karls, 1985).

Kisthardt and Rapp (1992) describe case managers as consultants who listen to what is requested by individuals, couples, or families; help them find their own strengths and inner resources; and plan and implement ways to address the barriers and challenges. An important part of playing this role is supporting the choices and the vision of the persons who have requested consultation (Kisthardt and Rapp, 1992).

Illustration

Essie has come several times to the clothing closet in the community center where Marie works, bringing her two small daughters with her (one 2 and one 4 years old). On this visit Marie decides to try to learn more about Essie's circumstances, because she looks more tired and ill each time that she comes in. Offering her a cup of hot tea, Marie asks Essie if she minds chatting a while. Essie gratefully accepts the tea and sits down wearily on the sofa. During this conversation, Essie tells Marie that she and her children have been staying in an abandoned building with several other people whom she only knows casually. She says that they all use crack cocaine and are either stealing or selling sex to get the drug. When Marie asks Essie about her health, Essie says that she is also addicted to crack but wishes she could get clean and sober. She is currently connected to a man who is her pimp and her drug supplier, and he has been physically abusive to her on numerous occasions. Crying, she says that she thinks constantly about going into detox, but she does not know how to arrange for her children being cared for while she is in treatment. She is terrified of losing her children to the child welfare system, because this is what happened to three of her older children. Marie asks her if she has any ideas about what to do, and she just shakes her head "no." Essie seems to feel defeated and stuck.

Think About It #47

1. What are Essie's strengths?
2. What are Essie's challenges?
3. What would be your first steps if you were Marie?
4. How might a case manager help Essie to feel more powerful?

Cultural Competence

The concept of delivering culturally competent services was explored in Chapter 3. Being aware of someone's world view and cultural context is extremely important in developing a service plan. It is a waste of time and energy to come up with a plan, no matter how good it looks on paper, if the steps cannot be implemented by the individual, couple, or family due to their beliefs, history, religion, or background.

Illustration

Margaret is a case manager in a community mental health center. George, a man diagnosed with bipolar disease who had been coming to the center to have his medications monitored, was hospitalized suddenly because of a manic episode. George called the center from the hospital and left a message for Margaret that he was very concerned about his elderly mother, who was visiting him from Bosnia. Margaret looked in George's case record and saw that he had several younger siblings in the area, so she decided that George's mother should just go stay with one of the other adult siblings temporarily. Feeling very pleased that she had such an easy solution, she went to the hospital to visit George. She greeted him by saying, "I know you called about your mother, but don't worry about a thing. I'll just arrange to have her moved into your brother's house right away." Margaret was taken aback by the look of horror and confusion on George's face. She said, "What's wrong? Do you prefer that she go stay with one of your sisters?" George replied with shock in his voice, "Oh, no. You just don't understand. We just don't pass our parents around from place to place like that. I'm the oldest son; I have responsibility for her! I would be ashamed for her to leave my house!" Margaret replied, "George, I am so sorry. I should have talked to you before I ever thought about making a suggestion. In most African-American families like mine, it would be commonplace for an elderly relative to stay in someone else's home in a situation like this one. I was not thinking about what Bosnian families might expect." George was then able to tell Margaret that he and his family had already talked about this and that they had decided that they would take turns staying with his mother at his house until he was discharged. He then told Margaret that when he had called her to say that he was concerned about his mother, he was not worried about her being alone. He had called because he just wanted to talk about how embarrassed he was that this hospitalization had happened while his mother was visiting.

Think About It #48

1. In the above illustration, what did Margaret do that was not culturally competent?
2. What did Margaret do to correct her mistakes?

Barriers in the Provision of Case Management

Case managers can fall into several pitfalls if they are not careful. The following are a few things to watch.

Biased Service Plans

One pitfall is the tendency to connect persons only with those services with which you are familiar or which your agency provides. It is very important that a helper offers all possible services and benefits and allow individuals and families to choose for themselves. Sometimes case managers develop service plans based on their own feelings about an agency, which may at times be based on misinformation.

Favoritism

A second mistake is spending more time or effort on persons whom you like or feel close to and slighting persons whom you experience as more difficult or demanding. It is vital that decisions about frequency of contact are based on the real needs of the person, not on whom you like the best! Good case management treats everyone equally, taking into account the needs of the person rather than the convenience or preference of the worker.

Turfism

A third common barrier is letting interagency conflict over turf get in the way of accessing necessary programs. Territorial battles among disciplines or agencies interfere with the coordination of services (Kane, 1985). Sometimes competition develops with other agencies and workers in other organizations may be hesitant to coordinate services or make referrals to your agency. An effective case manager will strive to work through these problems with the service delivery system in order to develop the best possible service plans for persons and families. (See the discussion of external case coordination.)

Dumping

When you turn someone away from your agency without facilitating a connection with another organization that will better meet his or her needs, or make a referral to another setting just because you do not want to work with a particular person, or simply put someone on a bus and send him or her elsewhere, it is called "dumping." Obviously, dumping is not an ethical practice. Stay with the service applicant or caller until you have assured that he or she gets assistance from somewhere. Similarly, do not let another agency dump on your agency when they actually have the responsibility and the resources to help the person seeking help.

Neglecting to Involve the Person Requesting Help

It can be a disaster when a case manager tries to solve problems or put services into place without the full participation and knowledge of the person or family. You will be wasting your time and frustrating the service recipient if the service plan is not developed and negotiated mutually.

Improving Your Case Management Abilities

The following guidelines can assist you to develop better case management and service coordination skills.

1. Maintain an up-to-date resource library in your agency so that you can match each person's needs with existing community services.
2. Work on developing relationships with key personnel in other agencies who can help you gain access to services and better cut through red tape.
3. Learn the eligibility requirements of entitlement programs and community-based services so that you can make appropriate referrals. Keep applications for other programs in your office so that individuals and families can more easily obtain them. Know these forms well enough to assist others in completing them.
4. Get in the habit of assessing someone's unmet needs without thinking of what services are currently available. If you tailor your service plans only to what is available, what you know about, or what your agency offers, then the case management plan is driven by the limits of your community's resources, not driven by the true needs of your constituency (Rose, 1992). Then document recurring requests and needs that cannot be met with current resources so that you can provide data to policy makers and agency administrators who may be able to develop new services or funding sources.

Results of Good Case Management

To summarize some of the important points in this chapter, the following, illustrate the benefits that occur when service management and case advocacy are performed well.

- When there is one efficient and trusted human services professional acting as a case manager, he or she can facilitate the delivery of appropriate services and radically cut down on the service recipients' frustration with medical and social services bureaucracies, red tape, and paperwork.

- Since a good human services worker tries to anticipate future difficulties and help the person or family plan for them, case management decreases the number of crises in the lives of those seeking our support and lessens the impact of those crises that do occur.
- Because a case manager orchestrates services and coordinates care with other agency personnel, he or she avoids wasteful duplication of effort.
- With one human services worker on whom a person or family can depend, there is a single point of contact and access to a complex service network.

CHAPTER TEN

Class Advocacy and Community Organization

This chapter introduces the concepts and activities of class advocacy and community organizing and discusses how human services helpers can intervene in social problems at different levels, such as organizations, communities, and governmental entities.

Introduction to Advocacy

Working toward the goal of planned change in organizations and communities, addressing the needs of groups of people, and helping to organize people who are feeling without power are activities that are often called "advocacy." As was explained in the previous chapter on case management, there are two types of advocacy: case and class. Effective human services helping requires a dual focus: (1) commitment to delivery of quality services to individuals, couples, and families (called "case advocacy"); and (2) commitment to advocacy on behalf of populations whose needs are not being adequately met by the current social and medical services systems (called "class advocacy") (Rose, 1992).

Frequently advocacy takes the form of striving to change or make more responsive a social service or medical care system. This activity differs from interagency coordination, which can be accomplished when the service delivery system is fairly responsive to needs. Class advocacy for change means that helpers are striving to make a dysfunctional system more functional so that a group of people is more fairly treated. Even when social and medical services seem to exist to address a particular problem or population, it is useful to think about these programs in terms of appropriateness, adequacy, and access (Rose, 1992). More information about each of these concepts appears below.

• **Appropriateness.** If a service is available, is it offered in an appropriate and culturally sensitive way? For example, day-care centers for the young children of migratory farm laborers may be available, but may have no Spanish-speaking employees or volunteers.

- **Adequacy.** Is there enough of that service to meet the need, or are there long waiting lists or unfair distribution of benefits? For instance, there is a program in your state to help persons with HIV infection purchase expensive medications, but there is a two-year waiting period for assistance.
- **Access.** Can people actually get to the agency or service, and do they know about it? In one town, an after-school program had been funded so that children whose parents had not yet returned from work when they came home from school could have a place to go and be supervised, but no outreach or announcements had occurred so that very few parents knew of this opportunity. In another community, a medical clinic for persons without insurance opened, but it was on the second floor of a building without a wheelchair ramp and with no elevator. One of the authors, in trying to locate case management services for a friend who lived in another city, met with automated voicemail systems in the first six agencies she called. Not knowing the extension numbers or the names of any case managers, and unwilling to leave confidential information indiscriminately, she was unable to gather any information about the agencies' services. After numerous tries, she was finally able to talk to a human being. If the services are present but too difficult to access, then that is an advocacy issue.

Not only do advocates consider whether social services systems are meeting the community's needs, they also take into account the environment and context of the individuals, couples, and families. Sometimes the barriers and difficulties people experience are due to societal pressures or the lack of responsiveness of organizations or governments. Often attempts at social change are necessary to address a social problem. Therefore, human service helpers sometimes target their efforts on behalf of communities or groups of people who are finding it difficult to gain access to needed resources, are not adequately represented in community political decisions, or whose rights are being neglected or violated in some way.

In the human services field, how one confronts a problem depends on the causes or sources of the issue. Problems frequently have more than one cause, and therefore there is more than one way to go about finding a solution. Since causes or sources of difficult situations are most often complex and varied, the solutions may be out of reach of the individuals who experience their effects. A widespread social problem may require a larger social solution. Therefore, while providing support to individuals, couples, and families is very important, energies and efforts should not be limited to working only on those levels. Often interventions involve governmental bodies, human services organizations, law enforcement agencies, or Congress, because the solutions to the bigger problems depend on the responses of those systems. For example, it is important to provide shelter to a woman who has been battered when she is running from her abusive partner, but it is equally important to advocate for the community awareness, law-enforcement responsiveness, and change in laws that will help protect all persons who are experiencing relationship violence.

The Link Between Helping Individuals and Advocacy

"Practice" and "policy" are two areas of helping that have traditionally been artificially separated. The policies of human services agencies, as well as the laws of localities, state, and nation, have a direct impact on the delivery of human services. Economic, political, and social environments directly influence the lives of the people who ask agencies for assistance and the organizations who serve them. Conversely, what is learned from those served and from the actions of helping can influence the development of policy. The issues of direct service provision and of advocacy are therefore intertwined.

Advocacy skills are as important for human services workers as are other skills that we have discussed in this book, such as active listening, crisis intervention, and case management. The same basic principles of working with individuals, couples, and families are important in advocacy: starting where people are, finding and validating strengths, supporting informed choice, being culturally competent, and planning *with* rather than *for* communities that may need assistance. Advocacy requires courage, conviction, patience, and optimism, because changing systems, policies, and practices that have been in place for a long time is not easy.

Illustration

In the town of Arcadia there are several emergency shelters for persons without a permanent place to stay. However, all of these shelters (such as Salvation Army, Family Shelter, the Women's Shelter, and the Men's Shelter) require that all guests leave each morning at 6:30 A.M. The shelters do not admit temporary residents again until 5:30 P.M. This means that all adults who do not have jobs, as well as the children, have to be on the streets each day, where warm places are sometimes scarce. Adults and children therefore spend their days in downtown parking garages, post offices, and libraries. Sheila, a case manager with a local information and referral service, often helps individuals and families find a place in one of the shelters overnight. From listening to their stories of being on the street during the days, she has come to be very concerned about this policy of the shelters in the community. With her supervisor's help, Sheila sent a letter to all human service organizations in Arcadia inviting representatives from each agency to come to a meeting to discuss this problem. After discussing the problem at this meeting, the participants agreed on the following initial plan of action: (1) contact the shelter directors directly to ask them to consider opening up part of their facilities for persons who needed shelter during the days, and (2) contact some local foundations and corporations about whether they would be receptive to a funding proposal to establish a day-time drop-in center for persons without homes. They also agreed to meet every two weeks until some solutions to this problem had been put into place.

Think About It #49

1. Have you ever been involved in a problem, issue, or situation that could have or should have been addressed through advocacy efforts for a cause? What was the situation? How could it have been addressed through advocacy efforts?
2. What is your reaction to Sheila's advocacy efforts? What are your thoughts and feelings about being involved in advocacy efforts like the one described above?

Advocacy Activities

Human services helpers are often called on to intervene in social problems at a variety of levels, which might include advocating for a group or community of persons, developing new programs, striving to change an agency policy, educating the public about an issue, or acting to influence policy or laws (Rothman, 1994). You may help to organize persons who have common concerns so that they can be more effective in bringing about change. When advocacy takes this form, it is often called "community organizing." This term refers to helping a group of persons (such as a neighborhood, town, or interest group) to identify their most important strengths and to formulate a plan to address their problematic situations.

There are many activities that make up human services advocacy. Almost anything that has social change as its goal can be considered advocacy. The following activities are examples of advocacy efforts in which a human services worker may be involved.

- Presenting testimony to a local funding source about the needs of the community and its residents.
- Presenting written or oral testimony to the state legislature about a proposed bill that would be detrimental to the residents of the community.
- Working with a lobbyist from a large state agency to craft or change a piece of legislation.
- Talking directly to members of Congress or their staff members about legislation and issues that directly effect the population whom you serve.
- Providing information and technical assistance to a legislator on your agency or its service recipients.
- Developing or participating in coalitions that are formed to affect social change.
- Helping with a voter registration drive, perhaps with your agency's service recipients and their families, so that your agency or interest group can have a greater voice in the political process.

- Writing to legislators or the president about a national issue, like welfare reform, to state your views, concerns, and recommendations.
- Helping a local agency or support group to launch a letter-writing campaign to legislators about a particular important issue.
- Organizing or participating in demonstrations to bring attention to a certain issue or to protest a particular policy.

Illustration

Sally is a human services worker in the county Department of Social Services. Sally cares deeply about her community and participates as a private citizen in a variety of ways. She feels that she can perhaps make a small difference in the quality of life for citizens by taking a stand when and where she can. For example, this year she attended a "Take back the night" rally against sexual assault; she marched in the Gay Pride march; she participated in a candlelight vigil for the victims of relationship violence; she walked for the March of Dimes fundraiser; she attended the celebration of Martin Luther King Jr.'s birthday; and she attended a local ceremony for World AIDS Day. She has written to city officials protesting a hike in bus fare, believing that it hurts poor people. She has written to several county officials about her concerns regarding the lack of safe places for teenagers to gather after school. She has encouraged her Sunday school class to volunteer once a month in a soup kitchen.

Empowerment in Advocacy

Communities frequently face overwhelming barriers and difficulties, such as poverty, gang activity, lack of financial and social service resources, unemployment, drug or alcohol use, teenage parenthood, or substandard housing. These issues are real and should be addressed. However, the mistake that most people make when they approach a community, neighborhood, or group of people is that they only recognize and talk about the problems and neglect to discover and acknowledge the strengths. The principles of "empowerment" and "strengths perspective" are terribly important in this type of human services work.

Remember that the empowerment philosophy says that persons (1) have the right to decide for themselves and (2) are capable of managing their own lives, with the support of others. Community organizing increases the power of others who have traditionally had less power, so that the community can have a positive impact on their own current and future situations. Those persons who are most affected by a social problem have the rights and the power, not the advocates. Advocates are partners whose job is to help people learn to get along without them eventually.

In order to help an organization or group realize their own power, identify and acknowledge the capabilities, strengths, and successes that already exist. The strengths perspective strives to identify what a group has going for it and believes in the potential for growth in any group. This does not mean that advocates ignore problems and challenges; rather, helpers should balance their perspectives, identifying positives as well as negatives. Strengths are then used to begin to solve problems. In addition, it is important to build in successes and celebrate each victory, rather than contributing to people feeling helpless and like failures.

Illustration

A small group of staff members at the Bridgeville United Way became concerned about the neighborhood of Littleton, a predominately Latino area bordering downtown Bridgeville. For several months these planners brainstormed and researched the problems of Littleton and generated a comprehensive plan to address those situations. The United Way announced that there would be a public forum at the Littleton Community Center regarding the availability of funds for improving the neighborhood. The evening of the public hearing, the gymnasium was packed with interested and excited community residents. Members of the task force from the United Way spent 30 minutes at a microphone outlining their proposed plan to address teenage pregnancy, gang activity, illegal drug use, unemployment, the spread of sexually transmitted diseases, and the transmission of HIV. Their enthusiastic presentation was at first met with a sullen silence from the audience. Then a man raised his hand and stood up to ask for permission to speak. His comment was, "Are any of you Latino or Latina? Do any of you even speak Spanish? Do you have any idea who we are? Have you asked anybody from our neighborhood what we want to do with the money?" When he sat down many in the crowd cheered and applauded him. The planners who had offered their help and money were stunned and accused the people of Littleton of being ungrateful and uncooperative.

Think About It #50

1. Think about a cultural group or geographic community with which you identify. Now think about how you might feel and respond if an outsider, even one with good intentions, came into your group or community and immediately began pointing out the problems and presenting ready-made solutions.

2. What might be some more acceptable approaches for the well-meaning outsider to take?

Preparing for Advocacy and Community Organization

Human services workers are sometimes reluctant to become involved in advocacy for a cause because (1) they see these activities as diversions from "real" helping, (2) they view themselves as uninformed and ineffective, (3) they become frustrated by the complex agency or political processes, or (4) they see themselves as less powerful and influential than other types of workers or groups (Brieland, 1982). Perhaps some of you are feeling anxious about class advocacy and thinking, "I could never do that." There are several ways to begin to build confidence in your ability to be an advocate. The first thing to do is to learn about how advocacy is accomplished so that you feel more confident in your abilities and so that you can be ready to participate in advocacy activities when you have a chance. The second way to become more prepared is to ask the supervisors at your agency to begin to include you in advocacy opportunities that may arise so that you can learn on the job about how to be an effective advocate. Third, keep in your heart and mind the purpose of your participating in advocacy. Remember that many service recipients are vulnerable or lack power because they are ill, stigmatized, young, abused, under-educated, or frightened, and that they may depend on you to help them be represented fairly. If you do not make a stand for them, who will? Try to take courage from the fact that you are representing persons who are in danger of being neglected and who may not otherwise be able to get a voice in decision making about laws and policy.

Illustration

Gabriella is an outreach worker in a community nutrition program. Several Spanish-speaking adult male migrant farm laborers have hitchhiked into the office to ask for emergency food assistance for themselves and their families. Gabriella asks them for information about their circumstances and learns that they were recruited from Homestead, Florida, by a labor contractor (called a crew chief) and brought to the area to pick strawberries. They are living at one of the camps operated by the Sunny Slope farm. They have been in the community for a month without work, because the strawberry crops were damaged by late frost. The crew chief has been supplying them with food, but the prices are exorbitant, and they are all in great debt to him with no hope of income. Gabriella asks directions to the migrant camp and goes there the next day. There she finds over 30 men, women, and children, some of whom are sick and all of whom have very little food. She tries to interview the crew chief, who acts angry and hostile and orders her off the property.

Think About It #51

Brainstorm some possible next steps that Gabriella might consider taking. Whom might she talk to? To whom could she appeal for help? What organizations or civic groups might help her?

Beginning to Advocate

How can you begin to improve your advocacy and community organization skills? You may already have an important issue in your agency or community to work on. Where do you start?

1. Get to know the people in the neighborhood, community, or group. Talk to persons who are affected by the problem about the history of the situation and how they view the future. Do not assume that you understand their perspectives on the problem or that you can choose a solution that will work for them.

2. Be sure that you research and document the nature and extent of the problem (see "gathering data" below). You need to understand the situation and be able to communicate the scope of the problem to other human services workers, residents or consumers, potential sources for funds, and legislators. Always include members of the community in your advocacy efforts. Invite community members to speak on their own behalf. Also, be thoroughly prepared to speak on behalf of a group of people or about a particular situation if and when the need arises. When you speak about the situation, state the problem clearly, suggest at least one solution, and back your statement up with data and stories that illustrate the reality of the situation.

3. Use existing networks of interested persons or form coalitions to begin to formulate possible solutions. Be sure at every step to involve representatives from the populations that are most affected by the problem.

4. Consider using the media to publicize the community's position or proposals for change. Make sure that you are well informed and well prepared.

5. If you are representing a particular group, community, or organization, be sure that the persons involved have given you permission to do so and have endorsed your statement. If you are speaking on behalf of your own agency, make sure that your supervisors are behind you and know what you will be saying. Tell policy makers whom you are representing and that they have appointed you as spokesperson.

6. If you are trying to conduct legislative advocacy (such as introducing or stopping a bill in the legislature or a proposal in the county council), study the law-making process; learn about the proposals, laws, and rules affecting the population or community in which you are interested; and determine who can change the situation or decision.

Gathering Data for Advocacy

How can you learn more about a social problem in your community in order to educate your peers and lawmakers about it? One place to start is census reports, which show a trend over time (for example, the growing population of persons over age 80). You can also get crime reports, from law enforcement officials (for example, to show an increase in drunk driving, relationship violence, burglary, or gang activity). You may look at epidemiological data from the Centers of Disease Control and Prevention about incidence of health problems (for example, the increase in cases of HIV, herpes, or tuberculosis). You could survey agencies or community residents about their observations about the most important social problems in the community, search for scholarly literature on the topic, read recent media reports on the situation, see what has been happening in nearby or similar communities, or use the statements of politicians or celebrities who are supporting the cause. The purpose of collecting information on a social problem is twofold: (1) to understand the problem so that the proposed solutions are appropriate and (2) to have a better chance of convincing funders, politicians, and the public that the problem deserves attention (Netting, Kettner, and McMurtry, 1993).

Several points to be clear about as you do research about a problem are:

- Why has it been difficult to change this condition?
- How was this condition viewed and defined in the past?
- Why is the situation persistent or growing worse?
- What can be theorized about the origins and/or causes of this situation?
- What are the predominant public attitudes about this situation and the people affected by it?
- What is the general view by the people who are most affected by this problem?
- What resources are already available that may help?
- What solutions have been tried before, and what can be suggested as possible ways to address the problem?
- Is there sufficient attention paid to this difficulty by the public and political leadership?
- Do you need to create some attention for this situation? (Netting et al., 1993)

Tasks for the Politically Involved Helper

You may become more involved in many ways in the governmental and political processes that affect your community and the persons whom you serve. The following are examples of ways in which you may be able to get started.

1. Offer your expertise, data, experience, and assistance to legislators who are exploring solutions to problems about which you have information.
2. Develop relationships with legislative aides and other staff persons in the offices of local and state politicians so that you can call to register a concern about a particular issue. This includes your county council and mayor's offices.
3. Track important pieces of legislation, and follow where they are in the enactment process. You can do this by calling the state's legislative information line to request copies of bills or by asking where a bill is.
4. Help to form coalitions with other groups who have similar interests on a particular topic.
5. Furnish service recipients and agency personnel with the telephone numbers, addresses, and other information they may need in order to communicate with policy makers.
6. Pay attention to announcements about public hearings on issues about which you care, and attend those hearings. You may simply listen to get an idea of what the different opinions and positions are, or you may prepare and offer testimony of your own.

Illustration

Joe works as the transportation coordinator for a local Council on Aging. Recently, each time that he expresses a concern that the agency needs newer and safer vans in which to transport senior citizens to centers and doctor's appointments, he is told by his superiors that the budget had been cut by the county and the state, and that there is no money for new vans. With the support and permission of the agency director, Joe testifies before several bodies about the needs of his agency for safer transportation. He gives a brief statement to the State Legislative Committee on Aging when they have a public hearing. He submits written testimony to a hearing on the use of Social Services Block Grant money. He speaks at an open county council meeting about the need for funds. He writes a statement that his boss uses in a grant proposal that she is writing to the State Department of Transportation to request funds. When the local media become interested in their problem, Joe organizes some of the elderly persons who use the agency's services and helps them to make pleas for funds in newspaper articles and television interviews. With Joe's diligent help, the agency is able to raise the money from a variety of sources to purchase four new vans and to repair several others.

Advocating for Change in Another System

If you are going to try to advocate for change within another agency or system, learn as much as you can about the rules of that organization and how it operates. Try to determine what can and cannot be changed. Use your contacts and existing network to research possible solutions. Be flexible and willing to negotiate with the agency personnel in order to make headway in your goal of making the organization more responsive to persons who may need assistance.

Illustration

Lori is a human services student intern at a shelter for homeless women. Lee walks into the office, after having walked for miles, to request temporary shelter. Lee appears to be an African-American woman. She says that she has no friends or family, and no money or food, and nowhere to go. She says that she spent the night before in a public park in town. Lee tells Lori that she has just been released from prison, but the name of the prison she gives is the men's correctional center. Lori then questions Lee about this discrepancy. Lee tells Lori freely that she is a transgendered person and has been living as a woman for many years. She had begun hormone therapy but never had gender-altering surgery, so she has breasts and a penis. Lori asks her supervisor for a decision and is told that Lee is technically a man and cannot stay in the women's shelter. Lori then takes Lee to the men's shelter next door and explains the situation to two of the staff members there. The two men laugh in front of Lee and are obviously uncomfortable and frightened of her. Lori returns to the women's shelter with Lee and calls several other shelters, all of whom say no when they hear that she is transgendered. Finally, Lee shakes Lori's hand and says, "I appreciate all you've done, but honesty ain't doing us any good." Lee then leaves the shelter and begins walking back toward town.

Think About It #52

Lee is a service applicant who has fallen through the cracks. What could you do as an advocate about this situation? Where would you start?

Advocacy Within Your Own System

Advocacy for change within one's own agency can bring with it some ethnical dilemmas for the human services worker. The loyalty of an advocate is to the persons or groups who are being neglected by society or the service system; your complete loyalty cannot go to the organization that employs you. This conflict of values requires a strong sense of purpose and mission, as well as a strong identity as a human services advocate. Dilemmas faced by many helpers include questions such as: Will I jeopardize my job if I am too outspoken for the rights of those I serve? Is it safe to make waves in my agency? Workers in large organizations (such as hospitals, prisons, and bureaucratic agencies) often find themselves faced with rigid rules and restrictions that hamper their ability to provide empowering, appropriate services. These workers often must decide how to best serve those who need help within their own service provision system.

Illustration

Jill is an adoption worker in a local public agency; it is her job to interview prospective parents and perform the initial screening of their application. One of the state rules about adoptions of infants is that the prospective parents must be under 40 years of age. Jill has met twice (once in the office and once in the home) with a couple, the Holts, whom she feels should pass the screening and go on to be full applicants, but they are both 42 years old. They have been married for 20 years and are both college professors. They seem to be financially and emotionally stable. They have tried unsuccessfully for ten years to conceive and now wish to adopt an infant. Jill realizes that she can either reject their application or appeal to the state office for a waiver of the rule.

Think About It #53

1. Have you ever been in a situation where you have disagreed with your employer's policies because you did not think that they served the best interests of persons asking for help? What did you do?
2. Discuss with your fellow learners situations where agency policies might conflict with the wishes of service recipients. (For instance, a hospital administrator or physician who tells the discharge planner that someone must leave immediately, but the discharge planner

and the sick person both feel that discharge is too early). What difficulties are posed for the human services helper?

One of your most important tasks as an advocate within your own system is to determine what can be changed. For example, if you are restricted by federal guidelines, then you may have to exist within those parameters. However, if the decision about a certain policy has been made locally, you may be more successful in changing the way that things are currently done.

A human service worker should do everything in his or her power to make systems responsive to service recipients and their families. However, if it becomes too risky to be vocal with your bosses, you may need to consider other options for advocacy. Your alternatives include:

1. Finding an outside advocate (such as a social worker, attorney, clergy person, or someone in an agency designed to advocate on behalf of vulnerable populations) who will protest your agency's procedures and call attention to them.
2. Encouraging people to file grievances or to call legislators.
3. If you cannot live with the agency's policies, changing jobs.

Illustration

Toni had become very concerned about the HIV epidemic in her community and sought out employment that would allow her to have an impact on the growing problem. She was very pleased and excited to have secured a job with the local health department doing pre- and post-test counseling. This meant that she would be able to meet with persons who were requesting an HIV test, answer their questions, listen to their concerns and fears, and provide them with resources and safer behavior education. After a year of being with the health department in this capacity, Toni was upset to receive news of a drastic policy change. The state legislature had voted to abolish the anonymous testing policy and would now require each person seeking an HIV test to give his or her name, address, and Social Security number. Within weeks of enactment of this law, Toni noticed a dramatic drop in the number of persons seeking HIV testing from the health department. From talking to volunteers who answered a local AIDS Hotline, Toni learned that many hotline callers were asking for the addresses of health departments in neighboring states so that they could test for HIV anonymously. Toni was also told by advocates in a local HIV support organization that persons who were economically disadvantaged and likely to be without transportation or insurance were not getting tested at all. Toni launched a campaign of protest to her supervisors, agency commissioners, and legislators about the change of policy. After several months of being vocal about her concerns, she was told by her department head that she had to be silent about her concerns if she wanted to keep her job as a test counselor. Toni wondered if she should stay at the health department and continue to try to work more silently against the new law, or whether she should quit the job and seek employment elsewhere.

Think About It #54

1. What is your reaction to Toni's dilemma?
2. What would you do in her situation?
3. What do you think about the ultimatum from Toni's supervisors?
4. Are there ways in which Toni may be able to enlist outside advocates to help her change the agency?

PART FOUR

Applications to Special Areas

It is not possible to address all of the populations and social problems that human services helpers may face in their work. However, this section attempts to offer the beginner information on six issues that are commonly encountered: relationship violence (abuse or neglect by persons known to the victim), older persons, HIV infection, mental health issues (mental illnesses and addiction), developmental disabilities, and grief. The authors recognize that many other important issues have been omitted, such as persons without homes, migratory and seasonal farm laborers, persons receiving public financial assistance, and persons who have limited economic resources. Nor have we discussed important programs such as legal aid, criminal justice services, hospice, foster care, or adoptions. The list is endless. What we have tried to do is give the reader a taste of the variety of issues that may be addressed through a career in human services.

CHAPTER ELEVEN

Violence in Relationships and Families

It is likely that you, as a human services helper, will encounter violence within families or couples, or suspect that it is occurring. It is important that you be prepared for this situation. This chapter introduces you to a variety of abuse and neglect situations, including abuse of children, elderly persons, persons with disabilities, and partners. Physical, verbal, and sexual abuse; property destruction; neglect; and violation of rights are explored as well. Several theories concerning relationship violence are introduced.

The Human Services Worker in Relationship Violence Work

It is important for you as volunteers, intern students, and human services workers to have enough information about abuse and neglect so that you can alert your supervisors to possibly dangerous situations. You should never act alone, unless there is clearly a present emergency that requires that immediate action be taken. If you ever suspect that someone of any age or gender is being emotionally, physically, or sexually harmed, then your job is to document what you have heard or observed as clearly as you can and discuss the situation with someone in your agency who has more experience and authority. By law, someone in your agency must report any incidents of child or elder abuse or neglect.

After abuse or neglect has been identified, the role of the human services helper is often to provide emotional support to the victim or practical support to the family who is affected. You may also be called upon to provide case management to a family who has come to the attention of the formal service network.

Introduction to Relationship Violence

Violence can be defined as an act that has the intention to do harm or to hurt someone (Steinmetz, 1987). Violence perpetuated on someone whom the abuser knows well and with whom he or she has a relationship is often called "domestic violence" or "family violence." "Relationship violence" refers to all types of physical, sexual, and mental abuse of someone with whom there is a relationship. It is distinguished from abuse by a stranger or casual acquaintance. *Relationship violence is usually ongoing abuse over time, in a recurrent pattern, by someone with whom the victim has an intimate relationship,* such as a family member or domestic partner. Sometimes, however, there is only one incident of violence in a particular relationship.

Because relationship violence does occur within an intimate relationship, the dynamic is particularly painful. Ending the relationship may seem difficult or impossible. When victims of violence are children or persons who are ill or dependent, they may indeed be trapped and unable to escape the violence or the relationship. Even when the victim is a fully functional adult, the emotional, financial, and social bonds may be so strong that leaving the relationship does not seem like an option. Even if contact with the abuser ends, the relationship (father, sister, ex-lover, child, etc.) goes on. Sometimes the recipients of the abuse are in greater danger of being killed as they are leaving the relationship or after they have left, so even ending the relationship does not guarantee safety.

This relationship dynamic is very important to understand. You may believe that love and abuse cannot coexist. Regardless of your personal convictions, it is important to remember that abuse often happens in the context of an ongoing relationship. *It is this relationship dynamic that makes it possible for the abuse to continue.* If a stranger or casual acquaintance hit you, you would be likely to file a complaint and would certainly avoid that person in the future. What if your assailant is someone to whom you have pledged your lifelong devotion? What if your assailant is your parent, child, or lover? What do you do then?

Examples of abusive activities are: use of a gun or knife in violence or threats; threatening to kill someone if he or she leaves the relationship; breaking or burning another's possessions; withholding medication or money; belittling, name calling, and criticism; striking, slapping, or hitting someone; and yelling, shouting, and creating undue tension in the home.

The relationships in which violence occurs take varying forms: for example, the abuse of a child, an older person, a person with a disability, or a sick person by a caretaker. Abuse can happen between siblings, between parent and child, and between sexual partners.

Females are most often the victims of relationship violence, but men are not immune to this type of victimization. Elderly men who are frail or have dementia can be abused by a caregiver. Men with vulnerabilities such as an illness or disability can be abused by persons they trust. Fathers can be abused by their teenage sons. Gay men can be battered by their lovers. Heterosexual men can be abused by their female partners. Anyone can be the victim of relationship violence.

Unfortunately, family violence is not uncommon; it occurs in at least half of the homes in the United States at least once a year (Benedek, 1989). Women and children are more likely to be assaulted in their own homes than on the streets, even in the most violent of U.S. cities (Barnett, Miller-Perrin, and Perrin, 1997).

Weise and Daro (1995) estimate that 1271 children died as the result of child abuse or neglect in 1994 in the United States. Another million or more children are victims of physical abuse, sexual abuse, emotional abuse or neglect (Gelles and Cornell, 1985). There is also growing concern about the abuse, neglect and exploitation of elderly people and adults with disabilities. The prevalence of elder maltreatment is reported to be as high as 4 percent of all older persons (Cash and Valentine, 1988).

In 1992, of the 13,805 U.S. homicides (male and female) where the relationship between the offender and the victim was known, 15 percent were committed by a spouse, ex-spouse, boyfriend, or girlfriend, and 11 percent were committed by another relative. Of the 3454 murders of females during the same year, when the offender could be identified, 41 percent of these crimes were committed by a spouse or boyfriend and 15 percent by another relative (Barnett et al., 1997).

Sometimes the assumption is made that people abuse their family members because they are bigger and stronger than the victim. While size and strength may play a part, the most important factor is the fact that the abuser is willing to harm someone with whom he or she shares a relationship. For example, Mary Ann, a therapist, was told by a man in good health that he was being physically and emotionally abused by his wife. Mary Ann, puzzled by this because she knew that his wife relied on a wheelchair for mobility, asked how it was possible for him to be physically abused. The man answered, "This is the woman I love. This is the woman I live with. I do not expect for her to suddenly throw a lamp or a book at my head. When it happens, I am in my own home, relaxed and trusting, and I am unprepared."

Your definition of relationship violence should incorporate the possibility that any type of person has the potential to be in a relationship that is mentally or physically abusive. Otherwise, you may become blinded to the signs of relationship abuse.

Causes of Violence

It is widely believed that violence is a learned behavior. In other words, some people learn that it is OK to hurt another person. Families and individuals who are accustomed to violence or who believe that violence is an option may become abusive in stressful situations. Some families, couples, or individuals experience violence throughout the relationship, regardless of the external stress to which they are subjected. Families and romantic partnerships are prime situations for abuse, because everyone is emotionally invested and the power and resources can be unequally distributed.

The authors have a strong bias that relationship violence is never justified. Human beings do not have the right to hurt each other. Every person has the right to safety and health. No one deserves abuse, no matter what he or she has done or how obnoxious he or she has been acting. Helpers must take a strong stance against abuse in all forms.

Illustrations

1. *William was the only boy in a family of five siblings. William's divorced mother (Peggy) married his stepfather (John) when William was 6. John was loudly verbally abusive to William's mother and to all five of the children. He was physically abusive to Peggy. Peggy and the four girls were never aware that John was also sexually abusing William. The sexual abuse began when William was 7 and continued until he was 11, at which time the marriage ended. William did not know how to tell his mother, sisters, or teachers about the sexual abuse by his stepfather.*

2. *Dave was 21 when he married Sheila, a divorced mother of two. Sheila's two children lived with their father, who had custody of them. Shortly after Dave and Sheila were married, Sheila began showing extreme and irrational jealously, trying to control Dave's every move, screaming her suspicions and criticisms of him constantly and threatening to kill him if he ever cheated on her. Once in a fit of rage she burned Dave's clothes. The abuse escalated when Sheila pointed a loaded hunting rifle at Dave's head one evening. Dave was ashamed to admit to his friends and family that he was afraid of his wife.*

3. *Priscilla was a married mother of two who was in graduate school part-time. Her husband Brad was moody, unpredictable, critical, and unfaithful to her. He was verbally abusive, telling her that she was no good and was not smart enough to finish school. One day, after Priscilla stepped away from the ironing board to pick up one of her toddlers who was crying, Brad threw the hot iron at her head. Every time she thought about leaving, she thought, "What about the children?" and, "What about school?"*

4. *Judy and Regina had been lovers for several months before they decided to move in together. Judy had been displaying signs of irrational anger and jealousy, but Regina had felt that when Judy felt more secure about their relationship, Judy would stop that behavior. Judy's verbal abuse continued and worsened after they moved in, however, and she also became physically and sexually abusive of Regina. Regina was ashamed to call the police because she dreaded revealing that they were lesbians and that she was being victimized by another woman.*

5. *Esther was an 82-year-old woman who had suffered a series of strokes and could no longer walk or talk. She relied on her only daughter and son-in-law to provide her personal care in her home. Both of her caregivers routinely slapped her, treated her roughly when they turned her, let her lie in soiled linens, and neglected to give her medicine, food, and water. Even when other relatives visited, she was afraid to communicate her abuse and misery to them, because she assumed she would be placed in a nursing home if the abuse was discovered.*

Think About It #55

1. For each of the above illustrations, make notes about what you think the victim of violence in each situation might be experiencing. For example, what might be frightening to him or her? What emotions might the victim be feeling?

2. For each of those illustrations, jot down your own immediate thoughts and feelings about the persons and their situations.

3. Look at your responses to question 2, and identify areas where you are surprised at the intensity of your feelings or the nature of your thoughts. What biases or emotions surface for you in thinking about this topic?

4. Do you believe that relationship violence is ever justified?

5. Think about how your own experiences with abuse and neglect, as well as your attitudes about violence, may influence the way you respond to situations you may encounter as a helper.

Types of Abuse and Neglect

Abuse is a term usually applied to violent actions that are perpetuated on another person. Neglect usually refers to something that does *not* happen. An example of abuse would be beating a child. An example of neglect would be not feeding a child. Sometimes the boundaries between the two can be blurred, or both abuse and neglect can occur together.

The types of abuse and neglect can be categorized for a more detailed understanding: violation of rights, financial exploitation, destruction of property or pets, verbal or emotional abuse, physical abuse, physical neglect, and sexual abuse.

Violation of Rights

Violating someone's rights can include involuntary placement in a mental institution or nursing home or having a person who is older or someone who has a disability legally declared incompetent in order to use his or her resources. It can include denying someone the right to worship, the right to free speech, the right to privacy, the right to property ownership, or the right to make his or her own decisions.

Illustration

Anne is a young woman in a rural area who does not define herself as an abused woman because she is not being beaten. However, her husband keeps tabs on her every move. He requires her to ask permission to use the family car and to tell him where she is going; he keeps a log of the mileage and checks it every evening. He has actually traveled a few times to where she said she had gone is order to double-check the mileage. He put a block on the telephone that prevents her from making long-distance calls. He will not allow her to socialize with friends, to go to church, or to visit her mother. He does not allow her to have a checking account or credit cards or to use his. Anne wants to work outside the home, but he will not allow her to do so.

Financial Abuse or Exploitation

Financial abuse or exploitation is defined as taking advantage of someone financially, misusing resources, or controlling someone's access to money. It is most common with elder abuse and neglect, but often happens to abused partners as well. For example, a batterer may prevent a partner from getting or keeping a job, having a checking account, or managing his or her own money.

Illustration

Rose is an elderly woman who lives with her only son, who is an alcoholic. She feels dependent on him for her care and is afraid to live alone. He has forced her to transfer her savings and her house into his name. He also is spending her Social Security checks and her food stamps on himself. He has told Rose that he will put her out into the street if she does not do as he says.

Verbal/Mental/Emotional Abuse

Some violence is not physical but consists of the maltreatment of someone through words, threats, and criticisms. This type of abuse can occur alone or accompanying physical or sexual abuse. Many victims of violence report that the verbal and mental attacks hurt them worse and last longer than the bruises and broken bones. When a child experiences mental and verbal abuse, it can have a lifelong effect on that person's self-esteem and confidence.

This type of abuse includes humiliating, degrading, mocking, ignoring, withdrawing, provoking fear, following the partner or having him or her followed, intimidating through looks or gestures, displaying weapons, isolating him or her, punishing by delaying meeting someone's physical needs, game playing to frighten someone, convincing the victim he or she is at fault for the abuse, and name calling. Mental abuse can include making threats of murder, suicide, physical harm to oneself or others, or institutionalization.

Illustration

Sheri came to the battered woman's shelter after an experience with her boyfriend that was particularly frightening to her. He had taken her for a walk in the woods behind his cabin and showed her a large rectangular hole he had dug. He simply said to her, "That's your grave." She thought that he was possibly just trying to control her, but she was not sure that he was not planning to kill her soon. Although he had not recently hit her, she felt frightened and in danger.

Destruction of Property or Hurting Pets

Sometimes abuse takes the form of destroying things the victim values, in an effort by the abuser to prove the intensity of control or the legitimacy of the threats. For instance, one woman went outside one morning to get into her car to go to work and found her pet cat hanging by its neck from the rear-view mirror.

Her husband had left a note on the seat that said, "This could happen to you." One woman came home to find that her grandmother's china had been broken into bits by her abuser. Another woman experienced her batterer's showing up at her jobsite with a sledgehammer and proceeding to smash the windows out of her car. This type of violence is often the first step on a continuum of abuse and can be an important warning sign to heed.

Illustration

Rocky and Ron were lovers who were in couple counseling. The counselor was disturbed by Rocky's account of Ron's losing his temper one night and slashing the cushions of the sofa and many of Rocky's clothes. Suspecting that this might be the tip of the iceberg of violence, she asked Ron to leave the room while she talked with Rocky. She confronted Rocky directly about her suspicion that Ron may be hitting him; Rocky reluctantly admitted that he had suffered much sexual and physical abuse in the relationship but had been frightened to tell anyone.

Physical Abuse

Battering, or physical abuse, is what most people think of when they think of relationship violence. Certainly it has serious ramifications in terms of physical pain and injury and often death. Not only do the victims of physical abuse die as a result of abuse, the perpetrators of abuse are often killed by the victims. Deaths from physical abuse can be accidental or purposeful. For this reason, it is one of the most serious and dangerous forms of relationship violence.

Physical abuse can include hitting, pulling hair, burning, shoving, pinching, kicking, pounding, choking, depriving someone of sleep, stabbing, cutting, or throwing things. It includes being shot at, tied up, or locked up.

Illustration

Caryn was in an abusive relationship with her husband Sidney and had decided to stay because they were in the process of adopting a baby. During the most recent violent incident, he pounded her in the face with his fists while he was driving. He then opened the door on the passenger side and threw her out onto the highway. He stopped the car, dragged her over to a nearby highway overpass, and held her by the ankles upside down, over the traffic below.

Physical Neglect

If physical abuse is the commission of violence, physical neglect can be described as the omission of safety. This includes not providing needed personal care, food, medicine, comfort in extreme weather, shelter, a safe environment, and adequate supervision. The victim of physical neglect could be dehydrated, malnourished, left alone inappropriately, or denied medical care.

Illustration

Fred was a 30-year-old man who was very ill from HIV Disease. He was bed-bound, incontinent, and mentally confused. His only caregiver was his mother, who was alcoholic. Neighbors called the local AIDS Service Organization when they realized that Fred was being left alone in the house every evening while his mother drank in a local bar. He was taken by ambulance to the emergency room, where it was discovered that he had bedsores, was severely dehydrated, and had not been bathed in a long time.

Unintentional physical neglect.   Consider the motivation of the caregivers when you assess neglect. Sometimes a helper will mistake poverty or lack of resources for willful neglect. In situations where the caregiver is doing the best he or she can but needs better or more resources, then it is your job to assist the person in gaining access to needed support and resources.

Illustration

Workers from the child welfare agency were called to assess the possible neglect of three small children who lived with their grandmother, Delores, in a housing project in an inner-city area. The workers found that the children were playing in a vacant lot filled with garbage, where rats were evident. The children were clean, well, and healthy, but their clothes were tattered and patched. The grandmother seemed to love them and was adamant in her desire to care for them adequately. She explained, however, that they were all living on her Social Security check and that she could not afford to move to another neighborhood. The children were not removed from the home. Rather, the service plan included helping the grandmother to apply for food stamps, AFDC, day care, and a Section 8 housing subsidy.

Sexual Abuse

Sexual abuse is a type of physical abuse, but it often has the added dimension of being even more hidden and shameful for the victim. Sexual abuse includes forced or unwelcome sexual activity, forced prostitution, sexual activity punctuated by physical violence, calling someone abusive names during sexual activity, involving other persons in sexual encounters against the victim's will, involving animals in sexual activity, videotaping the activity against the victim's will, or forcing someone to participate in any deviant or aversive sexual activity.

A person of any gender, sexual orientation, income level, ethnic group, or age can be the victim of sexual assault in a relationship. Up until recently, rape was not recognized within legal marriage; however, violent and coercive sex is possible in any relationship. Now there are laws in many states to protect the victims of marital rape. Having a marriage license should not give anyone the right to sexually humiliate, harm, or control another person.

Illustration

Jenny's husband was often abusive during their sexual encounters. He would slap her or throw her around, call her "bitch" and "whore," and would enter her vaginally or anally in a way that would hurt and tear her. Occasionally he would bring in a third person to have sex with them, which made Jenny very uncomfortable and upset. However, she felt committed to him and to the marriage and was ashamed to tell her minister about her experiences. She also had no sense of whether this type of behavior was normal for other couples; she was embarrassed to ask any of the women that she knew. One day she walked into the bedroom and found her husband with his penis in the mouth of their 4-year-old daughter. She confronted him, which led to an explosive abusive incident. It was at that point that she decided to leave him at the first opportunity.

The Cycle of Violence

Lenore Walker (1979), through her counseling work with women who were battered, documented a three-part cycle of violence that represents a fairly predictable pattern of abusive behavior.

The tension-building phase is characterized by many minor incidents of abuse, which gradually build in severity. This phase is that state where everyone in the house is "walking on eggshells" or "waiting for the other shoe to drop." In other words, the tension is mounting, and the power struggle is evident to both the batterer and the victim. This phase can last for a day or a year or more, depending on the nature of the relationship and where the family or couple is on the violence continuum (discussed in detail later).

After the tension building comes the acute battering, the violent incident, the explosion resulting from the stored up tension and rage the batterer feels. It is impossible to predict the nature of the attack or its intensity. The abuser can be blind with fury and feeling unable to control his actions. For the batterer, this is often a welcome release from anger and tension.

The honeymoon phase follows the battering incident. The batterer exhibits calm, attentive, loving behavior. At this point the assailant is genuinely regretful for losing control and for hurting the other person. Some couples refer to this as "making up after a fight." Abused children experience this phase with much confusion, because it appears to them that their parent or caretaker is inconsistent and that they can not predict when they will receive affection or violence. The batterer may cry, apologize profusely, give presents, nurse the wounded person, pay close attention, and promise that the abuse will never happen again. Walker says that a woman at this point is facing her Prince Charming, the man of her dreams, who is pledging undying love for her. Since they love each other and are committed to each other, the relationship is reinforced, and she is able to forgive him.

The honeymoon portion of the cycle is temporary, of course. Unless the batterer begins to seek therapy that includes confrontation and behavior-modification the tension building inevitably begins again, and more battering incidents follow. Often this cycle is repeated over and over before the victim gets any sense of the fact that there is a pattern. By that time, many couples are even more bonded to each other through time together, social expectations, and financial commitments.

Illustration

Mari, a counselor for a shelter for women who are battered, visited Joan in the hospital the day of a violent incident between Joan and her husband. Mari found Joan determined to end her marriage. Joan had a broken jaw and a broken leg in traction and was frightened and angry. The next day, Mari returned to find the hospital room filled with flowers. Joan gestured to the floral arrangements with pride and with tears in her eyes and said, "They're all from my husband. He loves me so much." He had been visiting her in the hospital, and they had made plans to reconcile.

Sometimes victims of violence are aware on some level that a violent incident can be triggered or postponed through their behaviors. This is not to say that anyone ever asks for abuse or deserves it. Violence cannot be controlled, but sometimes it can be triggered or delayed. If a woman, for example, intuitively recognizes the tension-building phase, she may try to push the batterer into the acute battering in order to get it over with. She knows that she cannot stop it, that it is inevitable, but she thinks that she can perhaps manage it. By the same token, she may try to manage his anger long enough to get through a meal without violence, knowing that it may occur later that day. One woman who felt the tension-building phase intensifying decided to push her husband into the acute battering stage so that her cuts and bruises could be healed by the time her parents arrived at their house for Thanksgiving.

Education of Victims About the Cycle

The victims of an abusive relationship tend to minimize the violence or to lie to others about its existence or its intensity. This denial is often an unconscious way to protect the victim from the anxiety and threat of future harm. It also stems from genuine hope and love. "It's no big deal; I can cope; it will get better," are examples of messages victims send to themselves and to others. It is very important that the victims of relationship violence be asked to look at the patterns in their relationships; it is your job to introduce this in a loving, accepting, nonjudgmental way. Often when the victims of relationship abuse see the cycle drawn and hear it explained, they realize what they are experiencing and can make a more informed choice about how to survive. It is useful to teach victims about the cycle of violence so that they can make safety plans for the next incident and so the seed is planted for them to reach out for help again when the tension building starts.

The Continuum of Violence

It may be useful to see violence as occurring along a continuum of intensity and seriousness. (A continuum is a line that represents progression from one extreme to another.) An abusive relationship may start with less dangerous or life-threatening forms of violence, but as time progresses, the violence gets worse. At the far end of this continuum is death for one or both partners. Relationships that have progressed to intense abuse can often end in a murder-suicide. Also at the far end of this continuum, violence may occur more often, escalate rapidly with little buildup, and involve a shorter honeymoon period. It is useful for victims to realize that violence may worsen over time. Often victims have much hope in the improvement of the situation and believe the abuser's promises to change despite evidence to the contrary.

Victim-Blaming

Think about how often you have heard said about a battered woman, "Why doesn't she just leave?" Now think about how often you have heard said about a batterer, "Why doesn't he just stop?" Our society tends to be focused on the behavior of victims rather than on that of abusers. We therefore participate in the dynamic that perpetuates abuse. We generally blame victims for their circumstances, rather than affirming their survival skills and trying to facilitate their safety.

Victims of violence usually want the violence to end, but they want the relationship to continue. It is very difficult to make decisions about one without the other. It is important for helpers to listen closely to what the victims of relationship violence say about what is holding them in the relationship and why they are reluctant to end it.

Responses to Being Abused

The victims of violence can experience symptoms and conditions that result from being held hostage in their own homes and relationships. These symptoms, sometimes called "post-traumatic stress responses," can include: sleep disturbances, eating disorders, nightmares, uncontrollable flashbacks, depression, low energy, panic reactions, stress-related sickness (such as nausea, headaches, chest pains, asthma, feelings of choking, hyperventilation), and being hypervigilent (being easily startled or being very jumpy).

Regardless of the nature of the violence or the age and relationship of the victim, experiencing violence teaches a person certain lessons about life. Some of the messages that may be learned by victims of violence are:

- They learn to expect to be judged, criticized, blamed, and hurt by the person they trust and love.

- They learn to feel unlovable, demoralized, and worthless.
- They learn to distrust people and environments.
- They learn to feel guilty and to take too much responsibility for other people's feelings and actions.
- They learn to feel incompetent and helpless.
- They learn to feel that they deserve abuse.
- They learn that being violent is the way to get what one wants.
- They learn to live with terror, tension, and agitation.
- They learn to be paranoid, hypervigilent, and poised for defense.
- They learn to ignore physical and emotional pain.
- They learn to be cautious, immobilized, paralyzed, and unable to act.
- They learn that love means pain, abuse, and betrayal.

Myths About Relationship Violence

Many myths exist regarding the recipients of relationship violence, such as they are mentally ill or masochistic, or that battering only happens to people who are poor or to people of color, or that abuse only affects a small percentage of the population. It is also a myth that abuse is dependent on size or physical strength. Another myth is that alcohol or drug use causes violence. The most destructive myth is that persons who are abused have done something that justifies the abuse. No one deserves to be abused, no matter what he or she did or said.

The truth is that any person can be abused in a relationship if he or she happens to get into the wrong situation. No one, regardless of ethnicity or income level or age, is immune from this threat. It is a widespread problem in our society. Abuse and substance use are found together in some incidences, but there are abusive persons who do not use alcohol and drugs, and there are persons who drink or use drugs and are never abusive. An abuser may drink in order to beat someone, but alcohol does not chemically cause violence.

Types of Abuse and Neglect Victims

In addition to categorizing the types of relationship violence, abuse can be categorized by the type of victim. Child abuse, partner abuse, and elder abuse are the most common categories and are discussed in this chapter. Do not confine your awareness of relationship violence to these types, however; be aware that sometimes ongoing violence occurs between siblings or cousins, for example. In addition, be aware that some families are experiencing more than one type of abuse. Also, some of the distinctions are blurred; for example, when an elderly woman is being beaten by her husband/caregiver, is that elder abuse or partner abuse?

Child Abuse and Neglect

Child abuse (physical and sexual) and neglect are common in the United States. The most recent statistics available indicate that social service agencies received 31,400,000 reports of child maltreatment in 1994. This represents approximately 47 children out of every 1000 (Weise and Daro, 1995). "One-fifth to one-third of adult women have had a childhood sexual encounter with an adult male" (Benedek, 1989, p. 6).

The physical or sexual abuse of children and adolescents happens to males and females, and the perpetrators can be of either gender as well. Abusers can be parents or other relatives. Child abuse may take the form of obvious physical injury such as bruises, cuts, burns, and swollen or broken body parts. It can take the form of sexual assault. When sexual abuse is perpetrated by a trusted family caregiver, it is called "incest."

Sometimes the maltreatment can be seen as neglectful care: unclean body or hair; dirty or tattered clothing; malnourished appearance; and dress inappropriate for the weather. Child neglect can include not accessing medical care for the child, not feeding the child properly, allowing the child to live in an unsafe environment, or not allowing the child to go to school or have friends.

Frequently the physical or sexual abuse is not immediately obvious but is evidenced through the child's behavior. Children may be rebellious, including running away from home, or, conversely, extremely compliant. They may have many physical complaints; abdominal or pelvic pain is a common complaint among children who are sexually abused. Some appear chronically tired, experience sudden weight loss or gain, and/or have trouble with bed wetting. A child may have trouble interacting with others, have few friendships, or exhibit hostile, aggressive behavior. Sometimes parents who are abusing their children appear suspicious, mistrustful, and isolated; delay seeking medical attention for their children or seek treatment outside the neighborhood; are rigid and righteous; and believe in severe corporal (physical) punishment (Williams, 1980).

No easy explanation and no one reason for abuse or neglect of children by caregivers is sufficient. There are several issues that tend to coexist with child abuse, such as social isolation of the family, alcohol and/or drug abuse, unemployment, a larger than average family size, the stresses of poverty, and job dissatisfaction (Hamilton, 1988). Children in a unique position in the family (the youngest or oldest) and those with physical or emotional disabilities are also more likely to be abused (Sobsey, 1994; Valentine, 1990). Caregivers who abuse children are often unskilled at managing stress, have unrealistic expectations of children, and are accustomed to dealing with situations in a violent way.

Reporting suspected child abuse or neglect. Any time that child abuse or neglect is suspected, the helper can call the local child protection agency and report the suspicion. Professionals such as doctors, social workers, and teachers are required by law to report suspected abuse. Child protection agencies are required to respond in a timely fashion and determine the child's safety.

Helpers may worry that the report will endanger the relationship between them and the person with whom they are working, or they may question their ability to determine whether or not abuse has actually occurred. Reporting to a child welfare agency allows someone else trained in this area to make this determination.

The usual goal of child protection agencies is not to remove children from their families unless they are in clear danger; rather, it is to make sure that children live in a safe environment. Sometimes that may require that parents attend parenting classes, or it may mean that basic resources must be provided to the family. Every effort is usually made to leave the family intact; removing the children from the home is often considered to be a last resort.

If you are thinking of making a child abuse or neglect report, talk with your consultant or supervisor and discuss whether a report is indicated, who will make the report, and whether you will tell the caregivers about your intention to make a report. When you do make a report, tell the intake worker what you have observed and relate your concerns.

Although making a report to child protective services is serious, it is important to remember that a referral is in the child's best interests. It is also often in the parent's best interest as well; the vast majority of parents love their children and welcome opportunities to learn more effective ways to demonstrate their love. It is important for the helper to make the report in a way that reduces the risk that the relationship between the caregivers and the helper will be jeopardized. For example, if you suspect that a young child is being left unattended or is being hit too hard and too frequently, share your concern with the parents. Be as honest and as straightforward as possible. Then report your suspicions and actions to your agency supervisors.

If the abuse seems to be coming from one caregiver, give the "nonoffending" caregiver a chance to make the report herself or himself. Be clear that you will have to go to your supervisor with this regardless of what he or she decides, and that your agency will have to report the suspected abuse if this caregiver does not do so. You can offer to make the call in his or her presence, or be there when he or she makes the report.

Try to remember that the vast majority of parents love their children and want to be good parents. They may be under tremendous strain or not have the skills to parent effectively at this time. Your support as a helper may allow them to better express their love for their children and to protect them. This is one of the best times to practice being nonjudgmental and balance protection of the child with compassion for the parent.

Prevention of further abuse. Ideally, child abuse should be prevented rather than be treated after it has occurred. Frequently, parents benefit from having another adult to talk to in order to maintain a sense of balance and break the isolation. Support groups, parenting classes, telephone hotlines, and respite care are important interventions for parents.

Support groups provide adult conversation with others in a similar situation. Educational groups and parenting classes allow parents to problem solve alternative methods of discipline and punishment and to learn new ways of relating to their children. A hotline can put a parent in touch with support when he or she feels most out of control. Respite care can provide a needed break, especially when responsibilities increase during times of crisis.

Illustration

Mindy and Phillip Scott were married 14 months ago when they were both 18 years old. They have two sons; Aaron is 12 months old and Noah is 2 months old. Aaron was born with congenital blindness, and Noah was born without disabilities. The Scotts are involved in services provided by an agency that serves persons with visual impairments, offering a parent support program that links a trained volunteer helper with a parent who is requesting support. Barbara Garcia, the mother of a 3-year-old daughter with congenital blindness, was asked to call Mindy and provide her with information and whatever assistance she requested. Barbara quickly developed a relationship with Mindy and her boys. It was apparent to Barbara that Mindy was in over her head; she was a young mother with two babies, one of whom has special needs. Barbara had a lot of empathy for Mindy; she knew how difficult and exhausting it is to care for a family and meet your own needs, too. Barbara learned that Philip worked a full-time job as a stocker at a local grocery store chain during the late afternoon and evening hours. Mindy worked as a waitress during the 6:00 A.M. to 2:00 P.M. shift at a nearby cafe. Barbara was impressed with the care the children received until one afternoon when she stopped by to see Mindy and she found the two children home alone.

Think About It #56

1. List the feelings that Mindy may be having. Summarize her situation.
2. List the feelings that Philip may be having. Summarize his situation.
3. What would you have done if you were Barbara Garcia, the volunteer helper, and you found the two babies left alone in their home? What are the advantages of the decision you made? What are the disadvantages of the decision you made?

Elder Abuse and Neglect

Older adults may also be victims of an abusive or neglectful relationship. The number of suspected incidents of elder abuse and neglect by family members reported to Adult Protective Services in 1991 was 227,000; approximately 55 percent of these cases were confirmed to be maltreatment (Tatara, 1993). Two percent of elders are victims of violence each year in the United States. Adults who are abused are most likely over 65, female, ambulatory, and living with a relative other than a spouse (Cash and Valentine, 1987). Defining adult abuse and neglect is even more difficult than identifying it for children, because of the differences of opinion about who is responsible for adults and at what point someone else is responsible.

While there is no typical abuser of an elder, 81 percent of the reported perpetrators of elder abuse are relatives of the victim, 58 percent are women; 42 percent are adult children or spouses of the children, and 58 percent are the spouses of the elderly victim (Barnett et al., 1997).

There are many barriers to the discovery and reporting of elder abuse. Some of them are:

- Most abusive or neglectful families are socially isolated. They let few persons into their homes, and abuse may be one of several family secrets.
- The victims of violence or neglect may not be seen by the medical community due to the family's isolation and neglect of the elder.
- The recipients of maltreatment may not leave the home and may not interact with anyone outside the home. Unlike children, elders do not go to school, where injuries could be noticed.
- When helpers do see the family of elders, they may miss the cues of abuse and may assume that injuries, malnourishment, and confusion are due to illness or advanced age.
- Because older persons can bruise and break bones more easily, physical abuse can be explained by natural events, falls, and accidents. It may be logical to attribute injuries to poor balance, vision, coordination, or orientation.

Why would a neglected or abused elder be reluctant to report neglect or abuse? Reasons may include:

- Fear of being removed and placed in a nursing home.
- Unwillingness to send a family member to jail.
- Fear of losing the love and care of the perpetrator.
- Fear of retaliation.
- Shame, embarrassment, and a sense of failure as a parent.
- Being bedbound or housebound.
- Being very frail or confused.
- Fear that no support or services will be available.
- Depression and sense of helplessness and hopelessness.
- Fear of not being believed.
- Experience that violence is to be expected.
- Pride and independence.
- Hope that the maltreatment will stop on its own.

Reporting suspected elder abuse.   Most states now have laws against elder abuse, just as they have laws against child abuse and neglect. To report suspected abuse of an elder or impaired person, start with the adult protective services unit in your county or state government. Again, unless it is an emergency, share your concerns and suspicions with your supervisor before you make a report.

Signs and signals of elder neglect or abuse. The following situations may alert you to the possibility of elder maltreatment.

- Bruises, burns, or welts in various stages of healing; patterns that indicate a human hand or fist or the use of an instrument, rope, or cord.
- Wounds or puncture marks.
- Being left alone if sick, bedbound, confused, or immobile.
- Confusion or sleepiness that could be caused by overmedication.
- Complaints of pain or medications not being administered.
- Excessive weight loss, dehydration, no food in the house.
- Gap of time between injury or illness and the seeking of treatment.
- Evidence that the caretaker is using several different emergency rooms, clinics, hospitals, or doctors.
- Caregivers indicate that the elder is unwanted or the need for care is resented.

Illustration

Tanisha Motley, a young woman caring for her three children and her grandmother in a small apartment in low-income housing, attended a support group for caregivers. Ms. Motley worked full time at a textile factory. She had total financial and physical care of her three children, ages 2, 4, and 7. She expressed her anger at being expected to care for her grandmother, Eva Hart. Ms. Hart, age 67, was recovering from surgery to treat stomach cancer but appeared to be very sick and unable to care for herself. Prior to the illness, Ms. Hart had been working part time for a family caring for their home and children, and living in her own apartment. Her granddaughter described her as "mean" and "hard to live with" all of her life; she admitted that she had been beaten by her grandmother when she was growing up. Now, Ms. Motley was providing total care for her grandmother, the person who had abused her as a child. In the group, Ms. Motley indicated that she left her grandmother alone at home during the day, ignored Ms. Hart's calls for assistance when they were home, and occasionally hit Ms. Hart when Ms. Hart was particularly insistent.

Think About It #57

1. What do you think of Ms. Motley? What are your feelings about her?
2. What would you do, say, or feel if you were another caregiver attending the group?
3. What would you do if you were the facilitator of the support group?

Partner Abuse

You may encounter a situation where a person is the victim of abuse by his or her domestic or romantic partner. Your role as a helper would be to ensure immediate safety, refer to appropriate community resources, and provide emotional support.

The vast majority of relationship abuse of partners occurs in situations where the female is the victim. For that reason, it is often referred to as "wife beating" and the recipients of abuse often are called "battered women." Most services for adults who are abused are geared for women. When people talk about abuse of adults, they refer most often to women because statistically that is most often the situation. Men, of course, can be the victims of partner abuse as well. Men in heterosexual or homosexual relationships can be verbally abused, threatened, battered, and physically harmed by their partners.

When people have learned through society and their family that violence is the way to solve problems, they are more likely to engage in violent acts against their partner. Men who observed violence between their parents are more likely to beat their wives (Lystad, 1986).

Women who are being abused often find it difficult to reach out for help. Often they are not believed, they feel ashamed, and they are unwilling to risk the breakup of the couple or family. Men who are battered are also in a difficult situation because of shame and stigma and because the resources for them are fewer. The stigma and hidden nature of violence may also be greater for women who are gay. Regardless of the nature of the partnership, remember that it requires tremendous courage to come forward and confide in anyone, whether stranger or acquaintance, that abuse is present in the relationship.

Reactions to separation from the abuser.   When a battered woman leaves her partner and her home, she may feel grief over the loss of many important things. She may experience the loss of dreams and ideals; the loss of the marriage or partnership; the loss of the man she loves; the loss of home, belongings, and pets; the loss of privacy; the loss of respect and status accorded to her by others; and the loss of approval from her family or religion. She may feel lonely, sad, or angry. She may blame herself for the troubles in the relationship. She may feel like a failure as a wife, mother, lover, or woman. She may feel suicidal or homicidal. She may feel disappointment, fear, shame, and embarrassment. She may feel inadequate, vulnerable, or out of control.

Sometimes victims of relationship violence are further mistreated by law enforcement personnel, the court process, or by other agencies. Sometimes when they try to talk about their experiences, they are not believed or are ridiculed. Sometimes they feel that police officers, judges, or magistrates are blaming them for the abuse rather than blaming the abuser. Sometimes victims get messages from authorities such as, "This is a family matter," or, "This is a domestic argument," along with instructions to "Go home and make up," or, to "Try harder to get along." In these situations, human services workers and volunteers are called on to act as advocates for the victims of violence, striving to make law enforcement and social service systems more responsive and appropriate.

Characteristics of couples in which violence occurs. A few generalizations can be examined in an attempt to understand victims and perpetrators. However, remember that there are always exceptions, that each person and relationship is unique, and generalizations are just tools to summarize experiences. These characteristics are not meant to suggest that only certain kinds of persons either get abused or are abusive. They are not meant to provide an excuse to blame the victim. Being in a violent relationship is an accident, and there is no way to predict if someone will stumble into abuse or neglect.

Abusing and abused members of couples do seem to have some characteristics in common. Both victims and batterers in abusive relationships tend to feel trapped by the dynamic, do not see a way to stop it or leave, and feel helpless in the face of the violence. In addition, some victims and perpetrators tend to have low self-esteem; they tend to believe that they are worthless, incompetent, and unlovable. Both abused person and abuser tend to believe all the myths about battering and tend to be traditionalist in their beliefs about relationships, family, gender roles, and the home. Both may contemplate homicide or suicide. Finally, both abusers and abused persons tend to believe that they cannot survive alone.

Additional characteristics of women who are are battered. In addition to the tendencies that are listed above, the following characteristics have been noted in many abused women:

- They tend to accept responsibility for the abusers' actions and to believe that they have deserved the abuse.
- They tend to have severe stress reactions and physiological symptoms. They tend to have depression, anxiety, and panic.
- They tend to believe that no one from the outside can help.
- They tend to believe everything that the batterer says about them.
- They tend to feel drained, defeated, and without energy and optimism.
- They tend to have a deep sense of hopelessness, helplessness, despair, and powerlessness.
- They tend to minimize, hide, and lie about the abuse.
- They tend to protect the abuser.

Additional characteristics of batterers. Studies and observations of men who abuse their partners have demonstrated these general characteristics:

- They tend to blame others for their actions.
- They are often irrationally jealous, suspicious, and paranoid.
- They tend to have severe stress reactions, during which they may use substances or violence.
- They tend to use aggression to bolster self-esteem.
- They tend not to believe that violent behavior in the home should have negative consequences.
- They tend to believe in the privacy of the family, the validity of violence, and the supremacy of males.

- They tend to not be able to differentiate between emotions but to label and experience everything as anger.
- They are mostly not violent outside of the family or couple.
- They tend to have unrealistic expectations of others and a distorted view of the victim. For example, a man who is abusing a woman may expect her to read his mind, and he may not communicate directly with her.

Illustration

Caryn, who was mentioned previously, gave one of the authors a typed list about her reactions to her abuser, entitled, "Things I don't like about Sidney." Notice how matter-of-factly she treats the physical abuse and how controlling he has been. Here is her list.

What I don't like about Sidney

He beats me.
He nags.
He doesn't let me go shopping alone.
He doesn't give me any money.
He doesn't let me call or visit my mother.
He is jealous of everyone.
He doesn't let me clean house the way I like to.
He doesn't let me go out with the girls or play bingo.
He's using me for his satisfaction to meet his needs.
He listens in on my phone conversations.
He nags me about sex.
He accuses me of having sex with every man.
We don't spend any time together anymore.
He took all the ash trays out of the cars.
If he doesn't like someone he doesn't want me to be around them.
He tells me who I can and cannot ride in cars with.

Abuse of Child and Partner Together

It is common for child abuse and partner abuse to exist in the same home. When you are assessing child abuse or neglect, it is important to determine if the child's mother is also being battered by the abuser. If so, the plan must include ways to make her safe as well, or she will not be able to protect the children. Sometimes child welfare agencies and workers will be too quick to blame an abused mother for not protecting the children from abuse, without acknowledging the complexity of the danger in which she finds herself. She does share the responsibility for the safety and well-being of her child, but sometimes she is so consumed with just surviving that she feels powerless to protect anyone else. Often she is taking extreme measures to ensure the safety of her child but is unable to be effective because of the dominance of the abuser.

Any child who witnesses and lives with relationship violence, whether he or she is actually hit, is a traumatized child. First, when a mother is preoccupied with her own survival and in keeping the family together, she cannot have enough energy to focus on the children. Second, when a child is awakened at night by the sounds of his or her mother being thrown against the wall, that child is experiencing emotional trauma. The same can be said for living in constant strain and tension. Third, the child, who loves both the victim and the abuser, is torn in loyalties and is therefore at risk for being emotionally disturbed. Finally, their views of relationships are affected for the rest of their lives. Many children from violent homes learn that abuse equals love. For example, in one play group, a little boy said "I love you," to a little girl, who replied, "No you don't; you haven't hit me yet." The major message that children from violent homes get is that you never know when a trusted person is going to turn on you.

Services for women who are battered. The first programs and shelters for women who are battered and their children appeared in the early 1970s, partially as a result of the attention the growing women's movement paid to wives who are battered. Before that time, public awareness of the problem of the battering of women was very low. It was considered a private matter. Most communities now have shelters for women who are battered; these shelters serve as safe havens during a dangerous time. Support groups for women who are battered are also common.

Thanks to the attention focused on this issue over the last 25 years, when a woman who is battered is able to reach out for help, she is more likely now to be able to find a group of concerned helpers who will try to assist her to make a plan for time in a safe, confidential location. She can receive validation for the pain, confusion, and humiliation she is feeling. She can get a temporary separation from her abuser while she sorts out her options. She can get basic survival needs met for herself and her children, and she has access to counseling and peer support. By talking to other survivors of abuse, she may be able to feel less alone and more empowered toward her own safety. In some situations, she may have access to legal advice and assistance to live independently if she chooses to.

When a women enters a shelter, she faces many problematic issues. She may feel homeless and with fewer resources than she had before. She may feel overly controlled by other persons and by rules about behavior, which is what she felt from her abuser, but at least she was in control of her own home. She may fear losing her children to a child welfare system. She has the realistic fear of the batterer finding her and hurting her or her children further.

Asking for services takes much courage for a violence victim. Seeking help does not just mean seeking safety. It means having to disclose your abuse history to strangers. A woman is most likely to feel ashamed at the verbal, physical, and sexual abuse she has suffered.

What Do You Do?

As a human services worker, student, or volunteer, what steps can you take to help someone who is in an abusive relationship? As a helper, you should be prepared to intervene whenever and wherever you can. The following guidelines may be useful.

- When an adult or child talks to you about verbal, physical, or sexual abuse in the present or past, you should start with believing him or her. Listen to the stories without running from the pain, and communicate caring and a willingness to understand the experience. Many times, victims of violence have tried to tell someone about the abuse and have not been believed, or their experiences have been discounted or made light of.
- Determine whether you are dealing with an emergency; if immediate danger exists, then you may need to involve the police, ambulance, or emergency room. In these situations, always involve your supervisor in the decision making or inform him or her as soon as possible of your actions.
- Listen, listen, listen. Use active listening to be fully present with someone; help him or her to label and explore the painful or shameful feelings. The stories of abuse, neglect, and exploitation may be difficult for you to hear, but it is important that victims of violence get a chance to share their experiences, fears, and pain with someone who is neutral and nonjudgmental.
- Introduce the idea of having the right to be safe, to have privacy, to make one's own decisions, to have self-respect, and to pursue one's own goals. Firmly but gently insist that no one deserves or invites abuse and that violence is unacceptable. Support victims as they explore these new ways of viewing the world.
- Help victims develop plans to be safe in the future. (See section on safety plans below.)
- If you suspect that a child or older person is being neglected or abused, inform your supervisor immediately, and make a joint decision about reporting the possible abuse or neglect to your local authorities.
- As in all human services work, you must support the informed choice of the victim of violence if he or she is a competent adult. This may be difficult, especially when you are afraid for his or her safety, but there comes a time when you will be required to allow a victim of violence or neglect to make his or her decisions whether or not you agree.
- Explore options and help with constructive problem-solving and education. Help him or her to get information and resources, and educate him or her about the cycle of violence. Realize that people change when they are ready to do so and that you cannot control another person's behavior. Remember that you are responsible for supporting them, giving them resources, facilitating their safety, and then supporting their choices.
- Be sensitive to cultural, class, ethnic, and language differences in the way that people define themselves and their roles in the world.
- Respect the victim's need for choice, control, dignity, and power. If you set limits, be clear, honest, and fair.
- Do not make promises you cannot keep, and do what you promise to do. The victim of violence needs to be able to count on someone.
- When the victim is safe for the time being and able to consider the big picture, provide him or her with information on relationship violence, such as the cycle and continuum of violence. Be prepared to recommend (or lend) books that may help him or her to feel less alone and isolated and that will put his or her experiences in a larger perspective.

Developing a Safety Plan

One of the most useful services you can provide to a victim is to help him or her develop a plan to escape when the next violent incident occurs. Every one who is in a violent relationship needs to have a safety plan. The plan should be developed with the victim, using his or her sense of what might be possible. Questions to ask in the development of a safe plan are:

- What kinds of things have worked in the past to protect you and your children?
- What are you willing to try next time?
- What people or agencies could you turn to for help?
- Do you know how to get medical help? Legal help? Police protection?

The most dangerous time for a battered person is when separation has occurred or is imminent. If a batterer feels threatened with the loss of the victim, then homicide may be a real possibility. However, this is not the only time that a safe plan needs to be in place.

Possible elements of a safety plan:

- Have spare keys to car, house, and office hidden and accessible.
- Have cash and public transportation tokens stashed with your keys.
- Alert a trusted neighbor that there is violence in the house and that someone may be asking for help in calling the police. Tell the neighbor to call the police if it appears that violence is in progress.
- Have telephone numbers to trusted persons and agencies in your wallet, car, office, and stashed with a neighbor and/or friend.
- Be sure that you know how to get to the nearest law enforcement station; drive there if the abuser is following you.
- Keep a suitcase packed and hidden.
- Be sure that you know how to get to the nearest emergency room.
- Have copies of all important financial and family papers held with a trusted person; examples are birth certificates, Social Security numbers, legal papers.
- Teach children how to dial 911.

Illustration

Many victims of partner abuse are able to eventually leave the abusive relationship, are able to develop a stronger sense of self-worth, and are able to value and protect their own safety. People who are able to escape a violent relationship often refer to themselves as "survivors." An example of this new way of life is illustrated by the following poem, which was written by a group of women who were incarcerated because they had killed their batterers.

I WON'T

- *Live in an abusive situation ever again.*
- *Feel responsible for the abuse inflicted on me.*
- *Feel sorry for a man or for my reasonable actions.*
- *Dwell on the past.*
- *Be completely dependent on anyone again.*

- *Be held responsible for things that are not totally my responsibility.*
- *Make any more important decisions under the influence of sex.*
- *Ever put a man before myself again.*
- *Get involved with a substance abuser.*
- *Be manipulated.*
- *Come back to prison again because of abuse.*
- *Ignore my own needs/wants/desires/choices.*

Summary of Relationship Violence Points

Since relationship violence is an especially complex topic, perhaps it would be useful to recap the major points at this time. No matter what form the violence takes, relationship violence has the following characteristics:

- The perpetrator's belief in violence as a viable behavior.
- The coercion of the victim.
- The disrespect or disregard for the victim's rights, safety, well-being, and choice.
- Unequal power in the relationship.
- Violence that is recurrent, ongoing, and in a relationship over time.
- Violence that escalates over time.
- A cycle of tension, explosion, and peace.

Working with Abusers

Some helpers make the decision that they will not be able to provide services to the perpetuators of violence. However, others realize that the abusers must receive counseling if they are to have any hope of changing their behavior.

Many communities also have treatment programs for men who batter. In working with a perpetrator, the first priority is stopping the abuse or neglect. This must be non-negotiable. The perpetrator must be held accountable and responsible for his or her own actions.

The following are guidelines for helper behavior in working with an abusive or neglectful person.

- Provide support, empathy, and acceptance, while continually confronting him or her on the reality of the abuse. Separate the worth of the individual from the destructiveness of the behavior.
- Provide group counseling with other abusers where the facilitators provide an ethic of nonviolence and of taking responsibility for one's own actions.
- Teach anger management, assertiveness, anger control, and how to time oneself out.
- Help develop a "safety valve" plan for when he or she is feeling out of control. (For example, walking around the block, calling someone, going to a friend's house.)
- Help him or her confront denial and minimization.
- Develop a behavioral contract. (See the following example.)

Example of a Behavioral Contract

Effective immediately, the violent behavior toward _____ will stop. I will control my actions from this day forward. I will immediately identify and contact three persons whom I can call for support when I think or feel that I may be violent. If I think or feel that a violent reaction is coming, or if I begin to lose control, I will leave the house and call someone who will talk to me.
Signature and date: _____

Illustration

Joe Holliday, an African-American male, currently divorced, has in the past been violent in an intimate relationship and is working with other men in a similar situation who wish to change their violent behaviors. For several weeks, Joe has been acting as a helper with Leon Miller, an Anglo-American, who is struggling to learn acceptable ways to express his anger and to take more responsibility for his actions toward his partner. Joe received a call from Leon, who is in crisis because Leon and his wife have been verbally fighting and the intensity is escalating. Leon calls Joe for advice because he wants to avoid a physical fight. In this situation, Joe is able to listen to Leon's expressions of anger and loneliness and fear; provide immediate tips about stepping outside the argument, cooling down, and then discussing the disagreement rather than entering a physical fight; and problem solve with Leon exactly what he will do when they finish talking. Leon seems amenable now to intervention by professionals, also, and Joe provides Leon with the name and phone number of a professional working with the local organization serving men who batter. In addition, Joe leaves a number for Leon's wife to call for counseling. Before ending their conversation, Joe and Leon agree that they will meet the next day to talk about it.

In the situation above, Joe assessed the seriousness of the immediate situation as having potential for violence, but he decided that, with assistance, Leon would be able to choose another course of action. He and Leon already had a relationship, so establishing rapport had already been accomplished. Joe and Leon were able to identify the immediate problem: concern about physical violence. Leon dealt with his feelings by venting with Joe. Together, Leon and Joe explored alternatives to physical violence as a resolution to a disagreement and developed a plan that Leon felt sure he could implement. Joe followed up the next day to ensure that the plan had worked and to problem solve for future crises.

Wrapping It Up

It is likely that you, as a human services helper, may encounter an incident of family violence or suspect that family violence may be occurring. It is important that you be prepared for this situation by being informed about abuse and neglect issues. It is also important that you be prepared emotionally. Few issues for helpers dredge up the amount of pain, sadness, and anger that relationship violence does. We have rational and understandable rage against people who harm children or elders and who mistreat their loved ones. You will not be able to avoid the issue of violence in our society, no matter what type of work you do as a helper or what type of agency in which you work. Before beginning this type of work, clarify your own values and beliefs about issues such as marriage, sexual behavior, violence, dependency, relationships, parenting, power, and the discipline of children. As in all human services work, you must know what internal buttons will be pushed with you, so that you can be prepared for them, get support for yourself, and keep your own reactions out of the way of the helping relationship. If you offer help to the perpetrators of violence, you will really be called on to suspend your judgment in order to work with them effectively on (1) taking responsibility for their behavior and (2) stopping the violence.

Think About It #58

1. Call your local department of social services, and ask for the unit for child protective services or adult protective services. Ask them what the procedure is for reporting abuse or neglect and how the reporter's anonymity is protected. If you are working as a helper through an agency, check with them about their procedures for handling situations where child or adult abuse is suspected.
2. Locate the emergency phone number for victims of relationship violence in your community. Call and determine the appropriate steps for making a referral. Write the name of the contact person, phone number, and steps for referral in your referral list. Ask how many women are served each month or year and how many are turned away.
3. Search the telephone book and community resource directories for agencies or organizations that will provide support to victims of violence who are in gay relationships. If you can find none, ask the mainstream agencies for victims what their policies and practices are regarding this population.

CHAPTER TWELVE

Gerontology and Aging

Older adulthood is an important stage of life. In addition, this population continues to grow. Human service workers are often called on to work with older persons and their social networks. This chapter defines gerontology, aging, and ageism; explores myths, theories, and facts about growing older; looks at trends; and discusses settings, values, and roles in the provision of human services.

Introduction to Gerontology

The study of the physical, mental, and social processes of growing older is called "gerontology." The word is a combination of *geron* which means "aging" and *logos,* which means "knowledge." Gerontology, which is knowledge of the aging processes, is much like human services in that it is comprised of practitioners and researchers from a variety of disciplines. Gerontology is only about 50 years old as a formal area of study. "Geriatrics" is a subset of gerontology; it is the area of the field of medicine that pertains to the diseases and treatment of older persons (Hooyman and Kiyak, 1996).

Social Changes

When proportions of the United States population are examined, it is evident that there are more older persons than there were at the beginning of the century. In 1900, people over the age of 65 were 4 percent of the population in this country. In 1990, they were almost 13 percent (U.S. Bureau of the Census, 1991). As demographic trends go, this is a dramatic increase. This change is primarily because individuals in our society are living longer, possibly due to medical science and better access to health care and medications. The increase in the number of older persons is also attributed to advances in the treatment and control of infectious and childhood diseases, giving people an improved chance to achieve an old age (Hooyman and Kiyak, 1996).

The overall maximum life span has not changed; people still do not usually live much past the age of 100. The oldest known survivor Mme. Jeanne Calment of France was 123 years old (National Institute on Aging [NIA], 1996). But our overall life expectancy has changed over the years; more and more persons are approaching the age of 100. In 1900, the average life expectancy at birth in the United States was 47 years (NIA, 1996). The average life expectancy for babies born in 1993 is almost 80 for women and almost 73 for men (U.S. Bureau of the Census, 1993).

Because more persons are living longer, gerontology is an important area of study and of human service delivery. Probably, no matter what your specialty or setting is, you will be involved with older persons, if not as primary clients, as family members, caregivers, or significant persons in the lives of the persons who ask you for help.

Aging

Aging is the way humans move along the continuum of time. There are three important things to remember. First, "aging" describes the normal progressive changes that can be expected in cells, organs, biological systems, and in a total person. Second, aging is an individual process; no two people will age in the same way. Third, within one person, systems and organs age at different paces.

Aging can be defined and studied in a variety of ways, such as:

1. **Chronological**—number of years lived.
2. **Bureaucratic**—eligibility for social programs or services, such as age 65 for Social Security, age 60 for Older Americans Act services, age 55 for discounts in the stores, and age 50 for membership in the American Association of Retired Persons.
3. **Functioning**—mobility, ability to care for oneself, and overall health.
4. **Social**—roles and relationships in the social structure as people age.
5. **Biological**—physical changes to organs and cells.
6. **Psychological**—sensory and mental functioning, personality, attitude, and coping.
7. **Demographic**—the trends in the overall and aging populations.

Because the process of aging differs from person to person, it is hard to define a norm for aging. Gerontologists do know for the most part what is to be expected as humans age normally, and what is the result of disease or illness. Although general patterns of aging can be sketched, each individual's path may vary as a function of factors such as: heredity, occupation, environment, stress, health care, health maintenance, and level of activity.

When you consider how incredibly long the aging period can be, perhaps you can have some appreciation for how diverse this group can be. We may consider anyone from the age of 60 on to be old. Childhood and adolescence together are only about 21 years long, and we certainly consider children and teens to be unique individuals with unique personalities and needs. With at least a 40-year span, certainly older persons are very different from one another.

It is probably useful and meaningful to think of older persons as belonging in at least three separate chronological groups: the young old (approximately 60 to 74), the middle old (approximately 75 to 84), and the old old (usually over 85). Someone who is 92 will have very different perspectives, experiences, and needs than someone who is 62. These two persons may attend the same senior center, but they are in different generations.

The oldest old, those over the age of 85, is the most rapidly growing segment of our population. Seventy percent of this group are females who are widowed, divorced, or never married. Most are economically disadvantaged (Longino, 1988). Certainly these oldest old adults are more frail physically, less functional, and more impaired mentally. They make up more than half of the population in nursing homes (Hooyman and Kiyak, 1996).

Ageism

The concept of ageism is like other "isms" in our society (such as sexism, racism, classism). The word, coined by gerontologist Robert Butler, describes prejudice based on age (Butler, 1969). Ageism is prejudging an individual, a group of people, or a physical condition based on preconceived notions of age. Ageism causes us to assume things about persons based solely on their age. It also includes discrimination based on age.

Ageism is fueled by our society's worship of youth and physical beauty and by the predominance of the work ethic. If you are not young looking and are not working, then you are considered to be worth less. Our societal values sometimes lead us to see aging and retirement as an insult, an affront, rather than a natural occurrence. People in the United States often view the word *old* as undesirable: Some people lie about their ages for fear of being seen as "old."

Examples of ageism are:

- Assuming that older employees are not as valuable as younger workers in the workplace; forced retirement; not hiring someone based on his or her age; or forced job changes based on age rather than on performance.
- Assuming as a social service or medical care provider that older persons are not worthy of time, expense, or effort.
- Misdiagnosing an older person or failing to order medical tests because "it's just age" or "this can't be a problem at her age."
- Talking down to an older person, being patronizing, making decisions for him or her, treating an older person as less than human or undeserving of dignity.
- Subscribing to stereotypes and myths about old age; assuming that older persons are all alike.
- Media and commercial images of old age being predominately negative and focused on dysfunction and disease.

It would be useful to focus more on the positive role models of aging rather than on the negative views. This would help to give the old and young alike a better sense of the dignity and benefits of living long. Advocates are often dedicated to creating an age-irrelevant society, where people are judged based on their contributions and ability to love rather than on their appearance or age. With ageism, we all discriminate against our own futures. If we are very lucky, we will all be old eventually. Old age is the only minority that we all hope to join.

Think About It #59

1. Have you ever witnessed an act of ageism or other evidence of ageism? What was it?
2. How do you imagine you would feel if you had been treated differently because of your age?
3. Share with other learners what your experiences and feelings have been regarding ageism.

Myths About Aging

Some of the most common stereotypes about growing older are (Lowy, 1979):

- **All older people are alike.** All older persons think alike, act alike, or look alike. Examples of this are: "All old people are sweet," or, "all old people are crotchety and grouchy." A similar attitude is that people grow more like each other as they age. Actually, as we age we retain our unique personality and cultural characteristics, and we actually become less and less like each other as we add years to our lives. Personalities get more distinct and unique. Unless disease intervenes, our personalities probably do not change as we grow older (Costa, 1996).
- **Older people cannot learn.** ("You can't teach an old dog new tricks.") Older persons learn at a slower pace, but they do learn, and they do retain the information. The ability to learn will not stop unless there is a medical problem.
- **Older people are senile.** The common myth is that "senility" (a word mistakenly used to describe dementia, mental impairment, and memory loss) is inevitable. "Senility" is actually just a synonym for "old" and does not mean mental impairment. Mental deterioration is neither normal nor inevitable. The majority of the population do not develop dementia.
- **All older persons are in nursing homes.** At any one point in time, only about 5 or 6 percent of Americans over age 65 are in institutions. Approximately 85 percent are on their own and doing well, whereas only 15 percent are frail and at risk for disability or illness. Researchers have projected that for persons who turned 65 in 1990, only 43 percent will spend some time in a nursing home in their lives (Kemper and Murtaugh, 1991).
- **All older persons are sick and hopeless.** There are many myths of this nature, such as: older persons lose their sexuality, older persons cannot benefit from mental health services, and all older persons are physically impaired or ill.

Each person is different, of course, but in normal aging, generally none of the stereotypes listed on page 204 are true.

   • **Older persons become more religious as they age.** A Duke University longitudinal study found that people who are old and religious were young and religious (Blazer and Palmore, 1976). Spiritual attitudes tend to remain constant throughout life, but religious activities may decline in later adulthood due to mobility problems, transportation problems, sensory impairments, or health problems.

   • **Olden age is a "Golden Age."** There is a myth that retirement age is always tranquil and untroubled. This denies the reality that some older people face: decreased income, increased isolation, chronic illnesses and conditions, substandard medical care, and fear of losing one's independence and home. Some elders are faced with ageism, sexism, classism, racism, xenophobia, homophobia, or any combination of these elements of oppression and prejudice.

Think About It #60

1. The above myths are very common. Which ideas had you held about what aging would be like?
2. Discuss with fellow learners how media images and societal stereotypes can perpetuate these myths.
3. Watch a movie that portrays older people. Identify ways these films perpetuate myths about aging.

The Truth About Aging

The truth about growing older is that it is neither all good nor all bad. It can be a mixture of gains and declines. The later years are not necessarily a golden age with no difficulties; neither is this time a hellish existence with no joy. Aging is an individual event, as well as a condition that is heavily influenced by societal attitudes and support. Older people are widely different from each other, and their experiences cannot be generalized.

Social Theories of Aging

A theory is a particular view of the world, a way of putting on a particular set of lenses through which one might interpret perceptions and facts. Depending on the theory to which you subscribe, you will attach different meanings to the same situation. It is useful to be aware of a variety of theories and to try on these different lenses so that (1) you can judge for yourself what fits your perspective best, (2) you can be aware of different ways of interpreting aging, and (3) you can perhaps pick and choose parts of these theories to form your own beliefs. Theories are ideas, not facts. This means that no one view of the world is written in stone and that no one perception is absolutely right or wrong. To judge a theory, framework, or way of thinking, consider the extent to which it fits with facts and experience.

The following are some of the most common theories about the impact of growing older on a person's social functioning:

1. Disengagement theory, developed in 1961, says that people remove themselves from society as they age, and that society simultaneously eases away from them (Cumming and Henry, 1961). This theory states that older persons are inactive and withdraw from others; elders do not have to be active to be well adjusted. This view states that it is normal, adaptive, and natural for older persons and society to disengage from each other. For example, it is considered functional for society to transfer power and responsibility to younger persons, and for older persons to remove themselves from the workforce.

2. Activity theory (Havighurst, 1963), developed in the 1960s as a reaction to disengagement theory, says that old age is not much different from middle age. Persons who are active will be more satisfied and better adjusted than those who have disengaged. This theory is congruent with societal values about work and productivity.

3. Continuity theory (Neugarten, Havighurst, and Tobin, 1968) takes into account the course of an individual's life and states that personality and styles of coping do not change with age. The aging person changes roles but continues to adapt as he or she always has. Personality is the major factor in adjustment to old age; therefore, there is much diversity in the way that persons cope as they age. In this view, people as they age become more and more like what they were when they were young. Characteristics and personal beliefs become even more important and more pronounced.

4. The psychosocial model is based on a life cycle view and theories of human development. This perspective states that later adulthood is a life stage, distinct from but connected to other stages, and that human development continues until death (Erikson, 1982). In this perspective, there are two fundamental truths: (1) Life is a continuum of change and growth, and (2) each stage has its own unique problems, tasks, coping skills, crises, and potential. The successful resolution of the task of each stage can lead to satisfaction about one's life. The psychosocial task of older adulthood in this theory is called "integrity versus despair," meaning that a person who is facing the end of his or her life must come to terms with what he or she has been able to be and do throughout life. Integrity and despair are not rigid ends of a dichotomy but exist along a continuum.

5. Role theory (Cottrell, 1942) examines the fact that people hold a variety of roles and relationships throughout their lives, such as spouse, parent, lover, sibling, employee, and friend. Our roles determine our self-concepts and our self-images. Age alters roles, norms, standards, and expectations. Persons sometimes change roles involuntarily as well as voluntarily. For example, a woman of 75 who has defined herself primarily as a wife for 50 years will have a difficult adjustment upon widowhood, in addition to experiencing grief over the death of a significant person. A man of 65 who has defined himself for his adult life as primary breadwinner and as employee will perhaps have a difficult time adjusting to forced retirement. Successful role adjustment might be the older person being able to make the transition from worker to full-time grandparent and being satisfied and happy with his or her new life.

Think About It #61

1. As you can see, the theories discussed above focus attention on different aspects of aging; for example, social involvement, personal adjustment, or personality style. Which ideas help us to understand which aspect of aging best?
2. Which of the above theories fits best with your own view of aging?
3. How might human services work with older persons be influenced by a belief in each of the above theories?

Normal Biological Aging Versus Illness

Biological aging is a series of normal progressive changes with the passage of time. Aging is the same as growing: It starts with conception. Aging is normal and expected; disease (or pathology) is not. There are physical changes that most humans experience as they get older, such as graying hair, thinner skin, wrinkles in the skin, and bones that are less dense. However, many difficulties we tend to attribute to aging (such as mental confusion, loss of memory, and loss of mobility) are not normal changes. Rather, they are the result of illnesses that may be treatable and/or manageable.

It is important to distinguish "growing older" from disease. There is a widespread misperception that to age is to automatically grow frail and sick. There are normal, expected changes that occur with reaching the later years. However, there are two important facts to remember: (1) Humans age at different paces and in different ways, and (2) many problems are caused by illness and disease, not chronological age.

Of course, the normal, expected changes of growing older often cause problems with physical functioning and mental outlook. Older persons are sometimes struggling with normal changes, acute illness, and chronic conditions all at once. Just because a condition is normal does not mean that it cannot be treated or made more manageable. Advocates and caregivers must facilitate prompt and appropriate medical assessments for every change in function in an older person.

The normal physical changes of aging involve chromosomes, cells, organs, and the whole body. Individuals age at different rates, and systems within a single individual change at different rates as well. Since no two people age at exactly the same rate, it can be difficult to pinpoint what is "normal." The rate at which your body changes depends upon heredity, environment throughout life, stress factors, health habits and care, and preventive measures. Health maintenance and illness prevention include such things as good nutrition, adequate medical and dental care, avoidance of smoking, keeping one's weight down, exercising regularly, being moderate with alcohol and drugs, and remaining mentally involved and active.

Disease involves abnormal changes in tissues and biological systems. These changes have an impact on the functions or structure of an organ or system. Disease is related to aging in two ways. First, the older one gets, the more vulnerable or susceptible one is to disease. Secondly, disease can hasten, or speed up, the aging process. Aging does not cause disease, but disease can affect a person's experience of aging. Diseases that affect all age groups do occur more often in older adults. For example, one can get cancer at any age, including childhood. But the highest incidence of cancer of all types is in older adults. The longer one lives, the more chance one has of developing medical problems.

Normal aging is what happens to a human body when all goes well. All of the body systems are affected as we age, and the decline begins in the early years. None of the normal changes are drastic or fast. The decline is steady and gradual throughout life. For example, probably one's heart function declines at the same rate between ages 30 and 40 as between ages 70 and 80. Sometimes toward the far end of the lifespan, the gradual changes compound each other, and problems intensify. In the body as a whole, the effects are cumulative, so a person may get the impression that he or she is wearing out all at once. The body's systems affect each other and are interdependent, so a problem in one major system may affect everything else to some extent.

Ageism and Health

Ageism does not only have social ramifications. Ageism is also very dangerous in the realm of health care. It can have serious consequences for wellness, quality of life, and life expectancy. When caregivers, health-care providers, human service workers, and older persons themselves attribute physical or mental changes to normal aging (for example, "What do you expect at your age?," "You're just wearing out," or, "It's just old age"), then they are in danger of making a temporary, reversible, treatable condition permanent, irreversible, and untreatable. Appropriate and sensitive health care becomes more important, not less, as we age. If a physician, for example, ever takes the position that a condition should be ignored and not assessed or treated, based solely on the person's age, then it is time to change doctors. Sometimes older persons can be empowered to advocate for themselves in the medical system, but sometimes, if they are frightened or frail or have dementia, they need others to advocate for them.

One of the authors saw an original skit by a group of seniors where this issue was addressed humorously. Someone came into a doctor's office with a broken finger and was told that it was "just old age." Another complained of pain in her left knee. When the doctor said, "It's old age," she replied, "But my right knee is just as old, and it works fine!" Unfortunately, the real effects of ageism in the health care system are not funny.

Normal and Abnormal Changes of Aging

The following section outlines both the changes that are expected in human aging and those that may be the result of illnesses. In addition, some of the ramifications of these changes are explored (Hooyman and Kiyak, 1996; Rockstein and Sussman, 1979).

Eyesight. Normal changes in a person's eyes that can begin to occur as early as the 40s include a decrease in peripheral (side) vision, a decrease in the elasticity or flexibility of the lens, a decrease in the ability to see small print, decrease in depth and distance perception, and a decrease in the pupil's ability to change size and to adjust to changes in light and dark. As they age, persons also lose some ability to read and see things that are overhead, due to changes in the muscles that raise the eyeballs. Most persons become more farsighted as they age.

Abnormal changes in the eye can be caused by cataracts (a clouding of the lens), glaucoma (an increase in inner eye pressure), and diabetic retinopathy (deterioration of the retina of a person who has diabetes).

Hearing and ears. Some persons do not notice any loss of hearing as they age; however, a slight hearing loss may be normal in later years. Hearing loss with age in our society may actually be environmental (caused by exposure to loud noises throughout our lives). Persons living in societies that are not industrialized do not have as much hearing loss as we do. When President Clinton was prescribed hearing aids for both ears when he was 50 years old, he was told that his hearing loss was most likely to have been caused by a lifetime of rock concerts, political rallies, and being in crowded rooms.

The most common condition of normal aging is presbycusis (age-related hearing loss), which is a gradual hearing loss where it is more difficult to hear high-frequency pitches in voices or instruments. Since consonants tend to be higher frequency sounds and vowels are lower pitched, speech can be difficult to understand for someone with presbycusis, especially when there are distracting noises.

Another common hearing problem is tinnitus, or ringing in the ears. The high-pitched ringing sound is especially acute at night or in silence. Sometimes apparent hearing loss is simply due to a normal buildup of ear wax, which tends to accumulate more rapidly in some older people.

Hearing loss can be caused by injury or infection. At any sign of hearing loss, a person's hearing should be evaluated. A person needs to have a medical examination before purchasing a hearing aid. Not every hearing difficulty can be corrected with an aid; some require surgery.

Infections in or damage to the inner ear may cause dizziness and mobility problems. This is not an expected change of aging and should be treated.

Sometimes hearing impairment is mistaken for confusion or dementia because of the person's difficulty in understanding and responding to people. Hearing loss can be devastating when it leads to a person's withdrawal and social isolation.

Other senses. There is a normal loss in the intensity and sensitivity in the sense of touch, so it becomes more difficult to detect hot and cold sensations. This loss is especially evident in the fingers, palms, and lower extremities. Since it takes a few more seconds for heat to register, an older person can become burned more easily. In addition, because an older person may not be able to tell how cold it is, hypothermia (lowering of body temperature) is a danger for him or her.

The sense of smell becomes less sensitive also in normal aging, and the ability to distinguish one smell from another declines normally with age. This is because the olfactory nerve becomes less sensitive normally. This change could be dangerous, because an older person may not be able to readily smell gas or smoke.

In addition, over time, people normally lose some ability to judge taste differences, especially saltiness. Because the senses of smell and taste become less acute, the appetite and enjoyment of food can be adversely affected. In addition, an older person may increase the intake of salt because the taste is harder to distinguish, which may not be nutritionally wise. Abnormal changes in touch, smell, and taste may be caused by strokes and other nervous system disruptions.

Teeth and gums. If care of the teeth and gums has been adequate throughout the person's life, then he or she will probably not lose teeth. Teeth normally change color with time, becoming more yellowish.

Sometimes oral disease causes cavities or loss of teeth. The increased incidence of losing teeth is due not to normal aging but to historical differences in preventive dental care and in water fluoridation. Since jaws normally shrink a little in later years, dentures may need to be refitted. Dentures that are an inappropriate size will adversely affect chewing and digestion, as well as cause problems with speech.

Cancers of the lip, tongue, mouth, gum, pharynx, and salivary glands increase with age as well. These are not normal occurrences but are the result of disease.

Speech. Speech should not change at all with age. If it does, it is most likely due to a problem, such as ill-fitting dentures or the occurrence of a stroke. Aphasia, a condition that may occur as a result of a stroke, causes a person to lose the ability to remember or pronounce certain words. Some persons who have had strokes lose the ability to speak altogether.

Skin and hair. Normally, the skin of older adults is not as elastic as that of younger people, so it is more susceptible to bruises and cuts. Skin also heals more slowly. Skin gets drier as well, and wrinkles always occur to some degree. Wrinkles can be more pronounced if the person has been overexposed to the sun or wind throughout life. Also, the lighter the person's skin, the more likely it is that wrinkles will develop. Older persons may also develop dark spots, called age spots or liver spots, which are harmless.

Hair becomes drier, more brittle, and thinner and usually turns gray or white. Many men and women have a decrease in amount of hair with age. Due to hormone reduction, men may experience a receding hair line.

Of course, many skin conditions are illnesses and chronic conditions and should be assessed and treated. Skin cancer is more common in older age, due mostly to the cumulative effects of sun exposure.

Body composition. Generally, over time there is a decline in lean body mass and muscle tissue and an increase in the amount of fat. In the very old, there is a tendency toward weight loss.

Kidney and bladder function. The kidneys usually lose about half of their function as people age, declining in volume and weight. This means that it is more difficult for toxins, vitamins, chemicals, and medications to be cleaned from the body. This can sometimes lead to accidental overdosing of medications. Regardless of whether there is disease present, changes in the kidneys of older persons are substantial.

The bladder's capacity also decreases normally with age, but the sensation of needing to urinate is delayed. Sometimes this results in urinary incontinence (not making it to the bathroom in time). Incontinence can often be managed through exercise and learning to plan ahead for bathroom trips and to urinate more often.

There are pathological conditions that can cause kidney (renal) failure. The most extreme of these may necessitate dialysis, a mechanical system for cleansing toxins from the blood. In addition, urinary incontinence is greatly worsened by diseases of the central nervous system.

Skeletal system. Humans normally get shorter with age, as the skeletal system changes. Between the ages of 65 and 75, a person may be one and one-half inches shorter than middle-age height. At age 85, the person may have shrunk as much as three inches. The muscles in the spine get a little weaker, and the discs get thinner. There is also a slight decrease in the hardness of bones with age.

A common disease of the skeletal system is osteoporosis, a condition most often found in women who are postmenopausal. It is thought that osteoporosis in linked to a calcium deficiency or to hormonal changes. In this disease, the spine begins to collapse, and bones become very brittle, fracturing more easily. There is also a stooping effect due to the spinal collapse.

Respiratory system. The lungs' capacity to hold air decreases normally with age, but not drastically. Cilia, the hairlike filters in the airways of the lungs, are reduced in number; this diminishes the amount of oxygen available and reduces the efficiency in removing foreign matter from the lungs.

Diseases such as emphysema, asthma, and lung cancer are often but not always linked to a history of smoking or to an occupation in hazardous conditions.

Muscular system. Normal changes include decreased size, tone, flexibility, strength, elasticity, and endurance of muscles. There may be increased twitches and spasms in the muscles. These normal changes should not usually have an impact on mobility.

There are chronic and acute conditions that can affect mobility and functioning. Arthritis, osteoporosis, muscular dystrophy, multiple sclerosis, and strokes are all examples of conditions that may create problems. There is currently some increased attention in gerontology to the loss of muscle mass, which used to be seen as normal aging and is beginning to be viewed as a condition that may be abnormal and should be treated. Someone who experiences drastic changes in mobility or muscle control should be evaluated for an illness.

Central nervous system. Normal aging does not result in significant cognitive impairment. There are some mild declines in memory function, known as benign senescent forgetfulness.

As people age there is usually a slowing of impulses to the brain, which results in reduced reaction time and reflexes. There is also a higher threshold for pain and heat, which could cause injury, heat exhaustion, or burns. Older persons are less able than are younger persons to discriminate among levels of perceived pain. If nerves are severed in an accident or operation, it may take longer in the later years for them to grow back.

The brain reduces its weight by 10 percent in older ages. Brain cells normally are lost steadily from about the age of 30, and they are never replaced or regenerated. The overuse of drugs or alcohol, exposure to pollution, and/or smoking cigarettes increases the constant loss of brain cells. However, we tend to function with fewer brain cells than we have, so a moderate loss does not create a significant decline in brain function.

Changes in brain wave patterns may also affect the sleep of persons as they age; they have shorter sleep cycles and may be more easily awakened. Some older persons shift to a sleep rhythm that includes more daytime naps and fewer nighttime sleeping hours.

In normal aging, the ability to remember the meanings of words and symbols does not decline. Neither does recall and recognition. Aging normally does make it more difficult to retrieve material that was stored in memory years ago. If problem-solving skills are not practiced and sharpened, they may decline. It is hard to generalize about the intelligence of older persons or to compare it with the intelligence of younger persons, because people in the later years have not usually been exposed to the same educational and technological factors as have younger persons. However, a researcher who followed some older persons over time found that changes in intellectual functioning did not occur until they were in their 60s, and those changes were gradual and did not occur across the board (Schaie, 1995).

There are many abnormal, disease-related changes that can occur to the brain that can cause dementia, which is serious and often fatal neurological impairment. These illness include Alzheimer's disease, Parkinson's disease, and multi-infarct (many strokes) dementia. Although not part of normal aging, the incidence of dementia increases with age. This will be discussed further in the section on mental health and cognitive functioning.

Gastrointestinal system. Normally the esophagus changes slightly in function; a decrease in muscle contraction necessitates a longer time for food to be transmitted to the stomach. This may produce an early sensation of being full, resulting in inadequate nutrient intake. Also, because digestive juices in the stomach diminish after age 50, older persons may develop a chronic inflammation of the stomach lining and may be more susceptible to gastric ulcers. Due to normal changes in the intestines, chronic constipation and impaction may be present in many older persons.

Persons are also at greater risk as they age for colon and stomach cancer, which are not normal conditions of aging. Diverticulitis is a chronic condition in which pouches form in the intestines because the walls have become weak. Older persons are more susceptible to hernias and gall bladder disease. Chronic constipation is a common problem of some persons as they age, but it is not the normal result of growing older. It may be the result of overuse of laxatives, lack of exercise, stress, an unbalanced diet, an obstruction, or an underlying disease.

Cardiovascular system. With normal aging, the heart gets slightly larger, pumps out less blood, and loses some efficiency. By age 75, the heart has lost 35 percent of its function. The heart and blood vessels both lose some elasticity. The heart increases in size and needs to rest more often. There is also a normal slight increase in blood pressure. The blood flow decreases, and circulation is poorer.

There are several pathological states in the cardiovascular system that can be very dangerous. Heart disease and strokes are common in elderly persons. Hypertension, or high blood pressure, is not a normal condition and can be treated with medication and/or attention to diet. For this reason, it is very important for older persons to have their blood pressure checked regularly. Hypertension results from the heart pumping too hard. It is estimated that half the population over age 65 has hypertension. Sometimes hypertension is called "the silent killer" because the symptoms are masked or hidden. This is a serious and life-threatening condition that can cause kidney failure, heart disease, strokes, or other medical problems.

Another possible chronic condition in older adulthood is hypotension, or low blood pressure. Hypotension, which causes dizziness and faintness from exertion or standing rapidly, is a problem only when it leads to a fall.

Congestive heart failure is a backing up of fluid into the heart, lungs, or limbs due to impaired and decreased pumping function. Angina pectoris is pain in the chest when not enough blood flows to the heart. An irregular heart beat is also an abnormal condition that may have many causes.

Atherosclerosis, a building up of fat, plaque, or cholesterol on the inner walls of the blood vessels, is a serious condition that can limit blood flow, causing the vessels and arteries to become smaller and less elastic. Blood clots may therefore develop more easily, which can cause tissue to die. If there is persistent lack of blood supply to the heart, an infarct, or dead area, may occur. If heart tissue dies, it is called a heart attack. If brain tissue dies, it is called a stroke.

Reproductive system and sexuality. The reproductive systems of men and women reduce the production of hormones in later middle age, which may affect sleep patterns, moods, or weight. Sexual arousal may be slowed but usually does not stop. Men do not normally lose their ability to produce viable sperm. In women, the expected changes of aging include the slowing and then ceasing of ovulation and menstrual periods (called menopause).

If sexual dysfunction occurs, it is not necessarily a sign of normal aging but may be a sign of another problem, either physiological or emotional. Sometimes stress, illness, or medication will cause a decrease in sexual desire. As in all age groups, relationship difficulties, depression, and the lack of a trusted and available partner will cause sexual difficulties.

Immune system. The internal and external healing processes in general become slower in old age. A normal loss of antibodies makes it harder to fight infections.

Other abnormal conditions. Some diseases that can occur at any age, such as cancer and diabetes, are more common in the elderly. In addition, injuries due to falls and other accidents are problems for this age group. People over age 65 have the highest rates of injuries and death related to traffic accidents (as drivers, passengers, and pedestrians) of any other age group except teens. Although older persons drive fewer miles than younger people, they have more accidents per mile driven.

Functional Health

One of the most useful concepts in thinking about the health of the elderly is the concept of functioning. Activities of daily living (ADLs) are a way to measure the level of functioning and capacity. Examples of ADLs are: ability to bathe, feed, and dress oneself; ability to get to and use the toilet; getting in and out of a chair or bed; managing one's own medications; and performing tasks such as shopping, cooking, and housework (Jett and Branch, 1981).

Persons over 85 years old are much more likely to be unable to perform the activities of daily living than those who are aged 65 to 84, due to the increase of disease, normal changes of aging, and chronic conditions.

Mental Health and Cognitive Functioning

Suicide is a concern for this age group. It is estimated that as many as a quarter of suicides in this country occur in persons over the age of 65 (Koenig and Blazer, 1992). The highest rates of suicide in the United States are in older white males, especially over the age of 85 (Belsky, 1984). It is not clear whether this is due to depression, illness, role changes, or lack of social support.

As was stated earlier, normal aging does not result in significant cognitive impairment. However, there are several diseases and conditions that can cause dementia, which is deteriorating mental and neurological functioning. Dementia is also sometimes called organic brain syndrome or disorder, or senile dementia. Dementias usually consist of impaired recent memory, comprehension, attention span, learning, recall, judgment, control of emotions, and orientation to time, place, and person.

Dementias are considered either to be *reversible* or curable, meaning that they can be improved with treatment, or *irreversible* or incurable, meaning that treatment will not have an impact. Reversible dementia can be caused by medications, drugs, alcohol, malnutrition, hormonal disorders, toxins, brain trauma, diabetes, infections, brain tumors, neurosyphilis, and depression. Irreversible dementia can be caused by a variety of illnesses, such as Alzheimer's disease, multi-infarct, Huntington's chorea, Pick's disease, and Parkinson's disease (Hooyman and Kiyak, 1996).

Alzheimer's disease is the most common cause of dementia. A person of any age can have Alzheimer's disease, but the rate goes up steadily with age. Less than 2 percent of the population under 60 have Alzheimer's, but over 20 percent of the population over 80 have it (Evans et al., 1989).

"Successful" Aging

There have been certain characteristics or activities identified with increased satisfaction in older adulthood. Gerontologists have referred to this satisfaction as "successful" aging. We must be careful with this language so that we do not imply that a person can "fail" at growing older; any way that a person faces older adulthood is valid for that person. With that caveat, let us examine what is usually considered to be signs of well-adjusted aging (Clark and Anderson, 1967).

- **A positive attitude and outlook.** Persons who are better adjusted do not tend to take an overly negative view of their situation. As in all ages, focusing on strengths, benefits, and blessings tend to make persons feel more satisfied.
- **Social connection.** As is the case at all ages in the human life cycle, people do better when they have other people in their lives who are positive influences. The happier and healthier elders remain congenial and gregarious; they are interested in others and accept friendship and support from persons they care for.
- **Personal independence and appropriate interdependence.** Most well-adjusted elders have relatives, partners, and friends who love and support them at the same time that those older persons strive for independence. They value and hold dear their ability to function as much as possible on their own and not be a burden to others.
- **Being realistic.** Although the more satisfied elders value independence, they also tend to be realistic about their limitations. Consequently, they adjust their lives based on whether total independence is actually possible. They take

into account their own needs for health and safety as they balance these realities with their desire to be on their own.

• **Being resilient.** The more satisfied persons are those who can bounce back from losses, illnesses, and disappointments. This does not mean that they are unrealistic or that they do not feel fear or grief. Rather, it means that they are resilient. "Resilience" means that they can determine what can be changed and what cannot be changed and can readjust and learn to cope in different ways when it is necessary to do so. Persons who are able to adjust more easily tend to plan ahead for retirement and for possible physical difficulties or medical crises.

• **Figuring out one's life.** One of the important tasks of older adulthood is reviewing and thinking about the meaning of one's life. One of the ways that older persons may do this is to talk to others about their life experiences. This process is often called "life review." When an elder can take the time and find the support to think about and talk about the accomplishments and disappointments in life, he or she has a better chance of being adjusted.

Think About It #62

1. Share examples with fellow learners of people whom you have known who seem to have some of the characteristics of "successful" aging.

2. Do you think that the above list describing well-adjusted elders could apply to other ages and life stages? Why or why not?

The Role of Oppression

Any discussion of "successful" or "well-adjusted" aging must take into account the fact that people do not always have control over financial resources, access to care, or attitudes of others. In studying demographic trends in the United States one can see that on the whole persons who are poor, less educated, African-American, Native-American, or Latino have a lower health functioning and do not live as many years as do those who are middle class, educated, and white (Markides and Mindel, 1987). This fact seems to be linked to the availability of medical care throughout one's life, which is in turn linked to one's income, education, and ethnicity. Advocacy for the elderly must therefore include working toward equality and social justice issues across the lifespans of all persons.

There are many situations that may make an elder's access to services for physical and mental health maintenance more difficult. Sometimes older adults who live in more isolated rural areas are at greater risk because social services and medical care are not as easily accessible. Sometimes elders who do not speak English or have learned it as a second language are not as able to explain to human services workers and medical providers what they are experiencing or what they choose. Gay older persons who grew up and became adults well before any liberation movements may be particularly embarrassed to discuss their sexual health or reluctant to admit that they are the "spouses" and therefore next of kin and primary caregivers for their same-gender lovers.

The plight of elders who are in minority groups is often described as "double jeopardy," meaning that one can simultaneously suffer discrimination for being older as well as a member of another minority group (including gender, race, ethnicity, poverty, or sexual orientation). Obviously, one can have multiple jeopardy as well if one is a member of several oppressed groups at once. In the double-jeopardy hypothesis, old age is more difficult when one is a member of more than one minority group and therefore bears an increased burden (Markides and Mindel, 1987).

Illustration

Betty walked into a drop-in center for homeless persons asking for shelter for the night. After listening to Betty for 30 minutes, Amanda, the case manager, realized that Betty was a person with multiple societal strikes against her. Not only was Betty African-American and female, but she explained that she was 60 years old, alcoholic, and without a home. She also said that she had received diagnoses of psychosis and AIDS. Amanda realized that she would need to connect Betty with many different sources of support and that they would need to have an ongoing relationship if possible.

Trends in Gerontology

It is likely that the older population in our country will continue to grow. By 2000, most of the older population will be females who live alone (Paradise, 1993). Many women reach old age with economic and social problems, as well as the physical issues that might arise with age or disease. The continuing trends toward fewer children per family will mean that there are fewer family members to provide support for persons who are functionally impaired. Because the futures of the Social Security system and health care financing are not secure and because political and economic trends are volatile, our society may have difficulty caring for those older persons who are in need.

One of the growing trends in our society is that older persons are more frequently becoming the caregivers for younger family members at a time in their lives when they may not have expected to re-enter that role. Because of the growing incidence of drug and alcohol use, homelessness, and HIV infection some grandparents are caring for their grandchildren when the parents are not available or have died (Minkler and Roe, 1993). In addition, many older persons are providing medical and personal care for their adult children who have end-stage HIV (Allers, 1990; Muschkin and Ellis, 1993), mental illness, or developmental disabilities (Valentine et al., 1998). These trends are receiving increasing attention in the field of gerontology. As a human services worker who cares about older persons, you will need to be prepared to identify and address many such family issues when older persons unexpectedly become family caregivers.

The HIV epidemic has affected the older population, just as it has affected younger persons. Over ten percent of the identified persons infected with HIV are over 50. However, because of general misconceptions about older persons not being sexual and not using drugs, many of these people are misdiagnosed by medical care providers. In addition, because HIV prevention messages do not usually include older persons, they tend to see themselves as not at risk for HIV transmission. When they are diagnosed with HIV, many of them feel the weight of HIV-related stigma and do not disclose to their social support networks that they have AIDS. Human service providers must be aware of the intersection of HIV and aging concerns and be diligent in their education and service planning efforts (Linsk, 1994).

Human Services Settings

Human services workers will be introduced to older persons in almost every setting imaginable. Every practitioner therefore needs basic information about the aging process and the experiences of older persons. Some common organizations where you may choose to concentrate in gerontological concerns are: councils on aging, senior centers, community mental health centers, mental health hospitals, acute care hospitals, nursing homes, dialysis clinics, retirement communities and other residences for older persons, rehabilitation facilities, home health agencies, and hospices. Of course, none of these types of agencies serve elders exclusively, but it is likely that a human services worker who wishes to serve older persons could get such an experience in these places.

Within community-based organizations for elders, many services are often offered to provide support for remaining in one's own home and community. Programs may include: telephone reassurance (calling each day to check on a frail older person), transportation, information and referral, home-delivered meals, congregate meals, recreation, job placement, legal services, shopping assistance, the provision of homemakers or chore workers to help in the home, support groups, friendly visiting, counseling, or protective services (in cases of caregiver abuse, exploitation, or neglect).

Because elders may be at risk for being home bound, may have some cognitive impairment due to depression or dementia, may be afraid of venturing out due to crime or bad weather, may be socially isolated, or may not be aware of financial, social, and medical services that may be available to them, it is especially important to provide aggressive outreach to this population. Organizations should consider public service announcements in newspapers and on radio and television to inform persons about benefits and programs. Another possible approach is to post fliers or leave brochures in medical clinics, hospitals, and physicians' offices to alert older persons and their family members that support may be available.

Applying Human Services Values to Practice

There are several underlying values (discussed in Chapter 2) that are especially applicable to working with older persons. They are: supporting choice, finding strengths first, respecting diversity, and building rapport.

Support Choice

Human services workers recognize that each individual, whatever his or her age, is a unique person with innate dignity and with unique characteristics, history, and dreams. Therefore, persons should be allowed to choose their own lifestyles as much as possible as they age (Lowy, 1979). This applies to choice about where and how to live as well as what to do with one's leisure time. Sometimes the recreational programs we offer to seniors do not at all reflect their talents, interests, experiences, or skills because we do not ask them what they want to do before we institute services for them. (For example, one retired businessman in a senior center said to one of the authors, "I came here to meet people, but I pray that no one will force me to make tacky things out of popsicle sticks.") Of course, human services helpers must balance the desires of older persons with the needs and limitations of their caregivers. Sometimes we make the mistake of assuming that friends, neighbors, or family members can take on the responsibilities for looking after the financial or medical needs of elders, when these informal support persons are feeling depressed, burdened, or unprepared.

Conflicts sometimes arise when the value of self-determination clashes with the value of protecting vulnerable persons. This situation can happen with any population but is commonly faced when working with older persons. How do we balance the safety of an older man or woman with his or her natural desire and need for autonomy? When and how do we decide to take away the right to choice? There often are no clearly identified right or wrong answers to these dilemmas; rather, there only is an agonizing decision to make between two difficult choices.

Illustration

Stella Jones is a woman of 80 who lives alone in a small mountainous town in Pennsylvania. She immigrated from Germany with her parents and siblings when she was 5 years old. All of Stella's family members are now deceased. For many years she has been attending a nearby senior center for meals and socializing. Stella has a reputation with staff and participants in the senior center for being hardheaded and sometimes rude to others. Sometimes when she attends the center, her clothes are not clean and she does not smell good. She is fiercely independent and is very attached to her large rambling home and several stray cats that come and go. In the winter she lives in only one room of the house, which she heats with a kerosene heater. After a fall on the ice one November, which resulted in a broken arm, Stella was hospitalized briefly. The hospital so-cial worker, Carol, became alarmed about Stella's isolation and lack of financial resources, and began proceedings to have Stella placed in a nursing home. Jenny, the senior center director, learned of these discharge plans as she visited Stella in the hospital and opposed the plans because Stella told her that she did not want to leave her home. Jenny felt that nursing home placement, while keeping Stella safer, would rob her of her prized independence and thus endan-ger her mental health.

Think About It #63

1. What is Carol, the hospital discharge planner, trying to accomplish for Stella?
2. What is Jenny, the director of the senior center, trying to accomplish for Stella?
3. What is your reaction to Stella's situation? What do you think should be done and why?

Find Strengths First

As in all areas of human service provision, the strengths perspective is vital. Per-sons often do have real problems and real struggles as they age, stemming from their health, their losses, their role changes, and their positions in society. It is im-portant to remember, however, that they are often high-functioning, wise, funny, loving persons with a lot of experience and much to teach us. It is important not to fall into the trap of expecting older adulthood to be miserable for everyone and thus only focusing on the challenges and worries that old age can bring.

Remember to search for and identify the strengths in the lives of elders and to use these strengths in service planning. For example, many older persons have strong social networks and kinship ties. Many older persons have sur-vived many financial, personal, and social hardships and have remarkable wis-dom, humor, and resilience to bring to bear on their current situations. Many older persons are energetic, involved in their communities, and eager to volun-teer for human services agencies or educational organizations.

Respect Diversity

Be careful not to let stereotypes or ageist attitudes creep into your decisions and actions. Remember that older persons in the United States are a heterogeneous and diverse group. Individuals will differ in cultural and religious back-grounds, education, financial situations, health, and functioning level. They will be different in gender, ethnicity, age, and sexual orientation. They will also bring a diversity of experiences and expectations to the helping relationship. They will have different opinions about privacy, dignity, and what information can be shared with strangers. Persons want to be seen as individuals, not as a "type" (Gastel, 1994).

Illustration

Mr. and Mrs. Kim, born in Seoul, Korea, in the 1920s, immigrated to the United States 20 years ago to live with their son, daughter-in-law, and two grandchildren. The family now lives together in a four bedroom house. Mrs. Kim's health has begun to decline. She has recently developed a serious heart condition and diabetes. Yo Jin, the Kim's 45-year-old daughter-in-law, is the primary care provider for Mrs. Kim, as is traditionally expected in the Korean culture. Yo Jin is also primarily responsible for looking after Mr. Kim, the children, and the maintenance of the household. Elyse, a visitor from the home health agency, notices that Yo Jin seems tired and overwhelmed by her many responsibilities. Elyse suggests to the family that services such as home health and adult day care could be appropriate to relieve some stress on Yo Jin. All family members agree that they do not wish assistance. Yo Jin tells Elyse that receiving help from outsiders is unacceptable.

Think About It #64

1. What do you notice about the Kim family that may be different than or similar to the way you or your family might make decisions?
2. What cultural or social factors might account for the differences?
3. As a helper, how might you respond to the Kim family's refusal of social and medical services?
4. How might you as a helper respond to Yo Jin's needs as the primary family caregiver?

Build Rapport

Be welcoming and inviting. Do all that you can to make an older person feel at home with you. Make your office comfortable and physically accessible. Be aware of possible struggles with mobility, hearing, or seeing, and find ways to compensate for those difficulties. Be attuned to whether the pace of the interview is appropriate, and check out frequently whether you are being understood. Do not ever be patronizing or call an older person by first name without permission. Treat each person with warmth and respect. If the person is able to communicate and understand you, speak directly with him or her, instead of ignoring the person in favor of younger family members or caregivers. When it is necessary to explore private or sensitive topics, do so gently, directly, and respectfully (Gastel, 1994).

Wrapping It Up

A human services worker should be aware of his or her own barriers, strengths, and preferences in choosing to work with older adults. (This is true for any specific topic, social problem, or population.) In order to best work with elders, you should be a person who does not give up on others easily and does not lose faith, even when someone appears to be quite ill or impaired. It helps if you believe that each person deserves affection and advocacy, even when he or she is suffering from dementia, is incapacitated, or is dying. An effective human services worker also needs patience and compassion to work with frail seniors. Often our "cure" goals are replaced with "care" goals. Instead of always working toward remarkable progress, sometimes a little bit of change or response is sought and rejoiced over. In addition, you should truly value older persons as contributors to our society, as persons who have something to teach us, and as persons who have earned the right to dignity. You must be able to appreciate wisdom and experience, and not judge a person's worth based on productivity or appearance. If you feel drawn to work with older adults, then it is an area of human services provision that is rewarding and meaningful.

CHAPTER THIRTEEN

HIV and AIDS

Human Immunodeficiency Virus (HIV) illness is a recent phenomenon that unfortunately continues to grow in its impact. Human services workers, students, and volunteers are more and more frequently called upon to provide competent and sensitive services and care to persons who are infected with HIV, as well as to their HIV-affected caregivers. Presently, many case managers in HIV service organizations are helpers with an Associate of Arts degree. Human services helpers also are involved in educating persons about safer behavior in order to prevent HIV infection. This chapter outlines some basic information about the disease and its impact on persons who are affected by it.

Introduction to the Basics About HIV

"HIV" stands for "Human Immunodeficiency Virus." HIV illness is a serious, infectious, life-threatening infection caused by this virus. Once a person is infected with HIV, the virus attacks the person's immune system, destroying cells that ordinarily fight infections and cancers. Although persons who have HIV illness can live for years in good health, at the time of this printing HIV still is considered to be eventually fatal.

Once certain symptoms or infections begin to appear, the disease is often called "AIDS," which stands for "Acquired Immunodeficiency Syndrome." The term AIDS was first coined in July of 1982 to describe a collection of unusual illnesses associated with impaired immunity in previously healthy adults. This occurred before HIV itself was discovered to be the cause of the immune system damage. AIDS is now considered to be the end stage of HIV illness or infection. "Acquired" means that the condition is not due to a genetic cause; the virus is acquired through transmission of specific bodily fluids (blood, semen, vaginal secretions, and breast milk). "Immunodeficiency" means that the body's defense system is not as able to adequately protect itself against certain illnesses. "Syndrome" means that the condition is a collection of symptoms and illnesses.

223

The full spectrum or continuum of the condition from beginning to end is called "HIV disease," "HIV infection," or "HIV illness." After infection with HIV, people often live for many years without any symptoms. When persons do develop symptomatic disease, or AIDS, they can still maintain good health and functioning for several years with appropriate medical care, social support, good nutrition, exercise, stress management, and a positive outlook. With more and more effective medications being developed, the time during which persons with HIV can live without symptoms is lengthening.

There are still many myths, even two decades after the beginning of the epidemic, about the risks of HIV transmission. The virus can only be transmitted when blood, semen, vaginal secretions, or breast milk from an infected person gets into the body of another person. HIV is not transmitted through tears, perspiration, saliva, or urine. Nor is it transmitted through casual contact, such as hugging, holding hands, or sharing food, beverages, dishes, or linens. When talking to people about preventing HIV transmission, it is helpful to evaluate their behavior for the likelihood of getting infected fluids into their bodies.

HIV most often is a sexually transmitted infection because blood, semen, and vaginal secretions carrying the virus can enter someone's body during sexual encounters. Infected blood can also enter someone's body during a blood transfusion or when people share needles or syringes to inject drugs. In addition, the virus can be transmitted from women to their babies during pregnancy. There is also a risk that a newborn baby can become infected by ingesting breast milk from an HIV-infected woman.

To significantly reduce the risk of HIV transmission during sexual intercourse, the use of latex or polyurethane condoms (male or female versions) is recommended. Other forms of erotic activity that do not involve vaginal, anal, or oral intercourse are also safer ways of being sexual. Persons who inject drugs (whether illegal, like heroine, or legal, like insulin) should either avoid sharing injecting equipment with others or should flush needles and syringes thoroughly, first with bleach and then with clean water, before using them. Health care workers should use "universal precautions" to avoid accidental exposure to infected blood as well. It is also recommended that women with HIV avoid breast-feeding their newborns to reduce the risk of transmission.

At the beginning of the HIV epidemic, from its preliminary identification in 1981 until the development of screening tests in 1985, there was no way to guarantee the safety of the nation's blood supply because there was no way to test donated blood for HIV or to destroy any virus that might be present. During that time many persons who received blood transfusions or Factor VIII injections (a blood clotting protein given to some persons with hemophilia) were infected with HIV because the blood products they received contained the virus. Since mid-1985, blood banks in the United States have been screening all donations for HIV and destroying blood that tests positive. Factor VIII is now heat-treated, which destroys any HIV that might be present. It is now extremely rare for someone to become infected with HIV from blood products.

HIV infection does not change a person's appearance at all until perhaps the last stages of the disease. The only way to know if someone is infected with HIV is to perform laboratory tests on their blood. There is no test for AIDS. Instead, the tests are used to detect *antibodies* to HIV. Whenever we become infected with a virus, our immune systems begin to manufacture antibodies to fight that invading virus. Antibodies are usually detectable in the blood a few weeks after the onset of an infection. This period of time is referred to as the "window period."

Some people are confused by the concept of a brief window of time between becoming infected and having detectable antibodies. Although *antibodies* are not present immediately after a person is infected with HIV, he or she is indeed carrying the *virus* and can transmit it to someone else. Some people have mistakenly thought that if a person can live for ten or more years without developing any HIV-related symptoms, then HIV antibodies could go undetected in the blood for that long as well. That is not at all the case; it typically takes eight to twelve weeks for antibodies to HIV to be detectable using the blood tests that have been developed for this purpose.

When tests show the presence of HIV antibodies, the test results are called "positive" and the blood and person are referred to as "HIV positive," "seropositive," or "HIV infected." When a test does not show the presence of HIV antibodies, the test, blood, and person are referred to as "HIV negative."

The basic procedure for HIV testing is as follows. A screening test called an "ELISA" (enzyme-linked immunosorbent assay) is first performed on a specimen of someone's blood. The ELISA is very sensitive and may occasionally give a false positive reading. For that reason, if the first ELISA test is positive, a second ELISA and another test called the Western Blot are done to confirm the first test result. If the Western Blot test is positive, then the blood is determined to be HIV infected. Because these tests are performed by a laboratory, it may take several days for a person to receive his or her test results, whether positive or negative. Viral load tests are now available to measure the level of HIV; these tests are sometimes used to determine whether newborns are infected.

It is important for people who are considering being tested to understand this three-test process because it is best to have the testing performed in a facility that follows this procedure. Health departments, clinics, and hospitals routinely perform confirmatory testing with a second ELISA and a Western Blot. However, some private physicians may only perform a single ELISA test and thereby risk diagnosing someone as HIV positive when that may not be the case.

There are two types of testing available: confidential and anonymous. "Confidential" means that the name and other identifying information about the person who is tested will be kept on record by the testing facility and may be sent on to state and federal officials who track the epidemic if the test is positive. Test results and the personal information associated with them are supposed to be protected to the greatest extent possible to prevent discrimination against those who are tested. "Anonymous" testing and reporting are done with code numbers rather than names. At an anonymous testing site one can have an HIV test and receive the results without giving a name; however, anonymous testing is not available in all states.

People generally receive some form of individual counseling before and after being tested. The purpose of pretest counseling is to help people identify ways to reduce the risk of becoming infected and to make an informed decision about whether or not to be tested. The purpose of post-test counseling is to provide people with emotional support and referral resources as needed once they receive their test results.

Recently, at-home HIV testing kits have become available. To use them, a person sends a small sample of his or her own blood to a laboratory through the mail and receives the results over the telephone. These at-home collection kits are processed using the same procedures as described above, so they are as accurate and reliable as the tests performed by health departments and other HIV testing sites. The test kits are also anonymous, but they lack the face-to-face counseling feature of the more traditional testing approach.

Another new development is oral testing for HIV. This method involves using a special collection device to absorb oral fluids, which are then tested for the presence of HIV antibodies in the same manner that blood is tested. Some facilities are now offering this option to persons who would prefer not to have blood drawn through a syringe or finger stick. As with the blood tests, it may take several days for the results to be available.

Deciding whether to seek HIV counseling and testing can be difficult and sometimes terrifying. When working with persons who are struggling to make this decision you should listen actively; offer them options for confidential and anonymous testing sites; answer their questions about safer sex and drug behavior, the testing process, and the illness; help them consider the possible ramifications of testing positive; assist them to prepare for the waiting period before the results are delivered; let them know the hopeful news about more effective treatments and longer life expectancy; and offer them community resources for supportive medical and social services.

As stated previously, "AIDS" often refers to the end stage of HIV disease, when unusual infections, cancers, or other illnesses begin to manifest themselves. Most AIDS-defining conditions are called "opportunistic illnesses" because they take advantage of the body's damaged immune system to take hold. In other words, these illnesses take the opportunity afforded by a depressed immune response. Examples of illnesses that can lead to an official diagnosis of AIDS include pneumocystis carinii pneumonia, Kaposi's sarcoma (a form of cancer that often appears as skin lesions), cytomegalovirus, tuberculosis, cryptosporidiosis, and toxoplasmosis.

While there is no foreseeable cure, many promising advances have been made during the past several years in the treatment of HIV and AIDS that can greatly add to the length and quality of a person's life. There are two basic categories of drugs in current use: antiretrovirals, which help block HIV reproduction; and prophylactic agents, which help prevent some opportunistic illnesses. A recent and promising development is the use of potent antiretroviral medications called "protease inhibitors." Most of the drugs that address HIV and opportunistic infections can cause side effects, and their use must be carefully monitored by a medical provider. Unfortunately, these medications and the ongoing medical care required to manage HIV disease are expensive and out of reach for most people unless they have private insurance or access to public health benefits. In addition, taking the drugs strictly as prescribed (called "adherence") is vital. Human services helpers working with persons with HIV must be aware of the need for access to the treatments and the importance of adherence to the strict dosing regimens.

Cultural Competence and HIV

There are several reasons to be concerned with cultural competence (see Chapter 3) when working with persons who are HIV positive. First, like all people, persons with HIV have their own cultural beliefs and experiences that influence their perceptions and expectations regarding illness, adherence to treatment, seeking medical or social assistance, and the meaning of their lives. Second, HIV disproportionately affects persons who were marginalized or stigmatized by society even before they became infected (for example, men who have sex with men, sex workers, injecting drug users, women of color, and persons who are economically disadvantaged and/or homeless). Third, in addition to the need to take divergent cultural backgrounds into account, you need to realize that people who test positive become members of an entirely new and different culture: They are suddenly in the stigmatized group of persons with HIV, and as such, they have a new and deeper level of discrimination with which to contend.

As in all helping work, it is best to approach persons, couples, families, and groups with respect and an open mind and without bias. Helpers should avoid jumping to any conclusions about persons based on preconceived ideas about their lives or experiences simply because of their diagnosis. It is important to recognize that persons with HIV are not necessarily similar in any way except that they have a life-threatening disease that is stigmatized. It is also important to expect individual differences within any given culture or group. For instance, you can probably see that it would be a serious mistake to assume that all injecting drug users are alike, that all women are alike, that all Latinos are alike, or that all gay men are alike. Similarly, it is not useful for a helper to respond as though all persons with HIV are like each other.

A culturally competent human services worker, student, or volunteer can also help to educate coworkers and colleagues about the importance of listening actively and nonjudgmentally to persons with HIV in order to facilitate appropriate service and treatment planning. Adherence to treatment is crucial for persons with HIV because failure to take medications as directed can reduce their effectiveness and result in drug resistance. It is therefore vital that human services and health care providers make sure to involve the person with HIV at every decision-making step; explain clearly the treatment options and side effects; listen to the person's concerns and fears; and negotiate a realistic medication plan that the person feels is manageable. For immigrants who may not clearly understand English or for persons who are hearing impaired interpreters should be readily accessible so that instructions and conversations are clear.

Cultural competence is also key in understanding the personal relationships of those affected by HIV. The epidemic has highlighted the need for us to accept a person's own definition of who is "family." Gay men, for example, who may have been rejected by their families of origin, often form strong alliances with friends whom they call their "gay family." They may have lovers or partners they consider to be their spouses, but gay couples are not always treated as legitimate spouses by medical and social service personnel. African-American and Hispanic families often have strong extended networks of persons who are considered to be family, even though those persons may not be related to them by birth or marriage. As human services helpers, we must respect the family of choice just as we would respect the family of origin.

Helping people with HIV and their support persons can be an enriching and meaningful experience if you are open to learning. In this work it is useful to be able to celebrate and accept differences and diversity and to be willing to actively educate yourself about cultural issues. The strengths perspective (see Chapters 1 and 2) is as important in this type of helping as in any other; be careful not only to look at deficits, but also to find the wisdom, functioning, and adaptation that are inherent in most persons and situations. Learn to recognize and put aside your own biases and assumptions; if you cannot identify them and keep them out of the helping relationship, they will adversely affect your ability to assist persons with HIV. Accept the person's definition of the problem and the solution, and accept the person's perception of what is meaningful and significant. Never develop a program, service plan, or intervention for individuals, couples, families, a group, or a community without fully involving them in the needs assessment, planning, implementation, and evaluation. Seek out ways to learn about persons who have different backgrounds and experiences than your own. You can begin by reading, attending cultural festivals, listening actively to what you are being told by the persons who are seeking assistance, and having consultants who can answer questions for you about how to be more culturally competent.

Common Concerns

While it is true that each person with HIV is a unique individual and that it is not helpful to approach anyone with stereotypes and opinions already formed, it also is true that persons who test positive for HIV often have certain responses in common. Each person who is told that he or she has HIV will experience a complex set of concerns and emotions. No matter what a person's previous experience, being confronted with a stigmatized and life-threatening illness generates some personal and interpersonal concerns for most people. Although everyone may not have the same feelings or express them in the same manner, they will have similar concerns. The following are some of the most common reactions we have observed while working with persons who have HIV infection and suggestions about how to respond to them.

Living with HIV

Almost from the beginning of the epidemic, persons infected with the virus have generally focused more on living than on dying. The phrases "person with AIDS," "person with HIV," "person living with HIV," or "person living with AIDS" have been preferred by those who are HIV-positive over terms like "victim," "sufferer," or "terminally ill." This is not a reaction that signifies the inability or unwillingness to confront the seriousness of the infection; rather, it is a recognition that persons with HIV need to remain optimistic to survive and thrive. Helpers are encouraged to do all they can to support this hopeful outlook.

Disclosure and Stigma

When persons learn of their infection, they are immediately faced with intensely difficult decisions about whom to tell, when to tell, and why. There are agonizing choices to make regarding the disclosure of one's HIV status because of realistic fears about how others will react to the news. Our society tends to blame persons with HIV for having the disease (Baker, 1992; Cadwell, 1994; Herek and Glunt, 1988; Laryea and Gien, 1993). Sometimes people lose social support, employment, housing, or health insurance when they disclose their infection; these losses occur at a time when people most need stability and safety in their lives.

Although these issues may become less problematic in the future, the stigma surrounding HIV still is of great concern at this time. Persons with HIV, as well as their caregivers, family members, and friends, often are frightened of rejection, betrayal, and discrimination. Not only do they anticipate being shunned or mistreated for having HIV, they also expect to be judged by others according to whatever behavior may have been associated with their becoming infected. They fear that people will assume that they have been injecting drugs or sexually indiscriminate. HIV-related stigma is often compounded by other sources of discrimination, such as sexism, racism, ageism, or homophobia. Sometimes people are more afraid of being seen as gay or addicted than they are of being identified as a person with HIV. It is therefore important to recognize the multiple layers of stigma (often called double or triple jeopardy) that some persons with HIV may experience.

After testing positive, people are frequently faced with enormous questions. "Will I be able to find a dentist who will accept me once he or she knows I have HIV?" "Will my parents kick me out?" "Will my lover desert me?" "Will sexual partners reject me?" "Will I have any friends left after they find out?" "Will people judge me?" Deciding when to disclose and to whom is a process that is repeated over and over throughout the course of the disease. Sometimes persons with HIV or their family members delay asking for assistance from medical or social service agencies because they fear being judged harshly or discriminated against. Indeed, such responses are sometimes encountered from agencies and helpers, which only serves to further discourage persons affected by HIV from seeking assistance.

Victim-blaming still is common toward HIV-positive persons, in part because the stigma attached to HIV is more extreme than for most other diseases. Think, for instance, about how our society treats people who develop lung cancer as a result of smoking cigarettes, or people who develop skin cancer as a result of sunbathing, or people who develop liver disease after prolonged use of alcohol. Do our public leaders and policy makers suggest that these people deserve to die because they brought their diseases on themselves? Are these persons seen as evil or immoral? Do we consider denying them the medical care they need or their basic civil rights? Usually not. Yet persons with HIV often are spoken of in these terms.

Think About It #65

1. What are some of the judgmental or stereotypical things you have heard said about persons with HIV infection? List as many as you can.
2. Think about how you would feel or respond if you were HIV positive and heard these things spoken. What effect might this have on your willingness to disclose your diagnosis or ask for help?

Emotional Turmoil

A wide range of feelings can surface at various times throughout the illness. Responses can range from shock, numbing, and shutting down, to anxiety and panic. People can become very sad, depressed, or withdrawn. Sometimes persons with HIV become angry or enraged, resentful, or blaming of themselves or others. Any response is natural under the circumstances. If the person's emotions are causing him or her to act in self-destructive ways or to harm others, then those feelings or actions must be confronted. Otherwise, you should strive to listen to, acknowledge, and reflect whatever the person is feeling.

Uncertainty

HIV is a condition that is often unpredictable, and people may feel as though they are on an emotional roller coaster. The physical manifestations of HIV disease are often unexpected and surprising as well; a person may feel fine one day and ill the next. Persons with HIV and their caregivers often feel out of control and frightened because they truly do not know what might happen next. People have expressed this as "waiting for the other shoe to drop" or "waiting for the axe to fall."

Stress and Trauma Responses

Learning that you have a life-threatening disease is an extraordinarily stressful event under any circumstance. Since HIV disease carries with it an added burden of stigma, receiving a positive test result or a diagnosis of AIDS often causes an emotional trauma. Since HIV disease can be a roller coaster of unexpected emotional and physical ups and downs, persons with HIV or their caregivers may experience a series of crises throughout the course of the illness. Receiving the results of tests used to measure one's viral load and immune function can lead to crisis and panic. Likewise, being given a new prescription, being diagnosed with an opportunistic illness, or being hospitalized can be a traumatic event. People may respond to these traumas in various ways, including: disturbances in concentration, memory, sleep, or eating; panic reactions or feeling completely out of control of one's emotions; increased risk taking or dysfunctional behavior. Sometimes a current crisis can trigger unresolved trauma or grief from the past and compound the person's feelings of panic and terror (Poindexter, 1997).

When a person experiences emotional responses such as these, he or she may need crisis intervention and/or a suicide assessment. He or she may also request support from or connection to other persons with HIV or other HIV-affected caregivers. Make sure you let the person know that the reactions he or she is having are normal and expected during these episodic crises. Above all, the person will need nonjudgmental listening, validation, and affirmation.

Seeking Control

Receiving a positive test result or becoming ill with AIDS often causes people to feel a loss of control over their lives. They may need to have more choices about how to manage their mental and physical health and more opportunities to maintain their independence and dignity. Ironically, control is often taken from them at the very moments when they need to feel in control. For example, sometimes helpers, friends, or family members begin making decisions for persons with HIV without consulting them about their needs or wishes. It is important to help people regain as much control as possible and to support their efforts to do so.

Illustration

Ed, a 40-year-old gay white man, became seriously ill with AIDS-related illnesses in 1985 and decided to return to his parent's home in a small rural community after having lived in a large city for 20 years. By the time he moved in with them he was bed bound. He simultaneously lost his health, job, home, social network, city, mobility, independence, and future. He felt completely helpless and out of control. After he began receiving disability checks from his job, he asked his mother to cash these checks for him so that he could keep the money in a plastic zippered bank bag in his bed. From that bag he paid for his own cigarettes, food, and medicines; paid part of the telephone and utility bills; and paid his private duty nurse. Even though he could not leave his bed, he found a way to pay his own way, maintain some degree of control, and preserve his dignity.

Concern with Physical and Mental Health

Naturally, learning that one is HIV-positive leads to an increased concern about how to manage stress, find appropriate medical and social services, and maintain one's physical and mental health. Difficult decisions must be made about accessing health care and support services, starting treatment, and entering clinical trials for new medications. Persons with HIV often need more information about nutrition, exercise, medical regimens, and alternative therapies. They often agonize over whether to stop drinking alcohol, using illicit drugs, or smoking tobacco. Sometimes health care providers do not adequately explain the reason for a treatment, prepare the person for possible side effects, or listen to the person's fears or cultural concerns. When this occurs, people often feel unprepared to follow the medical advice that is given. For people who are not accustomed to being connected to a mental health, medical, or social services organization, approaching these systems can be quite daunting.

Grief

As with any serious illness, persons with HIV and their loved ones experience all of the complex emotions associated with loss and grief (see Chapter 16). They typically grieve over both current and anticipated losses (such as job, income, home, friends, independence, mobility, functioning, ability to have children, and life itself). Sometimes there are real changes in their appearance, social network, and sexual life, all of which are mourned. Persons with HIV often experience the illnesses and deaths of friends, lovers, and support group members as well. Likewise, caregivers feel grief-stricken over the untimely illness and death of a friend or loved one and sometimes feel isolated; they may hide their grief because it does not feel safe to talk about the presence of HIV. Many families have multiple members who are infected. All of these situations can compound the grieving process and lead to feelings of intense sadness.

Concerns of the Spirit

Persons with HIV often contemplate the meaning and purpose of life and death and may examine the role of spirituality or religion in their lives. They also often exhibit extraordinary strength, wisdom, resilience, and adaptation in the face of intense stress and fear. They may speak to you about perceiving things and people differently since their diagnosis, of having a heightened sense of how special life is, of being more focused about their goals, or of being clearer about their priorities. You may also assist the person in gaining access to a spiritual advisor. As usual, the helper's own beliefs and biases should not interfere with the ability to listen to and support another person.

Specific Populations

It is worthwhile to note some of the special issues and concerns that exist for specific groups of persons with HIV. We will briefly introduce concerns of women, children, older persons, persons of color, substance users, and men who have sex with men. As always, we warn you against stereotyping anyone based on generalizations that you read. In order to be prepared to work with any of the populations discussed below, you must do further research on their unique concerns. Recognize also that the categories are not mutually exclusive: obviously, someone can belong to several of these groups (for example, an adolescent woman of color, or a gay man of color who has injected drugs). This list is also terribly incomplete; other groups have specific needs as well, such as persons who are deaf, persons who have developmental disabilities, persons who are incarcerated, and persons for whom English is a second language.

HIV and Women

Although women have been diagnosed with HIV since the disease was first recognized by medical practitioners and scientists in the early 1980s, the number of infected women has climbed steadily during the past few years. HIV is spreading almost six times more quickly among women than it is among men (CDC, 1994).

Several issues have compounded the problem of HIV for women. Because health care providers and the general public did not consider women to be at increased risk of infection early in the epidemic, many women with HIV did not get the testing, early medical treatment, or aggressive health maintenance care they should have. Women with HIV sometimes face another form of double jeopardy because they are often seen simply as vectors, vessels, or carriers of the disease, rather than as human beings with their own rights and needs (Amaro and Gornemann, 1991; Anderson, Landry, and Kerby, 1991). Public health officials and policy makers have frequently treated them as "Typhoid Mary" types who are blamed for infecting children and men. It is important to help dispel some of the myths and biases that are common regarding women with HIV and to educate others about their complex needs.

There were other policies and practices that have put women at a disadvantage. Earlier in the epidemic, women with HIV were excluded from clinical trials that tested potential new treatments for HIV, in part because of concerns over the unknown impact of experimental drugs on possible future pregnancies. In addition, the definition of AIDS used by the Centers for Disease Control and Prevention and by the Social Security disability programs did not include some of the illnesses that were specific to HIV-positive women (such as cervical cancer), with the result that women with symptomatic disease were not always eligible for disability programs when they became ill. With an improved understanding of the disease in women and strong political activism, both of these inequalities have begun to be addressed.

Although both men and women with HIV often struggle with decisions about sexuality and reproduction, decisions about pregnancy, abortion, and raising children often place a bigger burden on women who are infected. It is now recommended that all pregnant women be tested for HIV because there is strong evidence that the antiretroviral drug AZT (if taken by women during pregnancy and delivery and given to their infants after birth) greatly reduces the likelihood that a child will be born infected. However, imagine the impact of learning during a pregnancy that you have HIV and could pass it to your unborn child! Agonizing decisions must be made at that time about whether to continue the pregnancy and whether to start antiretroviral treatment.

In addition, many women with HIV are caring for their infected partners and/or children and tend to place their own medical and emotional needs behind those of other family members (Hackl, Somlai, Kelly, and Kalichman, 1997). A woman with a life-threatening illness like HIV also is faced with painful decisions about what might happen to her children if she dies, often having to develop guardianship and adoption plans while she is still able to do so (Hackl et al., 1997).

In our society, where sexism is all too often a barrier to equality in our personal and public lives, women are not necessarily equal partners in their erotic partnerships. This often means that women do not have the power to negotiate safer behavior with sexual partners, which can put them at greater risk for HIV, other sexually transmitted diseases, and unplanned pregnancies. Human services helpers can help women practice talking with their partners about the use of condoms and other ways to have sex besides unprotected vaginal, oral or anal intercourse. Be sure to assess the occurrence of relationship violence as well (see Chapter 11); sometimes women who are being abused may risk being beaten or killed if they try to insist on condom use.

Illustration

Darlene Brown is a 32-year-old African-American woman who was eight months pregnant when her husband Arthur became critically ill and was diagnosed with end-stage HIV disease. Darlene took an HIV test, as did their two-year-old, Fanta. Both of them tested positive. During the next six months, Darlene took care of Arthur (who was bed bound until he died) and cared for a newborn and a sometimes sick 2-year-old. Her biggest concerns were managing the family's medical care and keeping the family's diagnoses completely secret from their small town and from the rest of the family. She did not receive any social services or medical care for herself until Arthur died.

Think About It #66

1. What barriers did the Browns face as a family?
2. What struggles and challenges did Darlene face?

HIV and Children or Teens

Before blood screening began in 1985, the majority of children with HIV were infected through contaminated blood products. Since then, most children with HIV have acquired the infection from their birth mothers. It is estimated that 3000 children are born to HIV-infected mothers in this country each year (Heslin, 1993). Without perinatal antiretroviral treatment, approximately one-quarter of the children born to infected women develop HIV disease. Now, however, the use of AZT can reduce the risk of a newborn being infected to 1 in 12. When babies are born to women with HIV, they carry some of their mothers' antibodies to the virus in their blood and may test positive without actually being infected. Infants who are *not* infected will eventually lose their maternal antibodies to HIV within 6 to 18 months of age and will subsequently test negative; those infants who are infected will continue to remain antibody positive. Nevertheless, whether infected or not, these children are likely to lose their mothers sometime during childhood.

The HIV epidemic has given rise to the term "AIDS orphan," which usually refers to a child whose primary caregiver (most often the mother) has died from HIV (Forsyth, 1995; Levine, 1993). Approximately three-fourths of HIV-infected women have children, and the majority of these mothers have primary or sole responsibility for child care (Forsyth, 1995). With the rate of HIV infection increasing in women of childbearing age (15 to 44 years old), there is likely to be a corresponding increase in the number of AIDS orphans. The CDC (1994) estimates that from 1992 to 2000, 93,000 to 112,000 thousand uninfected children and 32,000 to 38,000 infected children will be born to women with HIV.

Disclosure issues are very important for families caring for children with HIV. Decisions about telling a day-care provider, school, or the parents of playmates are very difficult to make because of desires to protect the child and give him or her a normal life. In addition, caregivers must decide if and when to tell the child about the diagnosis and the nature of the illness.

HIV-negative children who have HIV-positive parents, siblings, or other relatives have special needs as well. These HIV-affected children may feel confused, frightened, or stigmatized. They may find themselves being moved to another home or city after the death of a parent. Any HIV-infected or affected child who is old enough to do so should be offered the chance to express fear, grief, or other feelings through reading, talking, drawing, or play therapy.

HIV-infected adolescents are another group with special needs. They perhaps are old enough to understand the seriousness of their illness but not necessarily old enough to make sense of everything that is happening. They also have difficult disclosure decisions to make with regard to school and friends, as well as the difficulties of negotiating safer drug and sexual behavior in the face of peer pressure. Adolescent women especially often have to face difficult decisions about reproduction and parenthood.

Children and teens with HIV have as many medical, social, and emotional needs as adults. One of their primary concerns often is the intense desire to live as normal a life as possible. Their problems and joys are the same as those of other kids, with a chronic and life-threatening illness added on. Some HIV-infected young people are in foster care or living with relatives and may feel unsettled, disrupted, and withdrawn or distrustful of adults. These children and teens may yearn for stability and normalcy.

The following guidelines may ease your way in working with a child or adolescent with HIV infection.

- Keep your language and explanations at an appropriate level for their developmental stage and language comprehension.
- Anticipate their possible distrust of you, withdrawal, and fear of rejection.
- Remember that young people have the same needs for affection, dignity, choice, and control as do adults.
- If a child or teen knows his or her HIV status, help him or her problem solve when and to whom this secret is safe to tell. You can help them create a safe list of persons they can talk to about their illness. The goal is to teach them about discretion without making them feel more ashamed.
- With children and teens, teach them about treating their blood carefully. Even if young children do not know they have HIV, they can understand that blood has "germs" and to be as careful of it as they are of urine or feces.
- With teenagers, reinforce safer drug and sex behavior, and help them practice how to negotiate with their peers and potential sexual partners.
- Children and teens who know that they have HIV or are very ill with AIDS often feel sad or depressed and need to talk about it. Offer age-appropriate books and other resources that can facilitate honest conversations about death or grief. Consider helping them make scrapbooks about their lives, draw pictures to express their concerns, or participate in play therapy.
- Whenever you are working with HIV-positive young people, help them retain as much choice and control as is possible and make their lives as normal as possible.

Illustration

Amy is a 16-year-old Caucasian woman who has been dating her boyfriend Andrew for over a year. She knows that Andrew has hemophilia and has decided to write her junior-year term paper on this illness. In the course of her research, she reads that many people with hemophilia have been infected with HIV during the years when the blood supply was not safe. Amy talks to Andrew about this, and he tells her that his mother never wanted him to get an HIV test because she did not think that she could deal with the results if they were positive. Amy and Andrew decide to go to the health department on their own and ask for tests. They go together for the tests and then ten days later for the results: They are both HIV-positive. Amy and Andrew both become so upset that they end their relationship. Amy feels scared to tell her parents and is convinced that she will die soon. She considers dropping out of school and running away. Feeling desperate, she uses a pay phone to call the local HIV information and crisis line.

Think About It #67

1. What are some of Amy's worries and concerns?
2. What type of services and support might be offered to her?

HIV and Older Persons

HIV disease is a growing public health concern in gerontology. From 1981 to 1989, 10 percent of all AIDS cases reported in this country were diagnosed in persons over the age of 50, and the proportion of cases reported in older adults rose sharply during this period, from 7 percent of the AIDS cases in 1981 to 11 percent in 1987 (Ship, Wolff, and Selik, 1991). As in all age groups, the incidence of HIV in women is steadily increasing; in persons over 50 who are diagnosed with AIDS, 12 percent are women. In persons over 65 with AIDS, 21 percent are women (CDC, 1995).

The incidence of HIV in older persons is probably even higher than re- ported, since they are less likely to be tested for HIV or diagnosed with AIDS than are young adults. This is most likely due to a mistaken belief, held by older persons (Dawson, Cynamon, and Fitti, 1987) and their physicians (Weiler, 1989; Ship et al., 1991), that the aging population is not at risk for HIV. There have been few media reports about older adults living with HIV or being at risk for infection, which contributes to a decreased awareness of the need to be tested or to practice safer sex and drug behavior (Hooyman and Kiyak, 1996). There may also exist an unfortunate (and very ageist) attitude that HIV prevention efforts or messages are wasted on elders (Riley, 1993).

Other myths about older persons contribute to our failure to offer them safer behavior education and HIV screening: We assume that older persons are not sexual or that there are no older injecting drug users or gay men. Obviously, persons over 50 acquire HIV in the same ways as do younger people: through contaminated blood products, unprotected sexual intercourse, and the sharing of syringes or needles that contain HIV-infected blood. Another mistaken belief is that the majority of infections in this population are because of blood transfu- sions. While older people make up a large portion of the persons who were in- fected in this way before blood screening began in 1985, the majority of older persons with HIV become infected through sexual activity, both heterosexual and homosexual (CDC, 1995).

The combination of ageism and HIV stigma is also an issue because it influ- ences this population's ability to access care. Older persons with HIV often feel out of place both in the AIDS-service network, which they perceive to be tai- lored to younger people, and in the aging network, which they see as unrecep- tive to persons with HIV. Consequently, persons with needs related both to aging and to HIV are often inadequately served (Emlet, 1993; Linsk, 1994).

Another reason that HIV is a concern for gerontology is that older family members often serve as informal caregivers for younger adults and children with HIV. One-half of persons with AIDS are being cared for by older relatives (Allers, 1990). These older caregivers are often caught between the care needs of their own partners, those of their adult children who are ill, and those of their grandchildren for whom they have parenting responsibilities (McKinlay, Skin- ner, Riley, and Zablotsky, 1993; Ogu and Wolfe, 1994). Older caregivers are likely to be at risk for chronic impairments related their own aging, which may be made worse by the lifting, carrying, emotional stress, anxiety, and other demands of caregiving (Emlet, 1996; Minkler and Roe, 1993). Sometimes older persons with HIV are cared for by friends, partners, or parents who are also old (Emlet, 1996). In addition, when children are orphaned by HIV, their care often falls to grandparents or great-grandparents through adoption or guardianship (Allers, 1990; Levine, 1995). Because the HIV epidemic has produced an unprecedented emergence of vulnerable, parentless children (Levine, 1990), more and more grandparent care can be anticipated (Boland, Czamiecki, and Haiken, 1992; Mellins and Ehrhardt, 1994).

Illustration

Bob was a 60-year-old gay man who was hospitalized after a heart attack when he tested positive for HIV. Upon hearing the news, he was concerned most about the care of his parents, who were in their 80s, who were both quite frail, and who had been relying on Bob for their care. Bob did not want to disclose his HIV diagnosis to them, because he did not want them to worry. Bob decided to continue to live with his parents and care for them. Bob died three years after this; both of his parents survived him.

Think About It #68

1. Have you given much thought in the past to how HIV affects older persons? Why or why not?
2. What are some of the barriers that you can identify in yourself that might make it difficult for you to talk to an older person about safer sex and HIV?

HIV and Persons of Color

Persons of color comprise approximately one-quarter of the U.S. population but shoulder a much larger share of the HIV epidemic. Since 1991, the majority of persons diagnosed with HIV and AIDS have been persons of color (CDC, 1995). The incidence of HIV disease among persons of color has continued to rise, especially among African-American and Latino persons who now account for nearly two-thirds of the AIDS cases reported each year (CDC, 1997).

Several social factors worsen the problem of HIV for people of color. The first is racism, which continues to hinder our society. Persons of African, Asian, Latino, or Native-American descent often experience prejudice and oppression. For people who have been marginalized because of their skin color, the stigma of HIV again creates a problem of double jeopardy. Many people of color have been badly treated by providers or agencies that were supposed to help them. It makes it more difficult for them to seek services and find providers whom they trust, especially if they have to disclose their HIV status, when they fear further mistreatment.

Another important factor is economic status. Both poverty and HIV disproportionately affect persons of color in this country, creating another form of multiple jeopardy. Lower economic status often means less access to services. The result is that many persons of color have difficulty getting health education, preventive services, medical care, and new treatments because they have fewer financial resources.

There are also cultural differences in the ways that illnesses are viewed and treated that can complicate HIV disease for people of color. Medical and social services agencies are often Eurocentric, that is, based on traditional white, Western ideas about what proper care should be. Although these systems usually are well intentioned, persons of color may be distrustful of them because of a perceived cultural bias. In addition, most mainstream providers are unfamiliar with, and may be hostile toward, the traditional medicines and healing practices used by other cultures. This can make some persons of color reluctant to use services from conventional medical and social services organizations. Whenever possible, agencies that provide HIV education, outreach, and/or support services to persons with HIV should strive to have trained volunteers, students, and workers available who are persons of color. Embracing diversity within an organization sends an important message about inclusion and commitment to all types of people.

Illustration

Cecilia is a 30-year-old Latina who was raised in Puerto Rico but had been living in Chicago for ten years. She was not aware that her male lover was infected with HIV until he died from AIDS. She then became very concerned about her own health. Terrified, yet determined to follow through on her decision to get tested, she contacted a local HIV clinic for an appointment. As soon as she learned of her positive test result, she wanted to return to Puerto Rico. She had a job with good benefits but decided that she needed her parents, sibling, and cousins more than she needed her current financial stability. She also felt that her family would know best what to feed her and what herbs would soothe her and help maintain her health. She sold her home and quit her job as quickly as possible and went home to Puerto Rico.

HIV and Substance Use

The social problems of substance use and HIV have been called "twin epidemics" because certain behaviors associated with drug use increase the risk of HIV infection. As has been stated, sharing needles or syringes contaminated with infected blood is a common way in which the virus is transmitted. It is illegal to purchase needles or syringes without a prescription in most states, so persons who inject drugs often find it necessary to share them. In addition, substance use is often a social activity involving family members or friends who routinely use drugs together. In these situations, people often feel pressured to share injection equipment rather than use their own.

Sharing needles and syringes is not the only drug-related risk factor for HIV transmission. Consuming alcohol or other drugs such as cocaine or crack can lower a person's sexual inhibitions and impair his or her judgment and memory. As a result, people who use substances and then have sex risk becoming infected with HIV when they often forget or choose not to use condoms. Similarly, some persons who are addicted sell sexual favors in order to pay the high costs of their drugs; in these circumstances, safer sexual practices are not always feasible. Because of the connection between HIV infection and drug use of all kinds, human services workers in the HIV field are sure to meet and work with persons who are addicted to substances or recovering from addiction.

The combination of addiction and HIV creates double jeopardy for people in several ways. First, drug use and alcoholism are often viewed with disdain, as is HIV, so persons with both conditions face a dual stigma that may make them reluctant to ask for help when they need it. Second, persons who use substances heavily tend to seek health care less often and may not get adequate nutrition. Third, there is some evidence that drug and alcohol use can suppress the immune system and can cause HIV disease to advance more rapidly (Stein and Hodge, 1995).

Working with persons who have both HIV and addiction can be challenging and complex. Addiction is a disease with its own set of behaviors, symptoms, and attitudes. To someone who is physically and psychologically addicted, the most important thing in life is the substance he or she must have in order to feel well and avoid withdrawal. When working with someone who is using drugs or alcohol, you must be especially careful to pay attention to boundaries (see Chapter 2), set limits, and ensure your own safety (see Chapter 6).

Illustration

Darryl was a 29-year-old African-American man who had left his small Southern town to live in New York City. He lived there for over a decade and worked steadily as a furniture mover. During breaks, he and the other men would routinely shoot heroin; they usually had one syringe between them. Darryl suddenly became very ill and was in the hospital when he was diagnosed with advanced HIV disease. He was very frightened and wanted to die "back home." He left his wife and two children in New York and went to live with his mother down South. After fighting off a serious depression for a month, he voluntarily entered a detox program and stopped using heroin and alcohol. He joined Narcotics Anonymous and eventually became a sponsor for someone else from the group. He sought and found medical care, a support group, buddy support, and case management services from an HIV services organization. After a few months he left his mother's apartment and moved into his own place.

HIV and Men Who Have Sex with Men

In the United States, gay men were the first group to be affected heavily by HIV (this was not the case in many other countries). This was due in part to sexual practices that increased the risk of infection. For instance, the virus is transmitted more readily through receptive anal intercourse than through vaginal or other sexual activities. As a result of the early link between HIV and gay men, HIV stigma became strongly associated with homophobia (an irrational fear or discomfort toward homosexual persons and practices) and heterosexism (a belief that heterosexual persons are of greater value). Although some progress has been made against these biases, the dual stigmas of homosexuality and HIV continue to create double jeopardy for men who test positive.

Men who identify themselves as "gay" or "bisexual" may be reluctant to openly disclose their sexual orientation (persons who choose not to disclose their sexual orientation are frequently referred to as being "closeted" or "in the closet"). Disclosure occurs along a continuum; a person may be "out of the closet" or "out" to his friends but not to his family, for example. In addition, not all men who have sex with men identify themselves as "gay" or "bisexual." Sometimes these men identify themselves as "straight" and have female partners. When they do have sexual contact with men, it may be hidden and/or anonymous; this sometimes happens in exchange for money or drugs, and sometimes because they do not want to be labeled as homosexual. Gay and bisexual men who are over the age of 50 lived half of their lives or more before the modern gay rights movement began in 1969 with the rebellion at the Stonewall Inn bar in New York City, so they are less accustomed to disclosing their sexual orientation than are younger men (Grossman, 1995). The important thing to remember is that not every man will be completely comfortable being open with you about his sexual orientation, behavior, or risk for HIV transmission. For men who have sex with other men, disclosing their HIV status can be agonizing because it may also mean having to disclose sexual behavior they have kept hidden from friends and family for years.

The phrase "the gay community" is often used to describe a specific culture of men who identify themselves as gay or bisexual or who are transsexual or transvestites. The gay culture is as diverse as any, so it would be unwise to assume anything about a man just because he may have sex with men or to expect a certain predictable behavior from the community. For example, you might expect that gay men would welcome with open arms someone in their community who has HIV, but that is not always the case. Gay and bisexual men who are HIV infected often face the same fears about being rejected and judged by their peers as do non-gay persons with HIV.

The gay community as a whole, however, has been extraordinarily insightful, courageous, and responsive throughout the HIV epidemic. Hit first and hardest, they rallied to offer culturally competent and innovative services for persons who were ill and created effective grassroots HIV prevention campaigns. Most of the early HIV services organizations were formed by gay men who were HIV-infected themselves or who had friends or partners who were. Many of these early visionary leaders have died from AIDS. Because of the tremendous toll the epidemic has taken on their community, gay men often experience tremendous anxiety, grief, and burn-out. The human services profession owes this community a tremendous thanks for stepping forward to help those in need when many human services providers were reluctant or slow to do so.

Working in the HIV Field

Human services workers, students, and volunteers may participate in a variety of services designed for persons with HIV or their caregivers. You may facilitate a support group, provide advocacy and case management, assist persons with applications for disability benefits or other financial aid, or work in a residential facility or hospice. In addition, you may provide counseling to persons who are considering an HIV test, educate other helpers or the public, or participate in services geared toward HIV prevention, risk reduction, and safer behavior. In addition, no matter what your work setting, from child day-care centers to nursing homes, you will most likely be involved with someone who is infected with or affected by HIV, whether you know it or not.

Generally, persons with HIV are quite concerned about (1) confidentiality and privacy, (2) being stigmatized and negatively judged, and (3) having access to competent and appropriate services. You can help tremendously by consistently responding with genuineness and respect toward anyone with HIV who requests your assistance. For example, you can assure them that you will respect their privacy to the greatest extent possible. You can avoid asking them how they think they might have gotten infected, since it is unlikely to have any bearing on the social services or medical care they need. You can support their informed choices. You can work diligently to combat stigma and discrimination in human services systems.

Working with persons with HIV and their caregivers is a difficult and challenging task that can be very meaningful and humbling. To be effective with this diverse and sometimes vulnerable population, you must be open and willing to learn from the persons who approach you for assistance. You must be accepting of persons who are different from you; be comfortable with health issues and facilities, dying, and death; and be sincere, committed, trustworthy, hardworking, and self-aware. You must also be willing to study the evolving nature of the disease and its treatment. It is also essential that you have your own network of social and professional support while you engage in this work.

It is important that you not blame the person with HIV for having the illness. Regardless of the person's behavior or your own attitudes, victim-blaming does not further the helping relationship and does not assist the person with HIV to confront the difficult issues in his or her life. Likewise, it is not useful to label a person as "in denial" or to try to move him or her into a different emotional state. You can be most helpful by listening to and accepting what a person is thinking and feeling without judgment. As a human services worker, it is important for you to help the person with HIV balance hope with reality, to confront difficult issues without destroying the optimism and enthusiasm that help all of us to fight and survive.

The Southeast AIDS Education and Training Center at Emory University (1991) developed the following list of the "silent hopes" of persons with HIV. These concerns are good examples of the types of thoughts and feelings that may be uppermost for persons with HIV who seek our help. Keep these in mind as you provide services and support to this population.

- **Please don't judge me.** I know that some people believe I deserve this illness because of my past behavior. Like all people, I have done the best I am able to do in my life with the circumstances that I have been given.
- **Please don't talk about me.** I know that you need to discuss my case with others who are involved in my care. I also know that HIV is an interesting topic of discussion at other times as well. Your talking compromises my confidentiality and could have serious consequences for me if word of my illness got around.
- **Please explain things to me.** I know you are busy, but please understand that the procedures and drugs you discuss are like a foreign language to me. It will help me accept and understand what I'm going through if I have your help.
- **Please understand how emotionally difficult this is for me.** I am too young to be thinking about the end of my life. I am worried about my young children. I am worried about my partner. I have kept a very big secret from many people, and this illness threatens to expose me.
- **Please acknowledge the people I consider my family.** I may have been apart from my biological family for many years, but I have a wonderful circle of people who are my 'family of choice.' I may have a partner to whom I am not married or a partner who is the same gender as I. Treat this person as my spouse, just as I do.
- **Please follow my lead with my biological family.** I may not have told my family about my illness. I may have reasons to avoid seeing them or to avoid their calls. Honor my wishes.
- **Please understand my need to be involved in my care.** So many things in my life now feel out of control. I want to be involved in making decisions about my care, and I want to be told what is going on. I am losing dignity all the time, and your respectful treatment can help me feel self-respect.
- **Please don't reject me.** I need love and acceptance now more than I ever have in my life. You will not become infected by listening to who I am, or holding my hand, or giving me a hug. Thank you for doing a very difficult job!

Making HIV Prevention Part of Your Work

You do not have to be an HIV worker to have a positive impact on this deadly social problem. Each and every human services worker, student, and volunteer should routinely talk to adolescents and adults of all ages about safer behavior in drug use and sexual activity. Persons who inject substances can either avoid sharing needles and syringes with others or they can flush them out with bleach and water before use. Persons who have sexual intercourse can use male or female condoms to prevent transmission. Although these practices seem simple, there are often obstacles that make it difficult for people to follow them. Therefore, you need to learn not only how to comfortably teach HIV prevention techniques, but also how to help people solve the problems they face when they try to practice safer behavior.

Try to make HIV prevention education part of your work and your agency's work. It is surprising and disturbing how few human services agencies include HIV education as part of their general routine. For example, human services helpers who work in clinics that address sexually transmitted diseases, contraception, and pregnancy do not always offer people information and counseling about safer behavior. Even workers in addictions counseling agencies do not always discuss HIV with service recipients. At the very least, each of us can strive to make condoms and HIV prevention brochures available to anyone who walks into our offices.

When you speak to someone about his or her sexual health and his or her drug use behavior, it is important for you to be calm, professional, and accepting. You should also be well informed and comfortable so that you do not appear to others to be embarrassed, unprepared, or unwilling to discuss the subject or to hear what they have to say. The following guidelines can perhaps ease your way in talking to someone about sex or drugs.

• **Approach.** Be sensitive to the person's level of comfort and openness. Be tentative and respectful in your approach. Try not to be abrupt or invasive in your introduction of the topic. Try to appear casual and at ease yourself; this helps to put the other person at ease (examples of how to begin a discussion appear below). Monitor the person's comfort level throughout, reflect and check out what you see and hear, and adjust your approach or end the conversation if necessary. Often people want to talk about safer sex and drug behavior but are embarrassed or afraid of our judgment. Invite individuals and couples to talk with you at anytime about HIV prevention. When people broach the subject with you, respond with acceptance and a willingness to talk honestly.

• **Language.** Observe how a person is speaking and what he or she is comprehending, and tailor your language to that individual. Be professional and matter-of-fact without using jargon or technical terms. Be accurate and appropriate in the words and phrases you use. Try not to use words that are clearly profane or slang, unless it becomes clear that the person asking for prevention information is unable to understand other words. If you feel that he or she is using graphic language in order to test you or shock you, or if you are truly uncomfortable with the terms being used, then state your objections or preferences firmly yet kindly, and set limits about what words and phrases the two of you will use. If you need to define or explain terms, do so simply and briefly. If someone doesn't understand something you have said, reword it.

Wrapping It Up

If you are interested in working in the HIV field as a human services worker or are already providing HIV-related services, you cannot rely on the brief introduction to the subject contained in this chapter. In order to offer effective prevention education, pretest and post-test counseling, crisis intervention, counseling, and case management, you will need further training in the biologi-

---

Examples of Questions Concerning Sex and Drug Histories

- "Do you mind if I ask you a few questions about your sexual health?"
- "Do you have sexual contact with men, women, or both?"
- "When you have sex, do you protect yourself against disease? If yes, what do you do?"
- "I wondered if you would tell me a little bit about what you've heard about safer sex."
- "Are you taking any drugs which involve needles or shooting up? If yes, do you share needles? If yes, do you protect yourself against disease? What do you do?"
- "Would you like to ask me anything about preventing HIV or other transmissible diseases?"

SOURCE: Adapted from materials developed by the Midwest AIDS Training and Education Center, University of Illinois at Chicago.

---

cal, psychological, medical, and social aspects of HIV. In addition, the landscape of the HIV field can change rapidly with new scientific and pharmaceutical discoveries, so it is imperative that you keep your knowledge of the disease up to date. HIV incidence patterns also vary from region to region and over time, so it may be useful for you to contact your state or local health department and ask to receive updated epidemiologic and demographic information.

Above all, you need to learn to speak intelligently about HIV and avoid some common mistakes in terminology that are confusing or offensive. For example, AIDS is not transmitted, HIV is. Likewise, there is no such thing as an "AIDS test." Be careful to avoid using terms such as "AIDS patient," "AIDS victim," "AIDS babies," or any other phrases that put the disease before the person. Remember to use the terms preferred by persons who live with HIV in order to show respect for their dignity and humanity.

Even if you do not work directly with children, teens, or adults who are HIV-infected or affected, you can still play an important role as an educator about transmission, safer behavior, discrimination, and bias. You can make a difference by confronting ignorance and prejudice against persons with HIV whenever you encounter it.

Think About It #69

1. List (by yourself or with your fellow learners) some ways in which you might become more informed about HIV or persons who are at risk or infected. (Examples might include getting brochures or statistics from an HIV service agency or health department, attending a class at the local Red Cross, reading a book about a specific population affected by HIV, or participating in a safer behavior workshop for the public.)

2. Choose one of the options that you or your group has developed, and promise yourself that you will act on your choice within the next two months.

CHAPTER FOURTEEN

Mental Health

Sometimes human services helpers provide practical and emotional support to persons who have chronic or acute mental illnesses or who are addicted to chemical substances. Human services helpers need some information about the nature of mental illness, including alcohol and drug dependence, and how these conditions might affect individuals, couples, and families. This chapter briefly introduces the concepts of mental health and mental illness, discusses some of the most common forms of mental illnesses, and considers the need for advocacy.

Introduction to Mental Health Services

There is an important message evident in the title of this chapter: "Mental health" is the term usually used rather than "mental illness" when describing this field of practice. To be mentally healthy is certainly a difficult concept to define; ideas about this probably vary widely. "Mental health" is generally thought to describe a state of functioning where individuals can operate independently, maintain relationships, and contribute to society. Sigmund Freud, a pioneer in research into the causes of mental disorders, said that mental health was the ability "to work and to love" (Freud, 1905). Sometimes, however, people are impaired by mental illnesses that, if untreated or unmedicated, greatly diminish their ability to function.

It is important that helpers never use diagnoses of mental illnesses or the label "mentally ill" to stereotype or dismiss the humanity of anyone. Always remember that persons who have illnesses are persons first.

What Is Mental Illness?

Mental illnesses are disorders of the brain or disruptions of emotions. Mental illnesses involve ways of thinking, feeling, perceiving, and behaving that are out of the normal or typical range. These disorders and the behaviors that accompany them are out of the control of the persons who have the mental illnesses.

They also can severely limit the person's ability to function. The term "mental illness" includes several categories of disorders, such as long-lasting depression, bipolar disorder (manic-depression), extreme anxiety or panic, schizophrenia, psychosis, and dementia. Mental illnesses can be temporary, like an episode of depression, or permanent, like schizophrenia.

Between 30 and 45 million Americans have mental disorders that interfere in some way with employment, school, relationships, or daily life (American Psychiatric Association [APA], 1988a). Eight to fourteen million people suffer from depression each year; 1.5 million have schizophrenia. Most mental illnesses can be lessened or controlled with medications, called "psychotropic drugs," which have been developed for this purpose. Some people who are diagnosed with a mental illness and are in treatment or on medication are able to manage their lives well, and other people never know. Others have chronic or acute mental illnesses that have not been diagnosed or treated, and their lives are severely affected by their loss of functioning.

Human services workers often work in residential facilities or community agencies designed to provide therapy and support services to persons who have mental illnesses. The main goals of services for persons with temporary or permanent mental or emotional disturbances are to provide relief from the symptoms and to maximize their functioning; in other words, to restore as much mental health as possible. Although medication and psychological therapy are the most common treatments offered to persons with mental health issues, human services helpers can play a valuable role in providing emotional support, practical assistance, advocacy, and service coordination.

Not only do human services helpers work in facilities that focus on mental health, they work with mental health issues in organizations where that is not the primary concern. Even if you are working in an agency that is not designed to offer services to this population, you are likely to meet persons with mental illnesses. Usually their mental condition does not have an impact on your ability to provide services. For example, if a man with schizophrenia comes to you for food, clothing, or temporary shelter, you will provide services to him as you would anyone else. However, sometimes the symptoms of mental illness may interfere with the helping relationship or may indicate the need for professional intervention. In those cases, consult with your supervisor immediately, discuss appropriate referrals, and decide on the action that your agency will take.

Common forms of mental illnesses are discussed here. The topic cannot possibly be fully covered in one chapter. However, to familiarize you with some of the most common forms of mental disorders, below we list and briefly describe depression, bipolar disorder (manic-depression), anxiety or panic, schizophrenia, psychosis, and dementia.

Depression

The most commonly diagnosed mental and emotional problem is depression. Depression can strike people of all ages, races, and economic levels. More women than men are diagnosed with this illness. Depression as a mental illness is much more serious than having a feeling of sadness or grief or being "blue." This illness is long lasting and deep and is often serious enough to require medication to treat the symptoms.

Characteristics of depression include: (1) changes in sleep patterns, either sleeping a great deal or being unable to sleep, including waking up in the middle of the night; (2) chronic lack of energy; (3) changes in eating patterns; (4) social withdrawal; (5) loss of sexual drive; (6) feelings of helplessness and hopelessness; (7) inability to work or to relate to others; or (8) serious thoughts of committing suicide or attempts to commit suicide (American Psychiatric Association, 1994).

Temporary feelings of being depressed are normal from time to time, especially at times of disappointment, grief, or crisis. However, when this feeling becomes particularly overwhelming or persists for long periods of time, medicine and counseling should be considered.

Illustration

Chris is a 23-year-old Native-American gay man who has just moved from the east coast to the west coast to enter graduate school in psychology. This is the first time that he has been away from his family, friends, and community. During his first semester he finds that the work is much harder than he anticipated, he is much lonelier than he expected to be, and his professors are more critical and distant than he had hoped. He finds that by December he is sleeping 12 hours each night, leaving his room infrequently, and eating very little. He is also finding it very difficult to motivate himself to go to class. Chris decides that over the holiday break he will volunteer at a local soup kitchen because he wants to focus on something other than his own troubles. He remembers that during high school he periodically went through similar dark moods when he felt that life was not worth living and that his mother always suggested that he "get out and help someone." While working at the soup kitchen, Chris meets and begins to date Pablo, another volunteer, and begins to feel that his new life might work out after all. However, after Pablo abruptly stops returning Chris's telephone calls, Chris plunges into an even deeper depression. He does not go to class at all during January, staying in his room and contemplating suicide.

Bipolar Disorder (Manic-Depression)

Persons who have bipolar disorder swing between two extremes or "poles" of moods and affect. This illness has been called "manic-depression" because periods of highs are interspersed with periods of lows. The depressions can be quite dark and dangerous, and the manic periods can be quite destructive as well. Persons who are in the manic phase of this illness can lose control of their behavior and go on shopping sprees or drinking binges or can have images of themselves that are grand and unrealistic. Persons with this disorder often cannot predict these extreme mood swings, may not be aware of having them, and are not in control of them. Manic-depressive disorder can usually be successfully controlled with appropriate medication.

Illustration

Lynn is a human services worker who provides support to children in foster care. Lynn's coworkers, friends, and family members are often puzzled and worried by abrupt and unexplainable changes in the way that Lynn behaves. Occasionally Lynn disappears for several days without telling anyone where she is going and without calling anyone during her absences. Lynn always returns after these disappearances in high spirits, having taken a spontaneous trip out of town. Ordinarily this would not be cause for concern, but during these vacations Lynn purchases an extraordinary amount of clothes, shoes, and jewelry, "maxing out" her credit cards and clearly living well above her income as a state employee. There are other episodes when Lynn withdraws into her apartment for days at a time, sitting in the dark, not showing up at her job, and not answering the telephone. Sometimes she does not answer the door when friends or family members come by to check on her. These depressive episodes seem to leave Lynn feeling completely despondent and helpless. Lynn has been warned several times by her supervisor, Diane, that she is in danger of losing her job because of her unreliable behavior. Diane has been documenting Lynn's patterns for several months in order to institute disciplinary action.

Anxiety Disorders

Anxiety or panic disorders represent an unusual reaction to fear. Being physically and mentally keyed up and alert is an appropriate physiological situation when one experiences environmental threats, because bodies and minds are being mobilized for possible action. However, when the anxiety or panic reaction is triggered in nonthreatening situations, occurs very frequently, or is more intense than the situation warrants, people may find their functioning severely impaired. For example, some persons with anxiety disorders cannot leave their homes, cannot be among people, or cannot fly on airplanes. Some people may be irrationally terrified of animals, heights, open spaces, or with whatever object or condition the anxiety has become associated.

Persons with anxiety disorders sometimes have frequent headaches, persistent fatigue, or chronic muscle tension. Persons who are in the midst of experiencing overwhelming panic have sudden feelings of terror that strike repeatedly and without warning. Anxiety in these situations may be experienced as a tight feeling in the chest, rapid heart beat, shortness of breath, dizziness, abdominal discomfort, sweating, and/or feeling very jumpy. Persons who are in the midst of panic episodes often feel that they are about to die.

A person who experiences repeated, intrusive, and uncontrollable thoughts or behaviors is referred to as "obsessive-compulsive." To cope with unwelcome thoughts, some people develop rituals, such as washing their hands many times; repeating words, phrases, or numbers constantly; checking the locks on doors and windows every few minutes; or checking the stove over and over. Others have intrusive thoughts about things they consider immoral, unacceptable, or painful.

When anxiety interferes with everyday living, or if one finds oneself in the emergency room because the symptoms have become so great that medical treatment is sought, it is a good idea to be examined by a medical professional. He or she may then refer the person to a psychiatrist or therapist, where treatment can begin. Effective medications are also available. Behavioral interventions, including relaxation exercises, have also been found to be successful in the treatment of anxiety disorders.

Illustration

Jack, 35 years old, is seized with feelings of terror and has extreme panic when he drives his car outside his immediate neighborhood. If he attempts to go anywhere outside his comfort zone, he feels his heart and pulse pounding rapidly, a tightening in his chest and difficulty breathing, and a sense of intense isolation and impending doom. During these unexpected attacks he feels trapped; he does not feel well enough to drive himself to a safer place, yet feels that the symptoms would be worse if he left the familiar confines of his car. Jack has learned to avoid these horribly frightening episodes by restricting his movements to a five-block radius around his apartment.

Schizophrenia

Schizophrenia, a disease that appears to be caused by genetics, usually becomes evident during the late teens and early twenties. Biological children of one parent who is schizophrenic have an 8 to 18 percent chance of developing the illness. An identical twin has about a 50 percent chance of developing schizophrenia if the other twin is schizophrenic (American Psychiatric Association, 1988b). Schizophrenia is not a direct result of the environment or poor parenting, as once thought; however, if one has a genetic predisposition to the illness, environmental factors may precipitate or trigger the onset. Initial symptoms include feelings of tenseness, lack of concentration, and social withdrawal. As the illness progresses, other symptoms develop, including unusual perceptions, peculiar behavior, talking nonsense or talking to oneself, or seeing or hearing things that others do not. In extreme cases, hallucinations occur, or the person hears disturbing messages or instructions from mysterious voices. There is a popular misconception that persons with schizophrenia have more than one personality; this is not the case. Medication and education enable many people with schizophrenia to lead normal lives. For others, the symptoms remain. Persons with schizophrenia and family members often find that having social support and accurate information about the illness is helpful.

Psychosis

Psychotic behavior is characterized by bizarre actions, such as dressing strangely; displaying unusual mannerisms; seeing or hearing things that other people do not observe; talking in nonsensical phrases; aggressive or suicidal behavior; and withdrawal from social contact and normal responsibilities. Psychotic behavior may be symptomatic of many different causes, such as a head injury, an emotional trauma, schizophrenia, and alcohol or drug use.

Dementia

Dementia, characterized by forgetfulness, confused thinking, and changes in affect, also has many causes and can be difficult to diagnose. Mild dementia frequently accompanies the grief process, so determining cause and seriousness of the symptoms can be difficult. Dementia may result from lack of flow of blood to the brain, Alzheimer's disease, medication, strokes, symptomatic HIV illness, and other causes. Being alert to the symptoms of dementia lets the human services helper react appropriately when the person they are working with forgets a name or does not return calls as promised. The helper who is unaware that a person has dementia might misconstrue the person's actions. Unrecognized dementia can create serious problems.

Illustration

Betty Moore, an African-American woman in her 60s, has joined a support group for people whose loved ones have AIDS. Her son Bobby, in his early 40s, has joined the group down the hall for people who have tested positive for HIV. After the first meeting that Ms. Moore and Bobby attended, Ms. Moore walked to the room where Bobby had been and found an empty room. Because Bobby was new to the group, other members did not question that he said nothing all evening, nor did they worry when he walked off by himself. After two tense hours, police officers found Bobby wandering several blocks away, unharmed but confused. Using the information on Bobby's driver's license, the police officers called his home, where his father, sister, and brother waited anxiously, and Bobby was returned quickly. Bobby's lack of care about his appearance, lack of response when spoken to, lack of emotion, and clumsy walk were signs of his having dementia, but the other group members were not aware of the need to watch and protect him.

Think About It #70

1. What do you feel and think about people who have mental illnesses?
2. Who do you know that has problems like those discussed in this chapter? Do you know their diagnosis? How do they act? What treatment or medications are they receiving? How does this affect your relationship with them?
3. Go to a library, and look at the reference book *Diagnostic Statistical Manual IV* (DSM-IV). Notice the wide range and types of diagnoses described in this manual for mental illnesses.

Alcohol or Drug Addiction

Human services helpers sometimes work with persons who are impaired in their functioning through their use of alcohol and/or other chemical substances. When a person must have alcohol or other drugs in order to feel normal or to feel that he or she can function, that person is "addicted" or "chemically dependent." Addiction is a complex problem that has physical, mental, and emotional elements. When people are addicted to a chemical, they become physically ill when they cannot ingest it when they need it and become psychologically uncomfortable when the substance is not readily available. The process of being without the substance is called "withdrawal"; being weaned off the substance gradually under medical supervision is called "detoxification." It can be physically dangerous to withdraw from a substance to which one is addicted; this is often called "going cold turkey." It can also be life threatening to take too much of the drug or drink too much alcohol; a person can die either from an overdose of a drug or from alcohol poisoning.

Someone who is addicted to a substance such as alcohol, heroine, or crack cocaine will go to great lengths to obtain it and have very little control over his or her behavior concerning the addiction. Even when persons have been through treatment, counseling, and detoxification, it is common to relapse, that is, revert back to using substances. Many persons who have stopped using the addictive substances and are clean and sober refer to being "in recovery" rather than "cured," due to the recognition that abstinence from these substances is an ongoing, life-long battle.

Levin (1987) defines alcoholism as "drinking which does serious harm of various sorts to the drinker; . . . continues despite its harmful consequences (p. 42). Alcoholism is reported more frequently with men than women, although the number of women diagnosed as alcoholic is increasing (Blume, 1988). Alcohol affects people differently; not everyone who drinks or gets drunk is alcoholic. Symptoms indicating alcohol consumption problems include (Nalty, 1997):

- Frequent use (and in increasingly larger amounts).
- Increasing fear and anxiety.
- Forgetting responsibilities (lateness for or absence from work or school; neglect of finances, interests, and family).
- Arguments over drinking and its behaviors.
- Promises to drink less or switching to a different or "milder" liquor.
- Blackouts (not remembering what happened during a period of drinking).

Use of legal or illegal drugs can cause a variety of difficulties. General guidelines to watch for are:

- Personality and behavior change including restlessness and boredom.
- Decreased interest in personal hygiene and appearance.
- An increased use of eye drops or breath mints.
- Withdrawal from the family.
- A new set of friends and acquaintances.

With the use of some drugs, signs may be more immediate or obvious: dilated pupils, redness of the eyes, needle marks on the arms (or wearing of long-sleeved shirts inappropriate to the weather), inability to focus the eyes, slurred speech; and presence of such drug paraphernalia as syringes and needles.

Many treatment programs exist, supported through both public and private funds; either inpatient or within the community; and focused on individuals, families, or in a group setting. Support and compassion for both the person seeking treatment and his or her loved ones are critical, because relapse is common and ongoing support and patience are therefore necessary.

For assistance in dealing with alcoholism or substance abuse, contact a physician, a clergy person, the local mental health center, a school counselor, or an alcohol and drug abuse program. Community support groups such as Alcoholics Anonymous (AA), Narcotics Anonymous (NA), Al-Anon (for the loved ones of persons with addictions), and Al-Ateen (for teenagers who love someone who is addicted) are easily accessible throughout the country.

Legal drugs. Sometimes people become addicted to or misuse medications that have been prescribed by physicians or that can be purchased over the counter at any pharmacy. Dependence or overuse of prescription or over-the-counter drugs is an especially common problem with persons as they age. Sometimes people become addicted to drugs that are supposed to help with weight loss, depression, pain control, or insomnia. These types of addictions are often as serious as dependence on alcohol or drugs (illegal or prescription), and withdrawal from these substances should be monitored by a physician.

Illustration

Agnes Hernandez, a Mexican-American woman in her 70s, attended a support group seeking assistance for her son, who has cancer. When she arrived, it was obvious that her voice was slurred and her attention wandered easily. Agnes later revealed to the group facilitator in an individual meeting that she was taking twice the level of medication prescribed for her depressive disorder. She had suffered depression and taken medication for it most of her life. Because of coping with her son's illness, she felt that she needed to take more medication to avoid becoming too depressed to care for him. With the support of her family, Agnes spoke with her doctor, who adjusted her medication, and the symptoms subsided.

Dual diagnosis. Sometimes human services helpers will work with persons who are struggling simultaneously with mental illnesses and chemical addiction. When chronic substance use is paired with mental illnesses, it is often referred to as a "dual diagnosis," meaning that there are two separate yet interconnected problems to address. Sometimes persons with mental illnesses use alcohol or other drugs to self-medicate, that is, to ease their symptoms and discomfort in the absence of prescription medications.

Your own recovery. You may be interested in working in the alcohol and drug field because you are recovering from alcohol or other substance use yourself or because you have seen a loved one struggle with these issues. Many addictions counseling and treatment centers seek employees or volunteers who are in recovery or have family members with addictions, because of the valuable knowledge that such persons possess about the complications of addiction, the effects on family members, and helpful support services. If you become a peer counselor in the addictions field, be sure that you take care of your own needs, get consistent supervision for your work with others, and make careful decisions about how much you wish to disclose of your past.

Learning more. If you are interested in learning more about how to provide support to persons who are in recovery from alcohol or drug addiction, there are many ways to gather information. Alcoholics Anonymous and Narcotics Anonymous groups periodically hold open meetings that members of the public can attend. Al-Anon meetings would also be an appropriate place for a human services helper to learn about the dynamics of being in a relationship with someone who is struggling with addiction. Information is also readily available through such publications as *Twelve Steps and Twelve Traditions* (1991), the handbook for Alcoholics Anonymous, first published in 1952 and now in its eighth printing. Jacquelyn Small (1990) provides additional suggestions for the "unlearned person" who is interested in "relating therapeutically" (p. 3) as well as appreciating the inherent wisdom that the human services helper possesses:

> Alcoholism is a problem that affects the whole person. And for you to be significant in the life of an alcoholic or drug user and his or her family, all you need is your own wholeness—your eyes, your ears, your brain, your heart, and your sensitivity—in other words, your ability to be fully human and naturally therapeutic (Small, 1990, p. 146).

Dependence on a chemical substance is so common that most United States residents have been affected through their own struggles with substances or the dependence of one of their parents, a partner, or another family member. Since substance use is pervasive, human services helpers are likely to encounter issues about addictions when working with someone who has come forward with other issues. In that situation, the human services helper who is aware of symptoms of the overuse of substances and the available resources may offer suggestions about sources of support for the person regarding recovery from addiction.

Think About It #71

1. What should human services helpers do if they suspect or are confronted with alcohol or drug use?
2. Have you been affected in some way by the abuse of alcohol or drugs? In what ways? How have these experiences influenced your attitudes toward people with alcoholism or other addictions?

Involuntary Commitment

When persons with mental disorders and/or chemical addiction are in danger of harming themselves or harming others, they are sometimes sent by a judge to a residential treatment center or hospital for a specified period of time. This process of being sent into treatment against one's will is called "involuntary commitment." Of course, it is possible to admit oneself into a treatment facility. In that case, the person can usually leave whenever he or she wishes. You may work in a residential facility where persons have been involuntarily sent, or you may work with family members who have decided that their addicted family member should be committed to an institution to avoid someone getting hurt or killed.

Deinstitutionalization and Early Discharge

An important recent trend in U.S. social policy is referred to as "deinstitutionalization," meaning that over the last few years a concerted effort has been made to maintain persons with mental illnesses in the community rather than in an institution. Of course, this is an admirable goal. This trend has given rise to a beneficial trend and an unfortunate one. The useful outgrowth of the trend of deinstitutionalization is that new services such as community mental health centers, outpatient clinics, and partial hospitalization have been developed. However, the deinstitutionalization policies have often led to discharging persons with mental illnesses from hospitals without the appropriate community support or without adequate planning. If persons with a severe mental illness are left without supervision to encourage them to take the important medications that can help them remain emotionally and mentally stable, they can become confused and dysfunctional and may end up on the street, without a permanent home and without mental health services or medical care. You have probably noticed men or women in public places who seem to be mentally ill and perhaps lacking the resources to keep warm or clean. Some aspects of the current problem of homelessness have resulted directly from policies and practices of deinstitutionalization.

Another trend results from shrinking resources for addictions treatments and from the shift from insurance to "managed care." Sometimes the period of time allowed for detoxification services or addictions treatment is so short that it cannot be effective. People with addictions are often discharged into the community well before they have completed the physical withdrawal process, are emotionally prepared to abstain from the substance, or have the community services in place to support their recovery process.

Stigma and Advocacy

Often mental illnesses and addictions carry a stigma. That is, persons with mental illnesses or persons who are addicted to drugs or alcohol are blamed, looked down upon, and seen as undesirable members of society. Even when a person with a mental illness is functioning well or a person is recovering from substance use, others may change their opinion of them when they learn about the diagnosis of mental illness or addiction. Individuals and families are understandably often reluctant to disclose the presence of mental illness or addictions, even to a human services helper or agency.

It is vital in both the mental health field and in the addictions field that we advocate for the rights of individuals and families. Discrimination can be a devastating side effect, resulting in not being able to obtain housing or employment. Helpers should work to secure the rights of all persons and to educate the public about mental illness and addiction. We should also advocate for adequate and appropriate resources to address these issues in our communities.

Remember to think of the person first and the diagnosis second. This is important when working with persons who have mental illnesses or addictions, because often their labels become more important than their rights as human beings. Regardless of what their behaviors are at any given time, they are still persons who are struggling to make it in the world. As always, educate those around you to respect persons with mental disorders or substance addictions.

Wrapping It Up

Human services workers, students, and volunteers are often called upon to provide emotional and practical support and case management to persons with chronic mental illnesses or with chemical addictions, as well as to their partners, friends, and family members. Being familiar with the symptoms of common mental illnesses, including substance use, allows the human services helper to understand potential reasons why another person is behaving in a certain way and to suggest that the person seek the opinion of a medical or therapeutic professional about possible interventions or medications that might make their lives easier. Understanding that the person's illness may cause him or her to withdraw from human services helpers just when he or she most needs assistance allows us to be more persistent about offering help. In other situations, the helper and the person seeking help might benefit from withdrawing from each other until a situation stabilizes. The key for the human services helper is to be aware and alert and to seek consultation with others when any confusion or concern arises.

In order to work with persons who have mental illnesses or who are struggling with substance use, you must have the ability to separate persons from behavior and to respect persons regardless of how challenging it is to be around them. You also should be able to treat persons with respect and acknowledge their rights and dignity. If you wish to work exclusively with persons with mental health or addictions issues or if you intend to work or volunteer in a facility that focuses on these areas, you need much more knowledge and extensive training. Many human services college programs offer courses or certificates in substance use counseling or mental health services. You may want to look into training programs in your community that could prepare you to be a helper in the mental health or addictions fields.

Think About It #72

1. Learn about the existing resources in your community. Using a telephone book or community resource directory, list some (a) community-based agencies, centers, and clubs that serve persons with mental health concerns; (b) residential facilities or hospitals that serve persons with mental health concerns; and (c) peer support groups, community-based agencies, and residential facilities or hospitals that serve persons seeking addictions counseling, detoxification, or long-term treatment.
2. Strive to be more observant of mental health and addictions issues in your area. Over the next few weeks, for example, notice persons in your community who may be homeless and have mental illnesses and/or substance use problems. Become more aware of drug activity (purchasing and using) on street corners and in parks.

CHAPTER FIFTEEN

Developmental Disabilities

In this chapter information is provided to help human services workers, students, and volunteers recognize and support persons with developmental disabilities over the life cycle; that is, from birth through childhood, adulthood, and old age. No matter what the human services setting, helpers are likely to work with people with disabilities and their families. They therefore need a basic understanding of the needs and vulnerabilities of this special population and their partners, parents, siblings, grandparents, and other family members who have unique challenges because they care for or about someone with a disability. Helpers also need to be aware of the unique strengths and capabilities of people with developmental disabilities.

Introduction to Developmental Disabilities

If you have known someone with a disability, whether the person was born with the condition or acquired it later, you probably already realize that people with disabilities are as different from one another as people without disabilities are. You may have noticed how people with disabilities are more like people without disabilities than they are different. Some advocates refer to people without disabilities as "TABs," people who are Temporarily Able Bodied. The term TAB emphasizes the possibility that anyone may acquire a disability at some point because of disease or accident.

The official definition of "developmental disability" refers to a specific severe, chronic disability of a person 5 years of age or older that:

1. Is attributable to a mental or physical impairment or combination of mental and physical impairments.
2. Is manifested before the person attains age 22.
3. Is likely to continue indefinitely.
4. Results in substantial functional limitations in three or more of the following areas of major life activity: self-care, language and communication, learning,

mobility, self-direction, capacity for independent living, or economic self-sufficiency (Developmental Disabilities and Bill of Rights Act of 1990 [Public Law 101-496]).

The definition of developmental disability depends on functioning rather than on a diagnosis. For example, a person may have a mild form of cerebral palsy but not meet the criteria for developmental disabilities. In contrast, a person may meet the criteria for having a developmental disability even though professional helpers are unable to determine a diagnosis or cause of the physical or intellectual limitations. Conditions such as epilepsy or hearing impairment, for examples, may or may not be severe enough to be considered as developmental disabilities.

Note that a developmental disability must have its onset prior to a person's 22nd birthday. This means that the disability affected the developmental outcome of a child or young adult. Some disabilities may occur during adulthood, from motorcycle or diving accidents, disease, or head trauma. If the disability occurs after the age of 22, the person may have a "disability" but not a "developmental disability." The definition of developmental disability also excludes children younger than 5 years old. Prior to age 5, a child may be considered "developmentally delayed."

Approximately 43 million people in the United States have a physical or mental disability (Johnson, 1993). Considering that people with disabilities have partners, parents, siblings, grandparents, neighbors, friends, and colleagues, it is easy to understand that the lives of many people are touched by developmental disabilities. The community and neighborhood, as well as family and friends, have a special role to play as people with developmental disabilities are being increasingly integrated in full community life. Therefore, social supports (such as employment, school, counseling, advocacy, and housing) play an important role in the quality of life for people with disabilities and their families.

Some people with disabilities experience oppression because of their disabilities; sometimes the stigma is compounded because the person may also be a member of another vulnerable population. We suggest that you begin your study of this field by examining your own attitudes about people with disabilities and confront whatever stereotypes and myths you hold.

The Developmental Disabilities Field

Helpers in the disabilities field may work in a variety of specialized settings that serve people with special needs, including the following:

- Hospital settings such as neonatal and prenatal intensive care units.
- Early intervention programs for children with special needs.
- Organizations serving special populations, such as the March of Dimes, Cerebral Palsy Association, Autism Association, or Craniofacial Society.

- Organizations serving all people with disabilities such as societies for protection and advocacy of people with disabilities, departments of mental retardation or disabilities, and special needs boards.
- School programs providing special education.
- Recreational programs, such as Special Olympics or Wings.
- Specific service programs to address the housing, employment, social, counseling, legal, or financial needs of people with developmental disabilities.
- Residential or supervised living programs for people with disabilities.
- Family-oriented programs that provide support to parents, siblings, grandparents, and other caregivers of people with disabilities.

You can expect to work with people with developmental disabilities no matter what setting you choose. For example, people with developmental disabilities request and receive services in all of the following settings:

- Corrections and criminal justice.
- Gerentology.
- Education and training.
- Schools.
- Medical and health care.
- Mental Health.
- Children and youth.
- Battered women and rape crisis networks.
- .Alcohol and substance abuse.
- Family services.
- Income maintenance.
- Community and neighborhood services.

People with disabilities face the same kinds of social problems that people without disabilities do. In many cases, people with disabilities are at increased risk for experiencing trauma (*i.e.,* assault) and social problems (*i.e.,* poverty.) People with disabilities are also often less prepared and receive less support than people without disabilities.

Think About It #73

1. Describe any experiences you have had with people with disabilities. Do you have a family member with a disability? How has this family member been treated? Have you had a neighbor with a disability? Was she or he a fully participating member of your community? In what ways?
2. If you have not had a meaningful experience with a person with a disability, why do you think this is?
3. Do you think you would like to work with people with developmental disabilities? Why or why not?

Language and Disabilities

Sometimes people use the term "handicapped" to refer to people with disabilities. The word "handicapped" originated when people with disabilities begged for money and food on street corners; thus, their *caps* were in their *hands* as they asked for charity. According to the *Social Work Dictionary* prepared by Robert Barker (1987), a handicap is "any physical or mental disadvantage that prevents or limits an individual's ability to function as others do" (p. 68). Therefore, although a person may have limitations due to a disability, the degree to which a person is "handicapped" is dependent upon the degree to which the environment can match the needs of that person (Simeonsson, 1986). For example, if an office building is totally accessible to people using wheelchairs, that is, there are elevators, accessible bathroom facilities, appropriate desks, and adequately spaced aisles, a person using a wheelchair is not "handicapped." The environment meets the individual's special needs. Scheer and Groce (1988) provide another excellent example. From the 17th century to the early 20th century, on the island of Martha's Vineyard in Massachusetts, a large number of people were born with profound deafness. As a result, the majority of the hearing population on this island became bilingual in English and sign language to accommodate those people with hearing impairments. In this community, those people who were deaf were not handicapped but could participate fully in community life.

Thus, because of the reasons discussed above, this book does not refer to people with disabilities as "handicapped." Instead, person-first language is used, referring to personhood first and disability second. Instead of saying "the Down syndrome child," say, "the child with Down syndrome." Instead of writing, "the retarded man," write, "the man who has mental retardation." Although these details seem trivial, language tends to shape and reflect attitudes. In some settings, people are frequently referred to by their diagnosis or disease instead of their name or their humanness. For example, have you ever heard, "I've got to check on the lung transplant in room 2C," or, "How's the amputee in room 306?" It is hard to even imagine a person behind these labels.

Respect and Disabilities

Another important point to think about when working with people with disabilities is the need to respect and honor strengths, capabilities, and rights to self-determination. For example, some people use wheelchairs for mobility. Please do not say that they are "confined" to wheelchairs or are "wheelchair bound." Most people who use wheelchairs to get around report that using the chair is liberating. The chair allows them to travel, be independent, and explore the world around them.

You will occasionally encounter people with visible or obvious disabilities (*i.e.,* blindness or use of a wheelchair, walker, or braces) whom you think might benefit from your help. No one likes to lose control, so before you help, make sure to ask if your help is welcomed. For example, if a person in leg braces drops one of her crutches and appears to be struggling to pick it up, rather than automatically assume that your help is appreciated, ask the woman if you may help in any way. She will let you know what, if anything, you could do. If a person who is blind seems perplexed and frustrated while at a street intersection, ask if you may assist in any way. You certainly wouldn't take the hand of a person without a disability and walk him or her across the street, would you? You should ask first.

Think About It #74

1. Close your eyes. What do you see when asked to visualize a "wheel-chair child?"
2. Now, close your eyes again and visualize a child who uses a wheel-chair for mobility.
3. Were your visualizations different? In what ways?
4. Try to list five reasons why a person using a wheelchair might not want you to push him or her down the street without having given you permission.

Types of Developmental Disabilities

Seven common diagnostic categories used to describe people with developmental disabilities will be discussed: (1) mental retardation; (2) cerebral palsy; (3) autism; (4) hearing problems; (5) epilepsy; (6) specific learning disabilities; (7) other conditions. Remember that just because a person receives one or more of these diagnoses, he or she may not have a developmental disability. That term is used to describe individuals who meet the definition discussed earlier.

Mental Retardation

The estimated prevalence of mental retardation in the United States is about 3 percent of the population (Drew, Hardman, and Logan, 1988). The definition of mental retardation has changed over the years. "Retardation" means "slowness in development or progress." A person with mental retardation learns more slowly than others. Until the 1960s, people with mental retardation were categorized as "morons," "imbeciles," and "idiots," depending on performance on intelligence tests (IQ scores) and the severity of the retardation. These words are now considered insulting and are no longer used. In addition to IQ, adaptive functioning is also considered to determine if an individual has mental retardation.

Since the 1960s, a new set of terms has been used to refer to degrees of mental retardation. The categories of mental retardation include borderline mental retardation (IQ 70–84); mild mental retardation (IQ 50–69); moderate mental retardation (IQ 35–49); severe mental retardation (IQ 20–34); and profound mental retardation (IQ below 20).

According to the 1992 definition of the American Association on Mental Retardation (AAMR), mental retardation refers to:

> substantial limitations in present functioning. It is characterized by significantly subaverage intellectual functioning, existing concurrently with related limitations in two or more of the following applicable adaptive skill areas: communication, self-care, home living, social skills, community use, self-direction, health and safety, functional academics, leisure, and work. Mental retardation manifests before age 18 (p.1).

Intellectual functioning is measured by the results of a standardized assessment designed to measure intelligence and resulting in an Intelligence Quotient (IQ). An average IQ is 100. Adaptive behavior is defined as an individual's effectiveness in meeting the standards of maturation, learning, personal independence, and/or social responsibility that are expected for his or her age level and cultural group, as determined by clinical assessment and, usually, standardized scales (Drew et al., 1988). Examples of adaptive functioning include drinking from a cup, cooperating by opening his or her mouth for feeding, sitting unsupported, imitating sounds or using gestures to communicate, and knowing familiar persons (Grossman, 1977).

According to Kevin DeWeaver (1995) this definition is congruent with trends in practice that include (1) moving away from labeling and highlighting the individual's strengths and functional limitations; (2) de-emphasizing the medical aspects of disabilities; (3) focusing on adaptive skills, not IQ; (4) providing for cultural and linguistic diversity in the assessment process; and (5) reflecting the ecological, contextual framework in which people live.

You might wonder why categorizing and labeling are necessary. First, determining whether a person functions intellectually significantly below average determines whether he or she is eligible for special services. These might include early intervention services, special school services, or the services of a department of mental retardation. Second, identifying the existence of mental retardation can help parents and helpers meet the special social and cognitive needs of individuals, that is, it may inform intervention and treatment. These might include specific accommodations at school or in vocational programs. Third, determining the existence of mental retardation may help provide information to determine an accurate diagnosis. Another question relevant to labeling is, "How can the diagnosis of mental retardation be used to enhance the quality of life of individuals and their families?" If a label detracts from a person's quality of life because it is used to justify limiting a person's options and rights, deny privileges and services, or demean or isolate, the diagnosis or label is being used inappropriately. If the diagnosis is used to increase a person's options and rights, open opportunities, or permit individualized services, then the label is being used appropriately.

Types and causes of mental retardation. There are numerous causes of mental retardation, including:

1. Low birth weight and inadequate gestational age. Preterm and premature infants are at significantly higher risk for mental retardation than are full term babies.

2. Chromosomal aberrations or genetic errors. An incorrect number or configuration of chromosomes in the body is considered to be a chromosomal aberration and may lead to mental retardation. Chromosomal aberrations may result in such diagnoses as Down syndrome, Turner syndrome, or Klinefelter syndrome. Genetic errors are those conditions that occur as a result of inheritance factors and involve specific genes. Such conditions include phenylketonuria (PKU), fragile X syndrome, or Tay-Sachs disease.

3. Maternal-fetal interaction factors such as Rh factor incompatibility.

4. Trauma during pregnancy or childbirth. Violence against the mother during her pregnancy may cause retardation in the child. It can also result from complicated birth trauma such as incorrect use of forceps. Retardation may also be caused in the child if the mother takes drugs, alcohol, medications, or poisons during pregnancy. Fetal alcohol syndrome (FAS) and fetal alcohol effects (FAE) are considered to account for the most common *preventable* cause of mental retardation in the United States today. Infants born affected prenatally by maternal ingestion of crack cocaine or other illegal substances have been the topic of many news stories in recent years. Methods of preventing and intervening with this population of infants and their mothers are very controversial. Babies born substance-affected benefit greatly from early intervention, and prognosis improves with consistent and responsive care.

5. Congenital infection or illness in the mother during pregnancy. For example, influenza, rubella (a form of measles), syphilis, and toxoplasmosis during pregnancy can cause mental retardation in a child. Maternal kidney disease, high blood pressure, and diabetes can also place a child at risk for mental retardation.

6. Malnutrition of the mother during pregnancy and poor prenatal care. In addition, malnutrition of the infant and poor infant stimulation also increase the risk of mental retardation. These conditions are often associated with poverty. Poverty is related to the majority of mild mental retardation. Circumstances related to poverty, such as children eating lead-based paint typical in older, substandard housing, are other examples of poverty-related causes of mental retardation.

7. Childhood infections and viruses. These include meningitis, encephalitis, and measles.

8. Child abuse and neglect. These are a significant cause of mental retardation. Physical assaults to an infant (such as shaken baby syndrome) and trauma to the head (traumatic brain injury) are examples of mental retardation caused by child abuse. Severe neglect of an infant's social and emotional needs is an example of a cause of mental retardation.

9. Unknown causes. Scientists are just now learning ways that the brain grows, develops, and is affected by its environment. Retts syndrome is an example of a type of developmental disability whose cause is unknown at this time.

Cerebral Palsy

Cerebral palsy is a general diagnosis that refers to a condition caused by damage to the brain either before or at birth. This brain damage causes problems with the muscle control centers of the brain. All three types of cerebral palsy are characterized by some degree of lack of control over the muscles of the body. In the *spastic* type of cerebral palsy, the individual typically moves stiffly and jerkily. In the *athetoid* form of cerebral palsy, the individual has involuntary and uncontrolled movements. In *ataxic* cerebral palsy, the individual's sense of balance and depth perception are impaired. Depending on the location and the extensiveness of the brain damage, symptoms can include "lack of balance, tremors, spasms, seizures, difficulty in walking, poor speech, poor control of the facial muscles, problems in seeing and hearing, and mental retardation" (Wikler, 1987, p. 424).

Baroff (1991) estimates that approximately 0.4 percent of the general population have some form of cerebral palsy. It is important to note that many people with cerebral palsy do not have mental retardation. It is dangerous to assume that physical limitations are equated with cognitive limitations.

Think About It #75

1. Watch one of the following movies that portray a person with cerebral palsy: "My Left Foot," "Gaby: A True Story," or "Test of Love: Annie Farrell."
2. Now answer the following questions about the movie:
   a. How were the main characters with cerebral palsy treated by their families in the early childhood years? Their adult years?
   b. How were the main characters with cerebral palsy treated by neighbors and other members of their community in their childhood years? Their adult years?
   c. Identify the strengths of the main characters with cerebral palsy.
   d. What could you offer the main character and his or her family if you were their helper?

Autism and Asberger Syndrome

Autistic disorder and Asberger syndrome are pervasive developmental disorders. According to the National Society for Children and Adults with Autism

(NSCA), autism is a severely incapacitating developmental disability that usually appears during the first three years of life. Autism is characterized by severe and pervasive impairment in reciprocal social interaction skills, communication skills, or the presence of stereotyped behavior, interests, and activities (Mesibov and Bourgondien, 1992). People with autism may appear to be in a world of their own. They may also demonstrate repetitive movements such as rocking, head banging, or hand twisting. An individual with an autistic disorder may insist that his or her environment and routine remain unchanged and may have unconventional ways of relating to people, objects, or events. Approximately 75 percent of children with autism also have mental retardation, typically in the moderate range (IQ 35–50) (DSM IV, 1994). Somewhere between 5 and 15 babies out of 10,000 births will have autism. Although no known agreed-upon cause of autism has been found at this time, scientists now agree that no factors in the psychological or emotional environment of the child have been shown to cause autism.

The definition of Asberger syndrome is as follows:

> The essential characteristics of Asberger's Syndrome are severe and sustained impairment in social interaction and the development of restrictive, repetitive patterns of behavior, interests, and activities. In contrast to autism, there are no significant delays in language development, cognitive development, or in the development of self-help skills (DSM IV, 1994).

Hearing Difficulties

An estimate of the prevalence rate for all hearing impairments in the United States in 1990 was approximately 16 million people. Schein and Delk (1974) conducted a National Census of the Deaf Population in 1970 and identified the following levels and definitions of hearing impairments. An individual with even the most profound hearing impairment is *not* considered to have a developmental disability. Hearing difficulties are included in this chapter because many children and adults with developmental disabilities also have hearing difficulties. *"Hearing impaired"* refers to any deviation from normal hearing, including deafness. *"Significant bilateral hearing loss"* refers to having significant hearing loss in both ears, with some difficulty understanding speech in the better ear. "Deaf" refers to the inability to hear and understand speech. *"Prevocationally deaf"* refers to onset of deafness before age 19. *"Prelingually deaf"* refers to the onset of deafness before age 3, before the acquisition of language.

Although about one-half of all hearing impairments have unknown causes, some of the known causes of deafness include heredity, maternal rubella, being born with the herpes virus, birth complications, and meningitis.

Debate and controversy continues about the best way to "treat" deafness. Some individuals who are deaf and their helpers and family members argue that hearing loss should be treated as a disability and should be corrected through whatever means possible (*i.e.,* cochlear implants). They also argue that children who are deaf benefit most from education that stresses lip reading and learning oral English. Other people argue passionately that people who are deaf do not have a disability. They are instead members of a subculture and are a linguistic minority. They argue that children who are deaf should learn American Sign Language (ASL) as soon as possible. Carol Padden and Tom Humphries, deaf linguists, write "Deaf people," emphasizing the upper-case D. They proclaim that the deaf share a culture rather than a mere medical condition (Dolnick, 1993).

If you chose to work with people with significant hearing impairments, you should learn ASL to facilitate communication. Most communities offer ASL classes at local community colleges or churches. If an individual with a hearing impairment comes to you or your agency for services, you should locate a translator to work with you and the client. Translators can be located through organizations in your community serving people with hearing impairments.

Epilepsy and Seizure Disorders

Epilepsy is a condition in which brain cells undergo abnormal electrical activity that causes disturbances in the individual's nervous system. A seizure occurs when there are too many discharges of electrical impulses from the nerve cells. Epilepsy is not a single disease or condition with a single cause. Generally, persons are considered to have epilepsy when they have experienced two or more seizures in the absence of external precipitating factors such as drug withdrawal or fever. Epilepsy and seizures are not contagious. Individuals who experience severe head trauma, stroke, and infections in the central nervous system are at the greatest risk for epilepsy. As is the case with some other medical conditions, some people with epilepsy have a developmental disability and others do not.

The Epilepsy Association of America estimates that 100,000 new cases of epilepsy are diagnosed each year. Approximately one percent of the U.S. population, or more than 2 million people, are currently diagnosed with some form of epilepsy. About 75 percent of the diagnoses of epilepsy take place before an individual is 18 years old.

Four different types of epilepsy have been identified: Tonic-clonic, absence (both are generalized), complex partial, and simple partial.

Generalized seizures affect both hemispheres of the brain and may lead to loss of consciousness, convulsions, and loss of memory. Two types of generalized seizures are the tonic-clonic and absence seizures. A *tonic-clonic* seizure (previously called "grand mal seizure") is a generalized seizure with loss of consciousness. The individual may cry out, fall, and lie rigid. The person's body may jerk, and he or she may lose bladder or bowel control. When the individual regains consciousness, he or she may feel stiff or sore. There may or may not be warning of the impending seizure. People experiencing this type of seizure do not usually remember the episode, may experience headache and drowsiness, and may take several days to return to normal. An *absence seizure* (previously called "petit mal seizure") is characterized by a sudden onset and blank stare.

These seizures last a short time but may occur many times a day, beginning and ending abruptly. The person is unaware of his or her surroundings and may not respond to others' attempts to communicate during the seizure.

Complex partial seizures sometimes affect the side of the brain near the ears (temporal lobes) but can occur in other parts of the brain as well. Individuals who experience a complex partial seizure appear to be in a trance accompanied by involuntary motor activities. They lose consciousness and have no control over these movements, which may include lip and tongue smacking, mimicry, hand movements, and repetitive speech or noises.

A simple partial seizure is characterized by uncontrollable movements such as arm or leg jerks while the individual is conscious. In this type of seizure, the part of the brain that controls vision, hearing, sensation , or memory is affected.

An "aura" is an unusual feeling experienced by many people with epilepsy prior to any type of seizure. The person may feel sick or apprehensive, have hallucinations, or notice a peculiar odor or taste. The individual retains memory of the sensation even if he or she eventually loses consciousness. The aura often serves as a warning that a seizure is about to take place, allowing the individual to move away from potential hazards before the onset of a major seizure (*A Woman's Guide to Coping with Disability*, 1994, pp. 135–136).

If the presence of epilepsy and seizures is *not* recognized and controlled early, a child runs increased risk for later learning disabilities, behavior problems, other types of seizures, and developing seizures that are more difficult to control (Epilepsy Foundation of America, 1989). Tonic-clonic seizures are usually easy to recognize; however, other types of seizures are more subtle. Sometimes people confuse the symptoms of seizures with other behaviors, such as substance use or attention-seeking behavior. If you suspect that a child may be having seizures, talk with your supervisor and the child's parents or guardian and arrange for an evaluation by a pediatrician. The Epilepsy Association of America (1989) offers the following signs indicating that a child is having seizures:

- Short attention blackouts that look like daydreaming.
- Sudden falls for no reason.
- Lack of response for brief periods.
- Dazed behavior.
- Unusual sleepiness and irritability when wakened from sleep.
- Head nodding.
- Rapid blinking.
- Frequent complaints from the child that things look, sound, taste, smell, or feel "funny."
- Sudden bending (bowing) movements by babies who are sitting down.
- Grabbing movements with both arms in babies lying on their backs.
- Sudden stomach pain followed by confusion and sleepiness.
- Muscle jerks of arms, legs, or body.
- Repeated movements that look out of place or unnatural.

Learning Disabilities

A learning disability is a "disorder in one or more of the basic psychological processes involved in understanding or using language, spoken or written, which may manifest itself in an imperfect ability to listen, think, speak, read, write, spell, or do mathematical calculations" (P.L. 94-142). Learning disabilities include:

1. **Spoken language:** delays, disorders, and differences in listening and speaking.
2. **Written language:** difficulties with reading, writing or spelling.
3. **Arithmetic:** difficulties in performing arithmetic functions or in comprehending basic mathematical concepts.
4. **Reasoning:** difficulty in organizing and integrating thoughts.

Students who have learning disabilities may exhibit a wide variety of characteristics. Some learning disabilities are more severe than others. In addition to the above academic challenges, people with learning disabilities may also experience hyperactivity, inattention, distractability, impulsivity, low tolerance for frustration, problems handling day to day social interactions and situations, and/or perceptual coordination problems.

The U.S. Department of Education (1987) reports that 4.73 percent of all school-aged children received special education services for learning disabilities in 1986–1987. This means that there were over 1.9 million children diagnosed with a learning disability in that school year (NICHCY, 1988).

Attention Deficit Hyperactivity Disorder.   Attention Deficit Hyperactivity Disorder is often referred to by its initials: ADHD. ADHD is a neurological disability that interferes with a person's ability to sustain focus or attention and to delay impulsive behavior.

The three main features of ADHD are inattention, impulsiveness, and hyperactivity. Children with ADHD have difficulty paying attention and remaining on task, especially in school. Children with ADHD are often impulsive and act before thinking about the consequences of their actions. They are often more restless and active than other people their age. The national organization Children and Adults with Attention Deficit Disorders (CHADD) lists the following characteristics typical of children with ADHD:

- Fidgeting with hands or feet.
- Difficulty remaining seated.
- Difficulty following through on instructions.
- Shifting from one uncompleted task to another.
- Difficulty playing quietly.
- Interrupting conversations and intruding into other children's games.
- Appearing not to be listening.
- Doing things that are dangerous without thinking about the consequences (1993, p. 1).

Approximately 3.5 million American children (as many as 5 percent of children under age 18) are estimated to have ADHD (Ingersoll, 1993). CHADD (1993) estimates that a significant percentage (perhaps as many as 50 percent) of children with ADHD are never properly diagnosed. Children do not necessarily outgrow ADHD. Not long ago, the national organization serving children with ADHD changed its name to "Children and Adults with Attention Deficit Disorders" for this reason. (CHADD can be reached at 499 NW 70th Avenue, Suite 109, Plantation, FL 33317.)

A diagnosis of ADHD should be made with great care and caution by a professional trained in this specialization. Appropriate diagnoses involve a multilevel assessment and a multilevel intervention plan. Paradoxically, the ADHD diagnosis appears to be given to children inappropriately and drug treatment therefore ordered too frequently. Many times the behavior of children who have experienced trauma exhibits characteristics of ADHD. Even if an appropriate diagnosis of ADHD is made and stimulants such as Ritalin can be very helpful, this form of intervention is not without its negative side effects. Frequently, environmental modifications to classrooms, such as increasing structure and mentoring a child with ADHD, are as effective in helping a child focus and slow down as medication (Hallowell, 1994).

Other Conditions

Several other sets of difficulties sometimes accompany developmental disabilities. Orthopedic problems, for example, involve difficulties in the functioning of muscles, bones, and joints. Visual impairment,which can include blindness, is another condition that affects more than 1.4 million Americans (DeWeaver, 1995). Head injuries also lead to developmental disabilities. Finally, numerous people have multiple disabilities. Some individuals are both deaf and blind. Others have both orthopedic difficulties and mental retardation. Very special means for assessing and treating people with multiple disabilities are required.

Social, Legal, and Ethical Issues

People with developmental disabilities and their families live within a larger community and societal context. Laws and policies protect people with disabilities from discrimination and ensure their rights as full citizens. People in our society also have certain attitudes toward people with developmental disabilities that affect their behaviors and decisions. Within a community context, people with developmental disabilities participate in the workforce, attend school and vocational training, marry, and have children. People with developmental disabilities also vote in elections, reside in community housing, and need social support.

Attitudes and Myths About
People with Developmental Disabilities

Negative attitudes and stereotypes about people with developmental disabilities are common. Take this opportunity to become more aware of your own opinions about people with developmental disabilities. You are encouraged to identify these stereotypes or biases, discuss them with others, and learn more about disabilities so that your attitudes are based on actualities rather than what you might imagine.

Livneh (1991) maintains that our cultural and social norms lead to the creation of negative attitudes toward people with disabilities. For example, our society stresses beauty, youth, ability, health, and wholeness. People with disabilities often do not measure up to normative standards of appearance. Our culture also highly values productivity, competitiveness, and achievement. Some people with disabilities are not capable of competing economically. Other negative attitudes toward people with disabilities might result because this culture generally supports the importance of independence and self-sufficiency. Some people with disabilities may never be totally self-sufficient.

Some negative attitudes toward people with disabilities are the results of the belief that people should be held responsible for their own circumstances. Thus, either people with disabilities or their parents are blamed for causing the disability. In this instance, the disability is considered to be a punishment for "sinfulness," the result of poor and inadequate parenting, or lack of motivation, poor work ethic, and laziness on the part of the person with a disability. The assumption is that if the person with a disability just wanted to, he or she could work hard and be able to overcome the disability (walk, do algebra, or work a 40-hour week, for examples).

A very interesting research study was conducted by Sobsey (1997) to determine the kinds of attitudes some people hold toward people using a wheelchair for mobility. In this study, 50 percent of the subjects (all undergraduate psychology students) were sent to an office where they were greeted by an able-bodied researcher who gave them a questionnaire to complete. After the questionnaire was completed, the subjects returned the questionnaire to the researcher. She placed it in a machine that scored the questionnaire. For half of the subjects, the questionnaire was scored, and the researcher thanked the subjects and told them that they could leave. For the other half of the subjects, the scoring machine destroyed the subjects' answers, and the researcher told the subjects that they would have to answer the questionnaire again.

The other 50 percent of the subjects were sent to an office where they were greeted by the same woman as in the above, but this time she was using a wheelchair. There were no other differences in the research project. These subjects returned their completed questionnaire to the researcher in the wheelchair. For half of the subjects, the questionnaires were scored with no problems; for the other half of the subjects, the scoring machine malfunctioned, and the subjects had to complete the questionnaire again.

In all four situations, subjects were asked at the end of the experience what their attitudes toward the researcher were. The researchers report some very interesting results. When things went smoothly, that is, when the scoring machine functioned well, subjects reported similar attitudes toward the researcher regardless of whether she used a wheelchair or not. However, in the situation where the scoring machine malfunctioned, subjects attributed blame to the researcher in the wheelchair. Attitudes were more negative, and some subjects made assumptions that suggested that the researcher in the wheelchair was incompetent. Attitudes toward the able-bodied researcher whose scoring machine malfunctioned were significantly more positive, and blame was attributed to the machine and not the person (Sobsey, 1997).

Dudley (1987) suggests that stigma must be lifted from people with mental retardation if they are to become fully participating members of our society. He identifies three misconceptions that have the "ominous effect of perpetuating social isolation and inferior status" of people with mental retardation. These include the misconceptions that people with mental retardation have little awareness or understanding of their disability; that people with mental retardation are indifferent to the language that is used when referring to them and their disabilities; and that people with mental retardation are not affected by the attitudes and stigma promoting practice. Contrary to the above myths, the vast majority of people with mental retardation are acutely aware of their limitations and the attitudes of people around them. In general the media does little to correct these misconceptions. Media coverage is often reduced to portraying "helpless victims" or "beggars" (as on fund-raising telethons) or "courageous cripples" (hero stories in which an individual overcomes his or her disability) (Dardick, 1993). Rarely are people with disabilities portrayed as ordinary people with ordinary joys and challenges. One notable exception was the television series, "Life Goes On," featuring Corky, a young man with Down syndrome.

Legal Rights and Entitlements

Human services helpers should be aware of the following three important pieces of legislation. First, Public Law 94-141, the "Education for All Handicapped Children Act" of 1975, mandated that all public school education be available to all children with disabilities. The purpose of this legislation is to:

> assure that all handicapped children have available to them . . . a free, appropriate public education which emphasizes special education and related services designed to meet their unique needs, to assure that the rights of handicapped children and their parents or guardians are protected, to assist States and localities to provide the education of all handicapped children, and to assess and assure the effectiveness of efforts to educate handicapped children.

This law entitles children with disabilities to a free appropriate public education just like the one to which children without disabilities are entitled. Prior to its passage, children with disabilities received no education, parents paid for private day school, or children with disabilities were institutionalized. If you attended school prior to 1975, you may recall that there were no or very few children with disabilities in either your classes or your school. Now, children are mainstreamed into public schools and receive education in the "least restrictive environment." "Mainstreaming" or "normalizing" refers to efforts to insure that a child with a disability be educated as much as possible with children who do not have disabilities. This means that children should not be segregated in separate schools or separate classrooms unless it is in the educational best interests of the child with a disability and is as "normal" an experience as the child can manage. P.L. 94-142 also mandates that schools develop an individualized education plan (IEP) for each child with a disability. The IEP is a statement of the unique needs and services to be provided to the child. The IEP is developed at a meeting that is attended by the parents or guardians, the child's teacher, a representative of the school who supervises special education programs, and the child if this is thought to be appropriate.

Second, Public Law 99-457, the "Education of the Handicapped Act Amendment" of 1986, extended public school education to children from birth to age 21 and recognized the need for early intervention for children under school age. This act acknowledged the fact that some children need extended education beyond age 18. The results of this legislation are: (1) enhanced development of infants and toddlers with disabilities; (2) minimized potential for developmental delay; (3) reduced costs associated with special education; (4) lessened likelihood for institutionalization; and (5) increased capacity of families to care for their children.

Third, Public Law 101-336, also known as the American with Disabilities Act (ADA) of 1990, gives civil rights protection to people with physical disabilities and mental illnesses. These protections are similar to those civil rights provided on the basis of race, sex, national origin, and religion. The ADA guarantees equal opportunity for people with disabilities in employment, public accommodations, transportation, state and local government services, and telecommunications.

Protection and Advocacy

In the 1960s, people with disabilities and their families and friends actively worked to change laws, policies, and attitudes such that children and adults could fully participate in their communities without discrimination (Asch and Mudrick, 1995). At the present time, each state in the union has an independent organization whose goals are to ensure the legal and human rights of people with developmental disabilities and advocate on the behalf of people with disabilities. Often these organizations are referred to as "protection and advocacy" systems for people with disabilities.

Illustration

Constance Fielder is a 28-year-old Caucasian woman with cerebral palsy. She has an undergraduate degree in psychology and a Master's Degree in Library Science. She uses a wheelchair for mobility and has a learning disability that im-

pairs her ability to do all but the most basic mathematical functions. She is very intellectually capable, has good communication skills, and earned a 3.0 GPA in her undergraduate studies and a 3.3 in her graduate studies. She has been out of work for the past 11 months. Her classmates have found jobs. Although she has been to six different interviews, she has not had a job offer. She thinks that her future employers are discriminating against her because of her disability.

Think About It #76

1. How can Constance determine if employers are discriminating against her because of her disabilities?

2. If you were working with Constance, how would you support her?

In addition, many communities have "self-advocacy" programs. Self-advocacy refers to self-help activities in which people with disabilities engage to better their own lives and advocate on their own behalf. Self-advocacy groups are designed to help people learn to act or speak for themselves. Examples of how self-advocacy encourages and teaches people with disabilities their rights include the following:

- I am not a client.
- I am not a resident.
- I am not a patient.
- I am a person.
- I am a citizen.
- I am an individual.
- I have value and worth.

Think About It #77

Explore your community's resources in one or more of the following ways:

1. Locate your state's protection and advocacy organization.
2. Determine if your community has a self-advocacy program.
3. Find out ways in which schools are complying with P.L. 94-152 or P.L. 99-457. Ask to see an IEP. Make sure that you promise confidentiality.
4. Go to a public place like a restaurant, an office building, or department store. Carefully observe with "new awareness" whether the facilities are accessible to people with disabilities. Is the facility wheelchair accessible? Are there telephone booths jutting from the walls that might pose a danger for people with visual impairments? Are there telephone devices for people with hearing impairments? Does your current place of employment or volunteer setting accommodate people with disabilities? How?
5. Notice advertisements on television, in the newspaper, or in magazines. Are people with disabilities included?

The national self-advocacy movement began in 1974 with the formation of People First in Oregon. (For further information from this group about starting a self-advocacy program, contact People First of Washington, P.O. Box 381, Tacoma, WA 98401.)

Developmental Disabilities over the Life Cycle

Sometimes people with developmental disabilities are thought of as infants and children. Often people forget that children with developmental disabilities grow up to be young adults, middle-aged adults, and older adults. People with developmental disabilities have different needs at different ages. Thinking about and responding to different needs of individuals with disabilities is taking a life-cycle approach. This section identifies some of the unique considerations that individuals with developmental disabilities might have during different periods of their lives.

Prenatal and Neonatal Periods

During the prenatal and neonatal periods of a child's development, special consideration is given to the prevention of developmental disabilities. For example, pregnant women are encouraged to take vitamins and receive early and consistent prenatal care. If a woman takes folic acid during her pregnancy, she will significantly reduce the risk of giving birth to a child with spina bifida. Fetuses are very sensitive to the mother's ingestion of drugs and alcohol during pregnancy. Alcohol usage during pregnancy is directly associated with a baby being born with fetal alcohol syndrome or fetal alcohol effects. Ingestion of crack cocaine, heroine, methedrine, and other substances also has an impact on infant health and development.

In addition, genetic testing of the fetus, amniocentesis, and chorionic villi sampling (CVS), are available during the prenatal period. With genetic counseling, parents can receive information about the fetus's genetic integrity. Amniocentesis and CVS can tell a parent if a child has Down syndrome, fragile X syndrome, or spina bifida, for example. Many women over age 40 receive genetic testing during their pregnancies because the risk for genetic abnormalities increases with age.

Developmental disabilities are also often diagnosed at birth or shortly thereafter. In most situations, early diagnosis is important so that children can receive early and intensive intervention. In the case of an enzyme deficiency like phenylketonuria (PKU), for example, a rigorous, prescribed diet can prevent disabilities.

Infancy and Early Childhood

During infancy and early childhood, children with developmental delays benefit from early intervention programs. These services often include physical therapy, occupational therapy, speech therapy, and early stimulation activities. These early intervention programs are often provided in the home by specialists called early interventionists. Services can also be offered at clinics or special day nurseries designed to meet the needs of young children with disabilities. Research has demonstrated that early intervention programs improve the language, cognitive, physical, and psychosocial development of children (Drew et al., 1988).

School-Age Children

The 5 to 12-year-old child with disabilities receives significant services through the school system. As you already know, public schools are required by law to provide relevant services to all children. Several alternatives are available for children being educated in the public school system. In the least restrictive setting, children with disabilities and without disabilities are fully integrated in all school activities and classes. For example, a child who is deaf participates fully in all activities and in all subjects, however the school district may provide the child with an American Sign Language (ASL) interpreter.

In some situations, a child with disabilities may be best served by being integrated with children without disabilities in all activities and classes except one. For example, a child with a learning disability that affects reading may go to a "resource" class with a specially trained teacher to be taught specific and individualized ways to learn to read.

In another circumstance, a child may not attend any academic subjects with classmates who do not have disabilities. The child may receive resource services for all academic subjects because of pervasive or severe learning disabilities or mental retardation. These classes may emphasize adaptive, functional, and social skills. For example, a child with moderate mental retardation may learn how to recognize money, pay for purchases, and receive the correct change while her classmates are learning long division or algebra. This student, however, is probably attending nonacademic subjects such as music, physical education, art, lunch, and recess with her classmates.

In the most restrictive school settings, children with disabilities are placed in a classroom separate from children without disabilities for the entire school day or, in some rare instances, they may attend a separate school for children with disabilities. Decisions to segregate a child from his or her classmates are made only under extreme circumstances. Sometimes a child has very special and unique needs that require individualized educational plans. Sometimes a child with a disability is violent or dangerous, and schools cannot take the risk of mainstreaming a child at a particular time.

Illustration

Aaron, a boy of 9 of Jewish heritage, was hit by a car while playing baseball in the street in front of his house. The accident caused severe traumatic injury to his brain. Although he has recovered most of his physical functioning, Aaron's behavior has been virtually uncontrollable. Despite the gentlest care, Aaron hits, bites, and kicks anyone who comes near him. Caregivers have had to wear protective gear such as protective vests to prevent injury. Aaron cannot safely ride the school bus without the driver and two adults supervising him. Despite many trials of medication and behavioral interventions, after six months Aaron is not ready to attend school with other children. The school decides that they could best meet Aaron's needs and the needs of the other children in the school by placing him in a separate classroom with his own teacher and teacher's aide. Aaron's needs will be reevaluated every three months to determine if a less restrictive environment is in the best interests of Aaron and his classmates.

The decision to isolate children with disabilities from other children must be made very carefully. Integration of children with and without disabilities benefits *all* children. Children without disabilities learn compassion, patience, and appreciation for difference. Children with disabilities benefit from the intellectual, social, and physical stimulation of mainstreaming. They also observe appropriate school behavior and are expected to behave in ways that allow for full community participation.

Adolescence

The teen years are difficult for most. During adolescence, young people are struggling with their identity and beginning to establish their independence. They are preoccupied with their social world, and friendships are extremely important. Most adolescents begin to think of romance, love, and intimacy during their adolescent years. They are also considering their futures. Teenagers with disabilities are no different. They are also facing these developmental tasks, but their disability may place some extra challenges in their paths.

Illustration

Ariel, a 16-year-old African-American girl with spina bifida, attends Evergreen High School. She is an average student and, except for physical education, participates fully in her school's classes and activities. Ariel expresses sadness and often worries about her future. Some of the kids at school are talking about moving out on their own after high school. Ariel would like to achieve this independence, but she doesn't see how this can be accomplished. She requires special daily medical care that her parents handle. When she thinks about getting a job, she wonders whether she can realistically accomplish her dream of being a veterinarian. Ariel is sad because she doesn't have close friends and has never been asked out for a date. Her classmates are nice to her, but no one has ever invited her to a party or to a dance. She thinks they probably don't even know that she likes to dance. While the other teenagers are with their friends, Ariel is often at home by herself or with her parents. Most of her class-mates have passed their driver's license tests, and either have their own cars or can borrow their parents. Ariel is unable to drive at this time. The cost of modi-fying a van for hand controls and a wheelchair lift is beyond her family's finan-cial resources. Ariel must ask one of her parents to drive her places. Even if she had a friend to offer to pick her up and take her to the mall, who would help her in and out of the car? And there wouldn't be room for her motorized wheelchair. Ariel wonders if she will ever leave her parents' house. She wonders if she will ever get married or even live on her own.

Early and Middle Adulthood

There are several special concerns for adults who have disabilities, such as where to live and work and how to form relationships and families. During adulthood, individuals in the United States typically maintain an independent household, have meaningful employment, and have intimate relationships with friends or a partner. Many adults also raise children. These are the developmen-tal tasks that face adults with disabilities, too.

Housing. Many adults with disabilities can live independently in apartments or houses in the community. Some adults with disabilities may need assistance or supervision. Several housing options for adults with disabilities fall on a contin-uum depending on how much assistance or supervision the person with the dis-ability needs and wants. The ability of the person with disabilities is only one factor that predicts the degree of supervision or assistance that is needed. Suc-cessful, independent community living also depends on the services available in the community such as "vocational opportunities, access to public transpor-tation and buildings, and adequate housing, medical services, and recreation and leisure" (Drew et al., 1988, p. 324).

1. "Independent living" refers to situations where people live by themselves or with others of their choosing. No formal supervision is used, however, per-sonal aides to help with bathing or toileting may be required.
2. Community-based residential living is a model for adults with disabilities who live in a group home or semi-independent apartment in a community or neighborhood. Typically, human service workers are hired to provide appro-priate supervision and assistance, depending on the needs of the adults with developmental disabilities. This may consist with help in paying bills, shop-ping, transportation, or medical care.
3. Institutional living refers to residential care that is segregated from the com-munity and often houses large groups of people with disabilities. All activi-ties are typically conducted in the same place. Residents are very closely supervised, and activities are governed by a system of explicit rules and regulations. Since the 1960s when it was found that many institutions were simply "warehousing" people with disabilities and not meeting their indi-vidual needs, deinstitutionalization of people with disabilities has been a pri-ority and the number of people in institutional care decreased.

Employment. Meaningful and interesting work and the capacity to be monetarily rewarded for this work is just as important for people with disabilities as it is for all people. Financial independence, that is, the ability to make decisions about how one spends his or her money, is a key marker of adulthood. Competitive employment for people with disabilities may be either full-time or part-time. Participating in competitive employment is often dependent on the quality of vocational training or education that a person with disabilities receives, as is true for people without obvious disabilities. Just as in the housing arena, some people with disabilities need more supervision in the workplace than others. Competitive employment can take three forms: (1) employment with no support services; (2) employment with time-limited support services; and (3) employment with ongoing support services (Drew et al., 1988). A second type of employment for people with disabilities is "sheltered employment." Adults with mental retardation occasionally participate in sheltered employment. The employees usually working under on-site supervision, typically work through contracts and with jobs broken down into small tasks. Rather than being paid minimum wage, employees working in sheltered workshops are usually paid on a piecework basis (Payne and Patton, 1981).

Sexuality, relationships, and parenting. Another important marker of adulthood is the development and maintenance of intimate relationships with other adults and/or the formation of a family of procreation. Most adults with developmental disabilities desire an intimate, sexual relationship. Many wish to be legally married and/or become parents. All people with developmental disabilities are entitled to the expression of consensual affection. According to Walter Fernald, in the early 1900s women with mild mental retardation were considered to be dangerous:

> They are certain to become sexual offenders and to spread venereal disease or to give birth to degenerate children. Their numerous progeny usually become public charges as diseased or neglected children, imbeciles, epileptics, juvenile delinquents or later on as adult paupers or criminals. The segregation of this class should be rapidly extended until all not adequately guarded at home are placed under strict sexual quarantine (Sloan and Stevens, 1976, pp. 76–77).

Involuntary sterilization is now considered a violation of a person's basic human rights. Now social services are available in many communities that provide support for parents with developmental disabilities.

Think About It #78

1. Write down some of your feelings, attitudes, and beliefs about providing adults with developmental disabilities full participation in adult activities such as employment, housing, sexuality, marriage, and parenthood.

2. Discuss your reactions in a small group. How easy would it be for you to provide services to adults with developmental disabilities in these areas?

Older Adults

The population is aging worldwide, and, in the United States, adults 65 years of age and older comprise 13 percent of the population. The number of people with developmental disabilities who are reaching old age is also increasing. Of all persons with disabilities, approximately 12 percent are over 65 (Anderson and Polister, 1993). Anderson and Polister (1993) estimate that between 200,000 and 500,000 older adults have disabilities in the United States, and the number is expected to double in the next 40 years. The life expectancy for individuals with developmental disabilities has dramatically improved over the past several decades and closely resembles that of the general population. In 1949, for example, persons with Down syndrome had an average life expectancy of nine years. In 1963, the life expectancy was 18.3. According to Adlin (1993), the average life expectancy for people with Down syndrome is now 55 years.

Older adults with developmental disabilities have very special needs; they seem to be at increased risk for psychiatric difficulties, nutritional deficiencies, and the typical challenges that face adults as they age, such as health and medical problems. For instance, Alzheimer's disease is associated with Down syndrome, and approximately 40 percent of persons with Down syndrome will develop the dementia symptoms of Alzheimer's disease (Cox and Parsons, 1994). Finding a nursing home or adult day-care center that will accept and appropriately care for an adult who is older and has mental retardation can be difficult.

The special medical needs of adults who are older and have developmental disabilities also challenge medical professionals who may have either specialized in developmental pediatrics (children) or geriatrics (older adults) but not "developmental geriatrics." Furthermore, these older adults may not have amassed the financial resources necessary to exercise control over their futures. They may no longer have family members alive to help care for them or monitor the quality of services they receive. The population of adults who are older and have developmental disabilities is a very vulnerable one. This is an area in the human services that will need sensitive and caring workers.

Developmental Disabilities and Families

Helpers have become increasingly involved in working with family members of individuals with developmental disabilities. Recognition that the family is the most important and consistent resource for a person with disabilities means that providing assistance and support to parents, siblings, and grandparents must occur over the course of the entire life of the individual with disabilities. Public law now mandates that the families of children with developmental disabilities be included as full-fledged members of the service team. Family members must participate in individualized educational plans (IEPs) and individualized family service plans (IFSPs) and give approval of these documents. This family involvement is called "family-centered service." The nine crucial elements of family-centered service are as follows (Key Elements of Family-Centered Care, National Center for Family-Centered Care, Association for the Care of Children's Health):

1. Recognize that the family is the constant in a child's life, while the services and systems and personnel within those systems fluctuate. Human service helpers should acknowledge and respect the fact that parents and other caregivers are the experts on the child with developmental disabilities. Parents and caregivers live, love, and care for the child 24 hours a day over the course of years. Helpers come in and out of a child's and a family's life. Listen carefully to family members. What they say they want from helpers is compassion, knowledge, respect, and support.

2. Facilitate parent–helper collaboration at all levels of service provision, including services to the individual child, program development and implementation, and policy formation. Family members should be involved in all aspects of services provided to people with developmental disabilities. In addition to being full team members in IEP and IFSP meetings, families should be invited to participate in the formulation of policies and laws, program development, and program evaluation. Be sure to ask yourself if a family member is represented on each one of the committees, task forces, or coordinating groups that your organization has. If these consumers are not represented, they should be.

3. Honor the racial, ethnic, cultural, and socioeconomic diversity of families. Different ethnic, cultural, and economic groups have differing attitudes, behaviors, and ideas about the treatment of people with developmental disabilities. For example, some groups of people are more tolerant of behavior that falls outside typical behavior than other groups. Ideas about the importance of independence and productivity differs. The willingness to seek help might differ.

4. Recognize family strengths and individuality, and respect different methods of coping. It is important to remember that all families have strengths and capabilities as well as struggles and challenges. Sometimes it is easier to find fault or blame families caring for a person with developmental disabilities rather than to find out about their special circumstances.

Illustration

Bob works as a teacher's aide in a classroom with a 9-year-old girl, Simone, who has been diagnosed with autism and mental retardation. Although it is important for Simone to take medication to control her seizures, she consistently comes to school in the morning without having done so. Simone's mother, a single parent with two other children, has been contacted. She says that she is aware of the importance of giving Simone her medication and promises to administer the medicine each morning. In less than a week, however, Simone is once again coming without having been medicated. Bob is at first critical of Simone's mother, seeing her as negligent, lazy, and a failure as a caregiver. Bob then comes up with a solution: He asked if the school could administer the medication when Simone arrives each morning. The school and Simone's mother agree to this plan. On the first day that Bob tries to give Simone her medication, Simone refuses to take it. The more Bob insists that she drink the liquid, the more agitated Simone becomes. Finally, she becomes so extremely aggressive and belligerent that she must be restrained. Over the course of the week, Simone's resistant behavior continues when anyone attempts to administer her medication. Bob now reflects that he understands why Simone's mother frequently sent Simone to school without having given her the medicine.

Think About It #79

1. What are some of the ways that Simone's mother coped with this difficult situation?
2. Identify some of the strengths that you recognize in Simone's mother.
3. If you were Bob, what might you recommend or try?
4. How would you work with the school staff, Simone, and/or her mother concerning medication?

---

5. Share with parents on a continuing basis and in a supportive manner complete and unbiased information. Sometimes helpers think that some information is too technical or painful for family members to hear and understand. Although the motivations may be understandable, family members have the right to be completely informed and read all records that pertain to their child with developmental disabilities and their family. How would you feel if your physician refused to tell you that you have a serious medical condition or that your son or daughter has a chronic illness? How would you react if your physician refused you access to your file?

Keeping families fully informed is often a challenge, however. You may need to find new words to explain medical terminology. You may need to educate yourself on a variety of diagnoses or medical procedures. You can also consider collecting written literature, audiotapes, and videotapes that you could give or lend to family members to help them better understand the disability that their family member has. The communication of this information must also be done in a compassionate and supportive way. Often information must be repeated several times because it is emotionally difficult to hear news that your child has problems. The important thing is to be patient, supportive, and available.

6. Encourage and facilitate family-to-family support and connection. Many communities have organizations that provide services to family members caring for children and adults with developmental disabilities. One valuable service is called family-to-family support. A new mother, for example, who has just received word that her newborn son has Down syndrome, can be matched with another mother who has an older child with Down syndrome. The support parent can provide the new parent with compassion, information, kindness, and understanding in ways that a human services helper cannot. Family-to-family support can also take the form of support groups for parents, grandparents, or siblings. Find out if your community has a family support organization or support groups for people with developmental disabilities and their families. The magazine *Exceptional Parent* is an excellent resource for families who have children with special needs. This magazine often matches parents of children with very rare disabilities or chronic illnesses with other parents who may live across the country.

7. Understand and incorporate the developmental needs of infancy, children, adolescents, adults, and their families into service delivery systems. Families have different needs as their child grows and develops. An often neglected period is the adulthood of a person with developmental disabilities.

8. Implement comprehensive policies and programs that provide emotional and financial support to meet the needs of families. Remember that the care of a person with developmental disabilities is life long. This puts emotional, financial, and practical stress on families. A human service worker can be very helpful in developing and accessing resources with families, thus assisting in reducing stress.

Think About It #80

1. Make a list of the needs of families when their son or daughter with developmental disabilities is in each of the following ages: infancy, childhood, adolescence, adulthood.
2. How might the lives of family members be affected when they care for a relative with a developmental disability?

---

9. Design accessible service delivery systems that are flexible, culturally competent, and responsive to family-identified needs. The only way to know the needs of a group of people is to ask each one. This means that you should feel free to ask family members what they want and need. You can ask how you can be helpful. You must of course listen carefully to their answers. For example, one human service worker was involved in helping a family with an adolescent daughter, Pamela, who had mental retardation and visual impairments. The family had multiple problems and needs. They were surviving on welfare benefits, living in a dangerous community in subsidized housing, without reliable transportation for medical appointments, and with little support from family or friends. The human service worker felt overwhelmed just thinking about the vast array of needs of this family. She asked Pamela's parents what they thought was their most pressing need at this time. They quickly answered that they most wanted a rubber mattress pad for Pam's bed. Pam had enuresis (wet her bed at night) and her soiled mattress made the entire apartment smell bad. The parents thought that their lives would be improved tremendously if they just had a mattress pad. The human services worker realized that she didn't have to solve all the problems of poverty and violence to help this family. First, she could be helpful by providing the family with what they asked for.

Think About It #81

1. What made the human services worker's question to Pam's parents important?
2. How do you think you might have felt after hearing from Pamela's parents that they could best be helped by having a mattress pad?
3. How would you have helped this family acquire a mattress pad in your community?

The unique needs and challenges of parents and caregivers, siblings, and grandparents of people with developmental disabilities will be addressed in the following sections.

Parents

In many ways, raising a child with developmental disabilities is like parenting someone with no special needs. Parents and other caregivers of children with disabilities, however, express that they also face unique challenges, stresses, and joys over the course of the family life cycle. Lynn Wikler (1981), a social work researcher, was interested in identifying and understanding the needs of parents caring for children with mental retardation and in how helpers perceived the needs of parents. Interestingly, Wikler found that helpers overestimated the needs of parents at the time the child was diagnosed with mental retardation and underestimated their needs at other times over the course of the life cycle. Parents reported that they needed services at ten critical periods before the child turned 21. Parents reported that predictable times of crises occur:

1. At the time of diagnosis of mental retardation.
2. When a younger sibling developmentally surpasses the child with mental retardation.
3. When there is serious discussion of possible placement of the child with mental retardation outside the home.

4. When the child with mental retardation exhibits exacerbated behavior or health problems.
5. When there is serious discussion of guardianship or long-term care for the child with mental retardation.
6. When the child should have begun walking but does not.
7. When the child should have begun talking but does not.
8. When the child with mental retardation is first enrolled in school.
9. At the onset of puberty (because of tension between physical appearance and mental/social ability).
10. When the person with mental retardation turns 21 (symbolic of independence).

These findings have implications for helpers. It is important to keep in mind that parents would like human service workers to be more available to help and support them throughout the time that they care for their children with special needs.

Gender. Although a vast body of literature has been written to address the experiences of parents caring for children with developmental disabilities, the majority of research conducted refers to mothers' experiences and needs. Most caregivers in the United States are women. Typically, mothers, daughters, and daughters-in-law provide the day-to-day care of people with special needs. When considering the consequences of deinstitutionalization and community-based care for people with disabilities, be aware that the responsibilities of caregiving primarily fall to women. Not only are mothers more likely to provide primary care for children and adolescents with disabilities, they are also more likely to provide the primary care for adults and older individuals with disabilities at the same time that they are experiencing the effects of aging. Some mothers report that they feel robbed of their retirement or quiet years as a result of this caregiving (Valentine et al., 1997).

Very little is written specifically about the unique feelings and experiences of fathers of children with disabilities (McConachie, 1982), although occasionally workshops and conference sessions are designed "for Dads only" (Meyers, 1986). Human services workers should be vigilant in efforts to include fathers in all aspects of services. Fathers should be invited to all meetings pertaining to their children with disabilities. Accommodations should be made for the work schedules of fathers just as accommodations are made for the work schedules of mothers (Valentine, 1988).

Parents of children with developmental disabilities also report specific sources of stress and support to their families. After interviewing 25 families, Valentine (1993) reports that the three most stressful relationships that parents experience are those with the child's school, the maternal extended family, and the paternal extended family. Significant sources of support included the father's employment, the maternal family, the church, the child's school, and other providers (*i.e.*, respite providers). It is interesting to note that for some families, school and the maternal extended family were significant stressors, while for other families they were major sources of support. Assessing potential sources of support and stress is an essential ingredient of working effectively with parents of children with disabilities. Parents who report being emotionally isolated from their family, friends, and community experience severe stress and exhaustion and report feeling overwhelmed (Valentine, 1993).

Realize that the stresses of caring for a son or daughter do not necessarily diminish as the years go by. Parents are life-long caregivers. When most parents launch their sons and daughters into independent living, many parents continue to provide day-to-day care of their children with developmental disabilities. McDermott, Valentine, Anderson, Gallup, and Thompson (1995) interviewed parents of adult sons and daughters with mental retardation to identify the burdens and gratifications that are experienced. Parental responses indicated that regardless of whether their adult child resided in the home with them or in an out-of-home placement, parents experienced social, emotional, and objective burdens. In a further analysis of this data, the researchers report, however, that mothers who are African-American report fewer burdens and a higher level of gratifications than do those who are white (Valentine et al., 1998).

Parents caring for sons and daughters with disabilities benefit from a wide range of services provided by human service workers in both the private and public sector. These services include case management or service coordination (refer to Chapter 9), parent-to-parent support, and respite care. Respite care provides day, night, or extended care and supervision for a person with disabilities during a caregiver's illness, absence, or an emergency, or when there is need for periodic relief of persons normally providing care. Respite care providers are specially trained to care for individuals with special needs in their homes, in the homes of the clients, or in a special facility.

Siblings

A person with a developmental disability does not live alone with his or her parents. Care and concern for a person with disabilities extends to other members of the family, including siblings and grandparents. Aunts, uncles, cousins, and other family members are affected as well. Siblings usually play a critical role in one's development, and sibling relationships are often strong and intense. The brothers and sisters of children with developmental disabilities have similar experiences and similar needs. They typically discuss the same general feelings, joys, and challenges (Powell and Ogle, 1993). But the experience of each sibling is also unique. Powell and Ogle (1993) suggest that one of the most important things that helpers can do is to listen to these brothers and sisters. Not only can much be learned from siblings, but helpers can provide them with an opportunity to share their thoughts and feelings and receive support and information. Shapiro (1983) suggests that human services workers ask siblings of children with developmental disabilities some of the following questions:

1. How did you first learn about your sibling's disability?
2. Do you have any special responsibilities to care for your sibling that are directly related to his or her disability?
3. Do you volunteer to help your brother or sister, or are you required to help?

4. Do you think the relationships between members in your family are closer or farther apart because of your sibling's disability?
5. Are you included in making plans for your brother or sister?
6. Do you ever worry about the time when you might be required to assume full responsibility for your sibling?
7. Does your sibling's disability ever affect your social life or your relationships with your friends or neighbors?
8. Does having a brother or sister with a disability affect your future goals or plans? In what ways?

Think About It #82

If you know a person who has or had a sibling with a disability, ask if he or she would be willing to be interviewed. Ask the interview questions above. (Be sure to use the interviewing skills you learned earlier in this book.) Take brief notes, and share your findings in a small group with other people who also conducted a similar interview.

Grandparents

The importance of intergenerational relationships has received increasing attention in the past several years. As noted earlier in this chapter, the extended family members of parents of a child with developmental disabilities are important sources of support and significant sources of stress. In addition, grandparents are increasingly more likely to be primary caregivers of children with disabilities now than in previous decades. More grandparents (especially grandmothers) are taking care of children born substance affected if it is determined that the parents are not capable of providing adequate care. Involving grandparents as potential supports acknowledges their importance in the lives of family members.

Interdisciplinary Practice in the Disabilities Field

No single discipline has the required breadth and depth of expertise and resources to alone provide adequate services to people with developmental disabilities (Drew et al., 1988). Meeting the needs of people with disabilities and their families requires expertise in medicine, social work, psychology, case management, nursing, education, physical therapy, occupational therapy, law, speech and language pathology, vocational rehabilitation, information technology, nutrition, audiology, psychiatry, behavioral specialists, and other human service workers. Thus, services to people with developmental disabilities are best delivered using an interdisciplinary approach. Andrews (1990) defines interdisciplinary practice as "the process by which the expertise of different categories of professionals is shared and coordinated to resolve the problems of clients" (p. 1479).

Interdisciplinary practice is group practice. According to Rokusek (1995), all members of an interdisciplinary team must share the ability to:

1. Understand a common professional language.
2. Loosen job and agency boundaries.
3. Understand the delivery system(s) available and remain open to all available resources.
4. Communicate openly and effectively to peers and others in and outside of the work environment.
5. Integrate their professional abilities and unique personal qualities into the team, and recognize the culture, values, traditions, knowledge, training, personal emotions, and experiences that the other members bring to the team.
6. Work well in teams and contribute towards building consensus (pp. 5–6).

In addition, Rokusek (1995) maintains that members of well-functioning interdisciplinary practice teams must have a commitment to their own discipline, respect for the expertise of team members from other disciplines, an ability to look at the whole person being served, and recognition that interdisciplinary practice ultimately benefits the consumer.

Interdisciplinary practice or teamwork is frequently used for serving individuals with developmental disabilities. Typically one person is designated to facilitate or lead the team. In medical settings, the team leader may be a physician.

Human service workers are important members of the interdisciplinary team. Frequently, human service workers are the team members who have the most contact with a person with developmental disabilities. Human service workers may work as care workers or aides in either residential or day-care settings. As direct-care workers, human service providers are in a unique position to provide other interdisciplinary team members with crucial information about a client's daily living needs.

Wrapping It Up

While there are challenges associated with working with adults or children with disabilites, there are also joys and rewards. Human services helpers in this field should acknowledge the special needs, considerations, and struggles of people with disabilities, while at the same time advocating for and supporting their rights to be treated as normal and to fully participate in community life.

You may have already decided that you wish to work with people with disabilities or their families. You may have experience in the workplace or as a family member, friend, or neighbor of a person with a developmental disability. On the other hand, perhaps you feel that you are not prepared to specialize in the disabilities fields. Maybe you feel nervous or uncomfortable around people with disabilities. Maybe you think that there is not much hope or progress possible for people with developmental disabilities, especially if the cognitive or physical impairments are severely limiting. Even if you hold these ideas, the majority of human services helpers work with people with disabilities during many points during their careers. It may be therefore useful for you to learn about this issue in order to be prepared to provide appropriate services. If you are unsure of your readiness to provide support or services to someone with retardation or cerebral palsy, for example, perhaps you could volunteer some time in a community center or residential care facility where you could get to know some people with those disabilities. It is always a rewarding experience to learn about people who may be different from oneself.

CHAPTER SIXTEEN

Loss and Grief

In most situations, helpers and the persons with whom they will be working have experienced a loss. Parents of a child with a disability experience the loss of their imagined child. The caretaker of a person with Alzheimer's loses a capable spouse, friend, or parent. The person with symptomatic HIV infection anticipates the loss of health and the potential loss of life. This chapter helps identify one's own losses and the losses of others. Education about the process and phases of grieving is an important component in both identification and subsequent normalization of feelings and reactions surrounding a loss. Throughout this chapter the reader has the opportunity to examine his or her own reactions and to gain skills for working with others who are grieving. Ways in which you can facilitate your own grieving processes and assist others in their grief are also suggested.

What Constitutes a Loss?

Life is a process of comings and goings, changes that enrich and irrevocably alter our lives. Babies are born; families move; children enter school; people graduate; promotions are awarded at work; grandparents die. Some changes are expected; others are complete surprises when they occur.

When conception, pregnancy, childbirth, and early infancy proceed without any complications, one tends to assume that his or her child will develop normally into a healthy, successful adult. The family faces an unexpected loss when the child who has developed "normally" begins regressing at about 10 months of age as a result of Rett's syndrome. Similarly, when a woman of 60 in apparent good health and enjoying her career, her grandchildren, and her newfound time together with her husband suddenly has a heart attack and dies, her family experiences an unexpected loss.

Some losses occur at a time that seems to be within the realm of expectations and appropriate for the age or stage of life, such as the death of a man of 83 who believes that he has lived a full life. Other losses are "out of time" (Neugarten and Hagestad, 1976), meaning unexpected. Examples are the end of a marriage or the death of a child.

Losses may occur from positive and desired events, such as childbirth and marriage. When one marries, the single life is lost, as the coupled life begins, and the grief that results from the transition can surprise people. When a child graduates from high school and goes on to college, even though this is an expected and developmentally appropriate event, the student and the parents may grieve.

The length of time over which a loss occurs varies, too. When a house burns down, the trauma occurs quickly. However, a couple dealing with infertility might experience a loss each month over a period of years, when, time after time, they lose the opportunity to achieve pregnancy.

Individuals may experience a loss, such as a leg being amputated. Couples grieve because of infertility or other shared losses. Families mourn the loss of a home when they move. But losses are not limited to individuals and their families. Communities, nations, and the earth also experience losses that result in grieving. A bombing in Oklahoma City, a typhoon in Bangladesh, a drought in central Africa, a nuclear accident in Chernobal, a flood in West Virginia, or an airplane crash in Charlotte can result in a grief response across neighborhoods, states, and nations.

Finally, grieving can occur prior to a loss, as well as after the identified loss happens. In "anticipatory," or "preparatory" grieving, the immediate loss is the change in one's expectations for the future. When a boy in the seventh grade learns that his family is moving to another state in the middle of the semester, he reacts to his anticipated changes by becoming sad at the thought of leaving his friends, angry at having no choice, and scared at the thought of entering another school and having to make new friends. A person with HIV infection thinks about immediate changes in health, relationship gains and losses as people around him or her learn about the diagnosis, changes in body image and bodily functioning, and ultimately, his or her mortality.

New losses can trigger previous memories as well. Grief about old losses can be reignited when a current event triggers old feelings. For example, when you hear about a friend who is very ill, it may trigger some of the sadness you experienced when your grandmother was ill. Hearing about a death in a coworker's family can bring up intense grief for you about the death of your brother.

Losses are not good or bad. They just are. Loss results from change, and change is inherent in living.

Think About It #83

1. List some of the losses you have experienced in your life.
2. Were these losses unexpected or expected? Sudden or long-term?
3. How do these experiences affect your life now?

The Grief Response

How do people respond to loss? Any change is a loss, creating disequilibrium and requiring adaptation by individuals, couples, families, or communities. That disequilibrium, both anticipating a loss and following one, is called *grieving* or *the grief response*. The grief response has some common characteristics across people and across types of loss. At the same time, each person grieves in an individual and unique fashion, depending on a variety of factors, such as the type of loss, the developmental stage of the person, and characteristics of the support systems.

Phases of Grief

Many writers talk about "phases" or "stages" of grief, suggesting that people experience common reactions to loss (Kubler-Ross, 1975; Lindemann, 1979; Sanders, 1989). Although common reactions can be identified, not every person experiences all phases or stages after each loss, and reactions do not occur in a particular order or by a specific time frame. However, identifying the most common phases allows the helper and the person experiencing the loss to normalize the grief reaction. It also enables a person to determine why he or she might be experiencing a particular feeling and thinking certain thoughts. What might seem crazy under other circumstances makes sense in this context. For instance, people who are grieving frequently think they see or hear the person who has died. Although this seems very unusual, it is apparently a normal part of the grieving process. The following discussion of phases of grief, an adaptation of John Schneider's (1984) theoretical framework for explaining grieving, provides one way to make sense out of the grieving process.

Think About It #84

Remembering early losses and your reactions to those losses may give you clues as to how you react currently and why. Think back to the first loss that you can remember in your life.

1. Put yourself back in the time and place of that first loss. What do your surroundings look like? Where are you? Who is around you? What smells, sounds, and other senses are you aware of? How old are you? What are you doing? What are your feelings?
2. Now, jot down words to help you remember the scene, or describe the scene on a tape recorder.
3. Do you have similar feelings now when you experience a loss? Do current losses remind you of this early loss or of other earlier losses?

Initial awareness. When a person first acknowledges that a loss has occurred, he or she may experience a rush of adrenalin or heightened energy and strength. One woman reports that she hit the pediatrician when he diagnosed her adorable 1-year-old daughter with muscular dystrophy. The impulse to hit someone, or, conversely, to run away, may move one to act in a way that others define as unacceptable. For instance, the teenager who went on vacation after hearing that her father had acute leukemia is criticized by other family members, but this is a predictable response for this stage in the grief process and her developmental stage.

Frequently, after the initial shock wears off, one finds oneself completely exhausted, unable to move. After the resolution of an altercation among a 19-year-old, her mother, and her parents-in-law about who was going to care for her infant, the mother of the 19-year-old, who had been the strong person in finding a solution, collapsed. She sank to the floor, unable to move; her muscles wouldn't work anymore. The popular media enjoy portraying this phase. For instance, in *Star Trek II: The Wrath of Kahn,* Admiral James T. Kirk discovers that Mr. Spock has entered the nuclear center of the ship. Kirk first attempts to rush in and save Spock. After he is restrained by other crew, he then places his hands on the wall of the nuclear center and slowly collapses to the ground, as if his legs no longer work (Sallin, 1982).

Limiting awareness. Because the intense emotions following a loss can be overwhelming and prevent one from functioning, people frequently enter a phase where their awareness of the loss and its ramifications is limited. In this stage people attempt to deny that the loss has occurred; bargain, promising to give up whatever they enjoy the most if the loss will just go away; and rage that such an unjust event could occur (Kubler-Ross, 1975). Sometimes people describe feeling numb. An individual may try to diminish the impact of the loss. Although some view denial as dysfunctional, this phase does allow people to accomplish necessary tasks.

People attempt to limit awareness in two ways, by denying the importance of the loss (diminishing the loss or letting go) and by assuming that if they work very hard, they can prevent the loss from happening. Denying the importance of the loss occurs in even the most obvious situations. For instance, a young woman, mother of two preschoolers, commented the week after her 27-year-old husband died of cancer, "Now I can take his ugly pictures off the wall." On the surface, this seems like an incredibly insensitive comment by a superficial person. In fact, diminishing her loss allowed her to move through an incredibly difficult time. She had to cope with two preschoolers, alone, in a new home in the country, after not having worked outside the home for four years, without the husband that she loved, and in the face of no insurance money to cover expenses. Had she actually acknowledged all that she had lost at that moment, she would surely have been immobilized, which would not have helped her children. A woman of 80 went so far as to pretend that her husband, dead for ten years, was not really gone permanently. She said, "I know he is dead, but it helps me get through the week to pretend that he is only on a sales trip."

The second way of limiting awareness, working hard to prevent the loss (holding on), also mobilizes one to accomplish necessary tasks. When parents receive a diagnosis of a disability for their young child, they frequently seek other assessments and diagnoses. They are motivated to try many different interventions to "cure" or "prevent" the disability. This energy allows them to complete necessary but seemingly overwhelming tasks, such as many visits to doctors and other specialists; acting as physical, speech, and occupational therapist with their child; becoming nurse rather than family member; taking the role of teacher; and facing the questions of friends, families, and the person at the checkout stand of the grocery store about why their child acts that way. In "Down to the Last Tear Drop," Tanya Tucker sings about holding on to behaviors during a relationship's ending, "I have been rearranging chairs in a ship that's going down." We frequently "rearrange chairs" in the face of overwhelming evidence that we can not prevent the loss; that it, in fact, has already occurred. A mother of a teenager with a chronic mental illness, when encouraged to join the Alliance for the Mentally Ill, responded, "That doesn't apply to me."

Awareness of loss. Eventually, a person who is working very hard to avoid thinking and feeling finds himself or herself depleted of energy and resources, unable any more to prevent the awareness of the magnitude of the loss. Denying the importance of the loss is no longer possible. At that point, the person experiences deep sadness. This phase is probably the time that most people think about when they hear the word "grieving." The impact can be seen and felt physically, emotionally, spiritually, and cognitively. The body feels broken. One's chest aches, as if from a broken heart. Aches and pains throughout the body are common during this period. Energy is depleted; even getting out of bed seems like an impossible chore.

Sadness is the overriding emotion. Feelings of hopelessness and helplessness pervade one's life. Sometimes others avoid the person who is grieving, because they have become tired of hearing about the loss during this period. During this phase, people are unable to think about the future and to plan ahead. They find themselves forgetting simple things. A physician stated that after his wife's death, he would go from one examining room to the other and not remember why he was there. People who are grieving frequently feel abandoned by whatever belief system they have been using to explain why things happen the way they do. This questioning of one's spiritual belief system may be another loss in itself.

Gaining perspective. A person may move back and forth between limiting awareness and having an intense awareness of loss. He or she may even intentionally set up this cycling back and forth. In "The Center of the Night," Jayne Blankenship (1984) describes this process as "taking out the loss"; after her husband's death, she would take out an object that reminded her of her husband, feel very sad, and then put it away, along with her feelings, until she was able to feel again. This allows one to tolerate experiencing the many intense emotions initially engendered by loss, such as sadness, anger, loneliness, and despair.

Eventually, though, hope and thinking about the future play a bigger and bigger role in one's thoughts, actions, and feelings. People begin gaining perspective: figuring out what was lost, experiencing appropriate sadness about the loss, and identifying what was gained. Although initially it seems crazy to think that a gain could follow from the death of a loved one, the birth of a child with a disability, the loss of a breast through cancer, the loss of a home, eventually people find themselves telling others about the wisdom, insight, courage, and strength they have gained through the experience. For example, in spite of the tragedy surrounding Magic Johnson's revelation that he is HIV positive, he and others have identified gains. Perhaps his experience has encouraged others to use preventive measures and will result in an increased public understanding for others who test positive. Duane, who is HIV positive, stated that he has learned to appreciate each day; he loves his wife, enjoys the children, is grateful for his home, and receives pleasure from working, despite his situation. Duane's wife is very sick from AIDS; they have been rejected by all family members because of their HIV status; the family has had to move frequently from one mobile home park to the other as neighbors find out about their illness; and he works as a day laborer. Yet he talks about the "blessings" in his life now because of finding out he has HIV infection.

This type of ongoing appreciation about what is important about life and relationships seems to be a common gain across individuals, regardless of the reality of their situation. People gain a greater wisdom, a greater tolerance for differences between people, a bigger way to view the world.

A woman whose son committed suicide at age 18 works as a grief counselor. She is able to offer hope for the future to other parents whose children have died because she has survived and made sense out of her own experience. Perhaps the greatest wisdom that emerges is that people do get through the grieving process somehow, and that others cannot provide a magical answer during the tough times. People do have that wisdom inside themselves; they only need a safe place and time to discover it.

Restitution and reconstruction. The previous phase, gaining perspective, is primarily an internal activity. The phase of restitution and reconstruction is external. People often seem to reconstruct their lives after a loss, regaining energy and developing a new direction for their life. Life does go on, although irrevocably changed. As part of reaching a new equilibrium, people who have experienced a loss frequently participate in a public expression or action related to that loss.

Individuals frequently experience guilt as one component of the grief process, whether or not they have any realistic reason for such feelings. They think through what part they played in creating the loss and what their relationship was like with what was lost. One common and useful way to respond, perhaps making restitution, is through working with others in a similar situation. One who has gained wisdom through experience has much to offer to others. People give in a number of ways, acting as volunteers in a medical setting helping other people with cancer, donating to a children's hospital, or organizing and managing a self-help group.

Transforming the loss. The final phase in this model consists of transforming the loss. This stage may result from the combination of losses that are experienced throughout life, rather than a specific loss. In this sense, transforming loss becomes one's life work, with the goal being personal growth (Kushner, 1986; Moustakas, 1961; Satir, Banmen, Gerber, and Gomori, 1991; Viorst, 1986). It seems, though, that when one has reached a plateau that makes sense, not just of one loss but of life, another loss occurs, and the grief process is experienced again. Perhaps the gain in aging is the knowledge that losses cannot be prevented, the losses cannot be subverted and life continues. Life satisfaction increases with age and is high in old age, when it seems that people would have lowered satisfaction due to their many losses (George, Okun, and Landerman, 1985). Perhaps the wisdom learned by living and knowing that we survive is the reason for the increased satisfaction.

Think About It #85

Think about a major loss in your life, and try to remember whether you experienced the phases of grief as discussed above. Identify the thoughts and feelings that characterized these stages in your experience. Use this knowledge to increase your empathy for others who are grieving.

The Grief Response

Grief reactions appear to happen in a spiraling motion. An individual may experience a particular phase over and over again but at a different place in his or her life. Reexperiencing a feeling of sadness years after one's father has died may not be regression or backsliding, but a view of the loss from a new and different perspective. The grieving results from realizing new losses resulting from an earlier event, and the grief process is reexperienced, not necessarily because the grief work was not done correctly the first time, but because of the realization of new losses. Crying at an important holiday three years after one's grandmother's death is not the result of not having grieved immediately after her death but of experiencing the sadness of Grandma's lack of presence right now.

   In addition, new losses or events may rekindle a sense of loss and therefore the grief process from previous losses, such as a parent who reexperiences strong feelings about his or her own child's death when he or she begins to work with parents in a similar situation. Commonly, when grieving a current loss, people remember other similar times, and feelings about current losses may be intensified as a result. People frequently find themselves wanting to talk about previous traumas in their lives while they are grieving a current loss.

Factors That Affect the Grief Response

People experience the grief process in their own individual way. Factors that may mediate, or influence, the grief process include individual characteristics, the nature of the particular loss, and external factors, primarily the level of support received.

Individual Factors

What people value the most, and therefore what they grieve the most when lost, may vary by gender. Also, males and females may express their grief differently. For instance, couples interviewed about their infertility routinely reported gender differences (Conway and Valentine, 1988). Men stated that everything would just be better if their wives would stop talking about the loss and get involved in other activities; women stated that they would feel better if their husbands would stop going off to play golf and would talk with them. One husband suggested that he and his wife experienced different losses. He lost his hoped-for child, someone with whom to play games, a person to carry on his genes. His wife experienced other losses, such as never wearing maternity clothes, or experiencing childbirth or breast feeding. These losses have social implications for women; when they were with friends, his wife was left out of conversations with other women about their experiences of pregnancy and childbirth.

Issues important to an individual also vary by their individual life-cycle stage and their family's stage in life-cycle development. Women who have diabetes, for example, may experience that loss differently according to their ages and developmental tasks. A 17-year-old is perhaps thinking about her future career and love relationships; the 70 year old is perhaps more concerned about her grandchildren's reactions to her illness.

Characteristics of the Loss

Individuals grieve differently depending on the type of loss experienced and the relationship between the individual and the loss. It appears that the most complex grief results from the death of a child, at any age, and the death of a partner when the couple is young. In addition, when the relationship with the loss is ambivalent, grief is more likely to be protracted with increased difficulty in making sense out of it. When the situation or illness is stigmatized (like suicide or HIV infection), and the bereaved person or family hides the circumstances of the death from potential sources of support, then the grief can be more complicated and painful.

A sudden loss may be more difficult, also. Young widows whose husbands die suddenly are less likely to remarry than those whose husbands died following a prolonged illness (Strobe and Strobe, 1987). When the loss slowly develops over a long period of time, an individual may not be able to identify just what the loss is or when it has occurred. A child of 4 has generalized developmental disabilities that have no apparent beginning or diagnosis but developed over a three-year period. Her mother described their process as, "We just slid into this. Each time something new developed, we just did what we had to do. What have we lost, and when do we grieve?" They have experienced and will continue to be on a roller coaster of emotions.

The grief process is altered when several losses occur together; although little is known about the reaction to multiple losses, people may put feelings on hold until they have the time to address them, thus lengthening the process. A young couple with two preschoolers separated the day after Christmas. The woman and the two children moved to another state. Two days later, the woman's father was burned severely and hospitalized for three months. She then entered graduate school. The immediate responsibilities required by the multiple life changes seemed to postpone her grief reaction to the separation and divorce.

When people are anticipating their own deaths, they may move through to an understanding of their life and death. Others who love them and anticipate their death have a much more difficult time achieving any sort of reconciliation or equilibrium until after the death has occurred. How can one give up hope for life for a loved one? Therefore, the grief process is mediated by whether the grief is anticipatory or not.

Think About It #86

Construct a time line of major events in your life. This may be done at one setting, taking approximately 30 minutes, or over a period of days. Draw a line across a large sheet of paper to represent your life. You may wish to indicate highs and lows in your life on the line. Record important dates and events over your life.

Which have been the most exciting times? the happiest? the hardest? the saddest? the most rewarding?

Overall, what feelings are you left with about your life after completing the life line?

Coping Strategies

How do people cope with the grief reaction? They can allow others to comfort them, honor the loss with a ceremony, work on accepting those things that cannot be changed, and turn to resources that help them make sense of their grief experiences.

Allowing Others to Help

Allowing others to assist with everyday activities and supporting others in their decisions to allow others to help is an important component of the healing process. Assistance might include respite care for a person with a disability or life-threatening illness or a meal that a neighbor delivers after the birth of a child. Allowing others to help is sometimes difficult. It often hurts one's pride and sense of independence. People in the United States frequently have the idea that all persons should be able to take care of themselves, all the time, alone. If someone cannot do so, then they are somehow diminished or disrespected.

A family with a teenager who had cystic fibrosis faced losing their home as the young woman became more ill and required constant care. A community-based program had the capacity to provide in-home care for her so that the parents could continue working and not lose their home. The parents felt extremely uncomfortable about going through the process of application and approval, however. Only people who could not care for themselves used "welfare," and this family was self-sufficient!

Social support systems, a crucial factor that enhances one's ability to move through a trauma and the resulting grief response, may include family members, friends, organizations in the community, public agencies, helpers, and even strangers who "have a heart." In some situations, family members and friends may not be able to help, because they are experiencing their own grief reaction to the loss. Parents of a child born with a chronic illness become aware that their parents are worried about their grandchild and feeling sad. A husband is asked by nursing staff to comfort his wife when their son died; no one thought to provide him with comfort, and all he could do was hit the wall with his fist. Friends and family may become tired of listening to a repetitive story.

Expanding one's support system through a volunteer, professional counselor, or support group also helps. Examples of support groups include Compassionate Friends, the National Association for the Mentally Ill, Parents without Partners, Parent Care, Parents Anonymous, Council on Adoptable Children, Resolve, and the many local groups that spring up when a need arises, such as Healing Wings, a support group for young widows; a support group for siblings of children with developmental disabilities; and Unconditional Support, a support group for people whose loved ones have AIDS. Professional counselors can be located through local hospice programs and community mental health centers, Family Service America, local chapters of the National Association of Social Workers, the Association for Marriage and Family Therapy, and the yellow pages of the local telephone directory.

Think About It #87

Think about someone whom you might help with loss and grief. Identify potential sources of community support for them. How might you assist the person in accessing those resources?

Ceremonies and Symbols

Ceremonies can help to make the grieving process easier. Funerals and memorial services are the traditional way in which ceremonies are used following a death. One 17-year-old said of her reaction to her aunt's death, "First, I felt numb; now I feel sad; I want to go the funeral so that I can cry." Many losses do not have established rituals or do not have ceremonies for important points in the grief process after the initial reaction. Therefore, people may need to develop their own ceremony when it seems that would help. A couple who experienced a miscarriage asked a minister to participate in their funeral ceremony for their child who died. A foster care worker conducted a funeral at the grave of a foster child's birth mother. The child had not been able to attend her mother's funeral two years previously, because the child had been burned critically in the same fire that killed her mother. The child needed a ceremony to celebrate her birth mother's life and a formal place to feel sad about her death. Because helpers have already moved through an experience, they may be able to suggest a ceremony that helped them, or problem solve another one appropriate for the current situation.

Acceptance

At times, acknowledging what can and can not be done is also healing. Helpers cannot make a child no longer have cerebral palsy, reverse the course of muscular dystrophy, or force someone to leave a violent relationship. Helpers cannot create a partner for an adult who is lonely and in a wheelchair, make a sister stop considering her brother with HIV as a hopeless sinner who deserves to be punished, or make the heartbreak go away when a friend dies. Giving up on the things you cannot change leaves you with what you can do: Be present, listen, touch, and care.

Media (Videotapes, Audio Tapes, and Books)

Others' words, music, and images frequently assist in the healing process. Books, audio tapes, and videotapes provide information about how to accomplish day-to-day activities. They provide suggestions for making sense out of questions about spirituality and impart a sense of universalism. For people who enjoy reading and who are at a place in the grief process where they can concentrate, books can be an excellent resource. Examples of resources are included at the end of this chapter.

Wrapping It Up

The human services helper is an important source of support at the time a loss occurs. He or she can be both a facilitator and a doer, listening while assisting in problem solving. Since loss is ever present in human lives, human services helpers will always be faced with persons who are grieving. Being familiar with the phases of grief can allow us to normalize and validate what others are experiencing. It is important that helpers face their own experiences with loss and grief so that they can be ready to assist others needing their support.

Think About It #88

As you have read this chapter and participated in the exercises, you have possibly felt many of the emotions about grief discussed as you remembered your own losses. At this point, you may wish to think about the gains you have experienced as a part of the losses experienced in your life. Find a quiet moment to relax, allow memories to wash over you, and then record the gains that come to you as you examine changes in you as a result of experiences in your life.

Resources for Helping Persons with Bereavement

Blume, J. (1972). *It's Not the End of the World*. New York: Bantam. A book for children whose parents are divorcing that normalizes the process for children, late elementary or early middle school age. Also helps parents think about their children's reactions.

Bunin, P. A. (1977). *Do You Think We Could Have Made it?* Pasadena, CA: Newaves. A short poignant book of poems for adults who are divorcing; powerful words and pictures.

Caine, L. (1988). *Being a Widow*. New York: Penguin Books. Her first book, *Widow,* recounted her experience as a widow; this one provides practical steps in surviving widowhood.

Greenfeld, J. (1986). *A Client Called Noah*. New York: Henry Holt and Company. The third in the series by the father of a child with special needs, Greenfeld eloquently presents his and his family's experiences honestly and, sometimes, painfully.

Krementz, J. (1983). *How It Feels When a Parent Dies*. New York: Alfred A. Knopf. This book captures in picture and short stories the experiences of teenagers following the death of a loved one, in this case a parent. Useful for adults whose parents died when they were young as well as for teenagers.

Kushner, H. S. (1981). *When Bad Things Happen to Good People*. New York: Avon Books. An excellent example of the types of books available to assist one struggling with questions of spirituality when trying to make sense out of a tremendous loss.

Monette, P. (1988). *Borrowed Time*. San Diego: Harcourt Brace Jovanovich. A beautiful story about love between partners, parents and children, and friends, told by a man whose lover died from AIDS.

Richter, E. (1986). *Losing Someone You Love: When a Brother or Sister Dies*. New York: G. P. Putnam's Sons. Descriptions of childrens' emotions and feelings when a sibling dies.

Audio tapes appeal in a different way. With the burgeoning of books on tape and other talking tapes, one can listen to words while driving or engaged in other activities. Tapes might be intentionally educational, such as "Humor and Healing," by Bernie Segal, Elizabeth Kubler-Ross's "Making the Most of the In Between"; or a volume of *Pediatrics* with Mel Levine describing attention deficit disorder. Tapes designed as entertainment may also be unintentionally educational, such as the segment about AIDS on "Twenty Years with NPR" and "It's Alright to Cry," by Rosie Greer on "Free to Be . . . You and Me." Songs frequently touch emotions; one woman stated she swore off country and western songs after her divorce; they all seemed to be about lost relationships, such as, "D-I-V-O-R-C-E," "All My Exes Live in Texas," and, "She Got the Goldmine, I Got the Shaft." Suzanne Vega universalizes violence in families with "Luka." And Dave Van Ronk playfully assists with raising children's self-esteem in "I'm Proud to be a Moose." People commonly associate certain music with people and experiences; one widow felt comforted whenever "The Wind Beneath My Wings" played on the radio, because she felt close to her deceased husband at those times.

With the easy availability of videotapes, people can bring into their homes a wide range of self-help tapes, sometimes offered free through chain video stores and libraries. Helpful videos specific for people who are grieving the death of a child or sibling include *Ordinary People* and *A Family in Grief: The Ameche Family*. *Trip to Bountiful* and *Driving Miss Daisy* gently portray interaction between caregivers and an older family member.

PART FIVE

Caring for the Caregiver

This final section wraps up this discussion of the essentials practicing in the human services field. Chapter 17 encourages helpers to care for themselves and discusses what to do if supervision, support, consultation, or referrals are needed. Chapter 18 reminds learners of some important points about helping. It is hoped that studying and thinking about the knowledge, values, and skills essential for human services helping have left you feeling more prepared, inspired, and confident.

CHAPTER SEVENTEEN

Taking Care of Yourself

In order to be useful to others, a helper should first be well, healthy, and mentally and physically balanced. If you are stretched past your limits, you may have nothing left to give. Similarly, if you are not trained or prepared to address a particular helping situation, you must refer the person elsewhere and/or get supervision and consultation for yourself. This chapter first addresses how to handle being "in over your head," and then how to take care of yourself, including recognizing and addressing burnout or stress.

In over Your Head

At some point, each of us finds ourselves "in over our heads," that is, in a situation where we need more information and/or someone else to step in. Every effective human services worker or volunteer must recognize his or her limits. It is unethical to try to participate in helping activities for which you are unprepared or untrained. Even when the type of helping is one with which you are comfortable, you will often want to talk the situation over with coworkers, supervisors, or consultants. When you or your agency is not the appropriate resource for the needs of the person seeking help, you must know how to make an appropriate referral. This section begins to address how you should proceed when you are in over your head.

Getting Supervision, Consultation, and Support

An important component of being a responsible and effective human services helper is using a consultant, either through professional supervision or peer support. Consulting another is logical when the helper is feeling intensely emotional as a result of his or her work, when the helper is uncertain about the next step in working with someone, or when someone requires interventions other than those the human services helper has been trained to offer or has the resources to provide. With a consultant, human services helpers are able to examine what has happened, what they are planning to do, and how they are feeling about their work. The consultant might be another human services helper who has agreed to act as consultant, a professional with an agency that coordinates helping activities, or a professional with whom the helper has contracted for support and feedback. The consultant need not be smarter or wiser than the helper to be of assistance; the most useful consultant is a person who is a good listener and a skilled problem solver.

Working with others during a crisis sometimes stirs up strong feelings in helpers, reminding them of their own current or past situations. For instance, a woman who has been battered at an earlier time in her life and who is now acting as a helper for other women who are battered may reexperience her fear, hypervigilance, or nightmares when working with another person in current danger. A parent whose child has died and who is now working with other parents grieving the death of a child may feel sad and angry, not because of the current work, but because past memories and feelings have been tapped. Talking these feelings over in a safe situation with a consultant allows helpers to make sense out of their own emotions and reactions and to rejuvenate themselves for future work. In addition, when a person does not seem to respond to your helping relationship and efforts but seems fragile or in need, talking with a consultant helps you problem solve whether or not to remain involved and how to proceed if involvement is continued.

Illustration

Margie Chang is a 54-year-old Chinese-American woman who was widowed five years ago. Her husband died of respiratory complications after he contracted a particularly virulent strain of influenza. Margie has been a cofacilitator of a support group for widows for the past three years. It pleases her to think of all the women that she has helped to face their grief. She knows she has touched the lives of many women, and it makes her feel useful. Usually, there

---

Think About It #89

Having a supportive and safe consultant or supervisor is very important for human services helpers. Identify at least one person with whom you would consult when you have questions or concerns about your own work as a human services helper.

1. In what ways can you best utilize this consultant's expertise?
2. Will you meet with this consultant regularly or only when you feel it is needed?
3. Do you anticipate feeling comfortable calling this person and relying on her or his judgement? Why or why not?
4. Will you feel free to call your consultant at home or only during certain hours in an office?

---

are about nine participants in the group each month. Three of the members have been involved in the group as long as Margie has. For the past three meetings, a new member, Lilly Schroeder, has been attending. Her husband died less than six months ago. Lilly tends to dominate the discussions each time. It seems to Margie that Lilly talks incessantly about her experiences, her husband, her children, her hobbies, and her dilemmas. Lilly seems to answer for everything and everybody. Margie finds it very difficult to warm up to Lilly. In fact, Margie has been thinking about not going to the next group meeting just to avoid Lilly. Margie realizes that she is angry at Lilly. She sees no sadness in Lilly's constant chattering and she misses the easy give-and-take the group discussions used to have.

Think About It #90

1. Focus on the feelings of Margie Chang. What do you think she is feeling about Lilly? What do you think are some of the reasons for her feelings?
2. Focus on the feelings of Lilly Schroeder. What do you think are her feelings in the group? Can you make sense of her behavior?
3. Do you think it is appropriate for Margie to talk with someone about her feelings? Who do you think she could talk to? What are some of the issues that you think Margie should raise during this consultation?
4. If you were consulting with Margie, what would you do or say?

Making Appropriate Referrals

People who are experiencing a serious change in their lives frequently feel confused, scared, angry, sad, overwhelmed, and in need of assistance with problem solving. The human services helper is often sought out at these times. Although human services workers, students, and volunteers have important characteristics and skills to offer, sometimes the person's situation warrants intervention by other professionals (such as psychiatrists, social workers, attorneys, physicians, pastoral counselors, therapists, psychologists, and psychiatric nurses). Examples of appropriate times to involve a professional include when: a person is experiencing sadness that does not abate; a person is thinking about taking his or her own life or harming someone else; panic attacks and nightmares persist after a frightening situation; abuse of alcohol and drugs occurs; or abuse and neglect are suspected. In these situations, the human services helper's job is to recognize when a referral is necessary, determine what the most appropriate community resources are, and make these connections in an effective way. Human services helpers are very valuable in these times, because they can provide information on resources, obtain the consent and permission of the person seeking help, and encourage the individual, couple, or family to seek other appropriate services.

In addition, sometimes a person seeking help will simply need services that your agency does not provide. If you are providing case management, for example, it is your job to arrange for a variety of services through a variety of organizations. The following important guidelines will help you make referrals appropriately.

1. Know the agency, organization, or program well before you refer someone to that program. Be somewhat familiar with their intake and referral procedures, their confidentiality and consent policies, and their eligibility rules. Check out whether the referral is appropriate before you refer an individual, couple, or family. Try to refer persons to a particular individual within that agency whom you know and trust.
2. Ease the way whenever and however you can. Try to make the connection with a new resource as painless as possible. You may offer to call the agency while the individual, couple, or family is in your presence.
3. Never promise or guarantee anything about another organization. Tell persons seeking help generally what they might expect, but do not pretend to be able to make predictions or decisions for another system.
4. Always follow up with people and with the organization or professional to whom you have referred them to make sure that the connection was made and that the referral was actually appropriate.
5. If another organization has been inappropriately unresponsive to those whom you have referred, you will need to advocate on their behalf. You may offer to step in and talk to someone in the other agency about their mistreatment of the persons you have referred.
6. Coordinate services and communication with the organizations who are working with those with whom you are providing services.
7. It is never ethical to attempt to dump a person seeking help into another service system because you do not want to deal with him or her. Never refuse a request for help without making a referral or developing alternatives. It is not acceptable simply to say, "Sorry, I can't help you." You must make an effort to secure help from somewhere.

Illustration

Kim is a case manager with a community-based AIDS-service organization. Bill Cullum calls the agency, seeking information about current treatments that are available or recommended. He says that he tested positive for HIV over a year ago and has been thinking recently about taking better care of his health. Kim makes an appointment to meet with Bill in the office so that she can better determine what resources might best meet his needs. When he arrives, she furnishes him with a list of clinical trial programs in their area, a list of infectious disease physicians, and some information published by various pharmaceutical companies about their products. She then asks him if he will sit down and talk with her for a little while, and he agrees. Kim realizes during their conversation that Bill has several concerns that he is interested in addressing, but that her agency does not directly provide these services. For example, he is interested in seeing a therapist so that he can begin to talk about his anger at his alcoholic fa-

ther. He is terrified about the prospect of depending on his parents for personal care when he gets seriously ill, because they were always unavailable, neglectful, and verbally abusive to him when he was a child and adolescent. Bill also wants to look into substance use treatment programs for himself, because he is concerned about his own heroin use and would like to get clean. Finally, he has not yet seen a doctor and needs help with connecting to an affordable and sensitive resource for ongoing medical care. After exploring with Bill his thoughts and feelings about available community resources and his ability to pay for services, Kim develops the following referral plan with him: (1) the community mental health center for counseling; (2) the county alcohol and drug abuse agency for help with addiction concerns; and (3) the federally funded HIV clinic for medical care. Kim obtains written consent from Bill to speak with each of these agencies on his behalf, gives him names and telephone numbers for contacts in each organization, and sets a time to follow up with him about how these services seem to fit for him. In addition, they agree to enter into an ongoing case management relationship. Kim promises to call him in a week if he has not called her.

Burnout and Stress Management

As we stated earlier, if basic needs are not met, it is less likely that helpers can meet the needs of others very well. It is therefore crucial that each helper be aware of his or her own emotional, physical, psychological, social, and recreational needs and whether they are being addressed. Having one's own source of strength may extend the length of time one can be an effective helper and postpone burnout. *Burn-out* is a common term for feeling that challenges outweigh one's resources. *Stress* means pressure or strain that we are not well prepared to handle. When you are burned out and your stress level is too high, you may feel sad, out of control, irritable, or extremely fatigued. You may act rigid or may dread hearing from those who are depending on you.

What Causes Burnout?

Human services workers and volunteers can begin to feel depleted for many reasons. Some of the possible triggers are:

1. You have not been paying enough time or attention to nourishing and nurturing yourself.
2. Your own crises, life stresses, or losses have become temporarily overwhelming to you.
3. You are not getting enough support or recognition for what you do.
4. You are feeling helpless and out of control due to serious or intense problems of those who have sought help from you or because of the lack of agency or community resources to address those needs.
5. You are having difficulty effectively managing your time. You are disorganized or trying to do too much too quickly.

6. You have either overestimated your own abilities or you have set expectations and standards that are too high for you. For example, do you expect that you will be able to fix every situation? Do you expect that you will succeed in every endeavor? Do you expect that you will never make mistakes?

Think About It #91

To test whether you may be feeling burned out, complete the following checklist. If you answer "yes" to a majority of these questions, you could probably use some help with stress management.

1. Are you frequently unreasonably upset, hostile, cynical, frustrated, dissatisfied, or irritable?

2. Are you often careless, accident prone, or forgetful? Are you seeing unusual signs of short-term memory loss or inattention, such as missing your subway stop, running a red light, forgetting to set your alarm, missing an appointment, or leaving the iron plugged in?

3. Are you withdrawing from key relationships in your life?

4. Are you feeling overextended, overcommitted, or overwhelmed?

5. Do you frequently feel physically tired and mentally drained?

6. Do your emotions seem to be all on the surface, surprising you at times? Do you have rapid or unexplained mood swings?

7. Do you have less time for relaxation or recreation than you would like?

8. Have you lost some of your sense of perspective? Do you catastrophize or exaggerate relatively minor incidents?

9. Are you suffering from physical complaints, such as headaches, insomnia, gastrointestinal problems, panic attacks, skin problems, or frequent colds? Have you experienced a change in sleep patterns, appetite, or weight?

10. Have you set unrealistic standards or goals for yourself?

11. Do you dread going to work or having contact with service recipients or coworkers?

12. Are you generally sad or depressed?

13. Do you seem to be working harder but accomplishing less?

14. Have you lost your ability to laugh at yourself?

15. Do activities you once enjoyed no longer interest you as much?

Stress Management

Being aware of and taking care of our own needs are especially important as helpers because many of the individuals and families we will work with are not taking care of themselves very well. If a mother is meeting the needs of a child with a traumatic brain injury, for example, it may be difficult for her to find the time to go to the doctor or dentist for a preventive check-up for herself. If a daughter is caring for her elderly and medically fragile mother, getting regular exercise may be extremely difficult or may be deemed a low priority. If helpers are to continue to meet the needs of others, however, they must take the time to stay emotionally, psychologically, socially, and physically healthy. Helpers can then be more effective and model the importance of self-care for others.

The following strategies can help you to manage your stress reactions and be more prepared to cope with whatever comes your way.

Know yourself. It is useful if you have insight into your feelings, habits, limits, strengths, and weaknesses. You should be introspective and observant about your own needs and stress triggers. For instance, are there certain types of behaviors in people that tend to agitate you? Are there certain times of the day where you are likely to feel overwhelmed with tasks? Are you not doing well right now with time management? Are there specific situations that push your buttons? Are there certain job duties that make you want to scream? Once you identify your stress triggers, take steps to anticipate difficulties. You may want to take a break, manage the work in a different way, or ask for help from co-workers, supervisors, friends, or family.

Get adequate support. Making sure that you have adequate social support is a very important feature of being an available and effective helper. Social supports can have a powerful, positive influence on one's ability to handle crisis and cope well with life's challenges. It helps to share burdens. Ensure that you are receiving the help you need as a provider of support for others. Helpers need to feel supported, valued, and validated. If you are not feeling that your supervisors, coworkers, friends, lover, and/or family members are behind you, you may be more susceptible to feeling depleted. If you are lacking support in one area, strive to make up for it through another source. For example, seek out or form a support group for helpers, utilize consultants, and form your own network of other helpers who can support and listen to you.

Be realistic. Evaluate whether your expectations and standards are realistic. Try to let go of your "superworker" or "savior" complex! Do your very best, but recognize that you, the people you serve, and organizations have limits. Admit your humanity, and accept the fact that we all occasionally make mistakes or get tired. Set realistic limits with yourself, your supervisor, your family and friends, and those whom you serve. Know that you cannot be everything for everybody.

Refresh yourself. Find stress management techniques that work for you and that are not dysfunctional or destructive, and begin to implement them in your life. These techniques do not have to be elaborate; they can be as simple as drinking hot herbal tea, taking a bath or shower, walking around the block, reading a magazine or short story, watching a movie, praying or meditating, hugging a loved one, petting your cat or dog, or calling a friend who makes you laugh. Try to avoid those activities that might increase your stress in the long run, like alcohol or drug use, overeating, gambling, or going on a shopping binge.

Exercises

To help assess your own healthy behaviors, complete the following "Checklist for Caregivers" and "Stress Management Plan of Action."

Part 1: Checklist for Caregivers (Bass, 1990)

Do you exercise regularly?	Yes	No
If so how often?	_____	
Do you have a regular volunteer activity that is gratifying?	Yes	No
Do you have a hobby or another leisure-time activity that you enjoy regularly?	Yes	No
If so, how often?	_____	
Do you smoke or drink alcoholic beverages?	Yes	No
If so, how many packs/drinks per day, and has it increased in recent months?	_____	
Have you gained or lost 10 pounds or more in recent months?	Yes	No
Do you eat balanced meals three times each day?	Yes	No
If not, how often do you eat balanced meals?	_____	
Do you often experience sleeplessness or anxiety?	Yes	No
If so, what are you doing about it?	_____	
Do you have a close relative or friend with whom you discuss problems and successes?	Yes	No
If not, do you talk with a professional or someone of your religious faith?	Yes	No
Do you make and keep preventive and necessary medical and dental appointments?	Yes	No
If not, what would help you do so?	_____	
What are your goals and what are you doing to achieve them?	_____	

Part 2: Personal Stress Management Plan of Action

1. I need to start _____

_____

_____

_____

_____ .

2. I need to stop _____

_____

_____

_____

_____ .

3. I need to continue _____

_____

_____

_____

_____ .

4. What will help me: _____

_____

_____

_____

_____ .

5. What will hinder me: _____

_____

_____

_____

_____ .

6. What I will do to increase my success: _____

_____

_____

_____

_____ .

CHAPTER EIGHTEEN

Putting It All Together

This chapter summarizes some of the highlights of the personal and professional journey toward becoming a human services helper.

What Is a Human Services Helper?

What does it mean to be a human services helper? You have learned that your experiences and life challenges have given you wisdom and that this wisdom can be helpful to others in similar situations. You have learned that often caring is just as important as curing. You have learned that human services helpers are important sources of strength and support for people in crisis or experiencing difficult times. You have learned that human services workers are important complements to professional helpers and can often make vital connections to other services.

You have also learned that although many people have a human services capacity to be helpful and have a special understanding of people, there are skills, attitudes, abilities, and guidelines that make each one of us better and more effective helpers. Effective human services helpers have the following qualities, among others:

1. Human services helpers have certain *attitudes and beliefs* that promote well-being in others, including:

   - **Seeing strengths first.** Seeing strengths means being primarily aware of the abilities in people rather than focusing solely on their limitations or deficiencies.
   - **Suspending judgment.** It is important for human services helpers to put away their biases and listen without criticism to the different ways that others cope with life challenges. This includes supporting the informed choice of those who ask for their support.
   - **Being honest and trustworthy.** The human services helper must be truthful in all interactions; the reliability and honesty of human services helpers

will solidify the relationship between the helper and the person seeking care and increase the likelihood that help will be accepted and effective. An effective helper is genuine and strives to use him- or herself in an honest and congruent way.

- **A commitment to empowerment.** The human services helper who can encourage others to take charge of their lives and become partners in service delivery systems will decrease feelings of hopelessness and helplessness in others and increase feelings of competence.

- **Being aware of diversity and culture.** This includes being aware of attitudes toward differences. Effective human services helpers are aware of their attitudes toward people, especially those people whom they perceive to be different in some way. Human services helpers are aware of stereotypical attitudes that they have toward people of other races, religions, ages, sexual orientation, or gender, and work toward preventing these attitudes from interfering with their roles as human services helpers.

- **Being aware of, appreciating, and respecting developmental stages.** Help is more effective if the human services helper has an understanding of typical individual and family development, both stages and tasks. Supporting and encouraging typical development is an important job for human services helpers.

2. Human services helpers have a system of *ethics and values* that promote the well-being of others, including:

- **Being aware of personal feelings and values.** Imposing personal values on others is not the role of the human services helper. Awareness of their personal belief systems will prevent human services workers from limiting or restricting the choices of others. When a human services helper becomes aware of feelings toward others that might get in the way of helping, it is important that these feelings be evaluated so that they do not interfere with the helping relationship.

- **Maintaining confidentiality and the privacy of others.** When people seeking help from agencies trust helpers with personal information, they are required to adhere to strict confidentiality guidelines for their protection.

3. Human services workers have a commitment to appropriate *awareness of, use of,* and *care of themselves.* This includes:

- **Taking care of themselves.** In order for human services helpers to be able to give, they must be well supported themselves and have their own basic needs met. This includes watching for burnout and successfully managing stress.

- **Using consultation and supevision.** Asking for and accepting support and assistance is an important feature of effective human services helping. Consultation and supervision should be available and used with regularity.

4. Although human services helpers already possess many helping qualities and abilities, other helping skills can be learned and refined.

- **Establishing and maintaining a helping relationship.** Effective helping occurs within a mutual professional relationship, established through certain worker behaviors and characteristics.
- **Listening** accurately and compassionately, both verbally and nonverbally. This includes labeling and validating feelings.
- **Focusing on feelings.** Being aware of his or her own emotions as well as the feelings of others is an important skill of the human services helper. Having the vocabulary and being able to communicate understanding is an important component of helping.
- **Communicating empathy.** Communicating understanding sends the message that the helper is consciously aware and accurately perceiving another's feelings. Empathic understanding is an imperative feature of human services helping.
- **Communicating positive regard and respect.** Helpers who are most effective have the ability to let service recipients know that they accept them for who they are and accord them dignity and worth as fellow human beings.
- **Starting where the person is.** A helper should always begin any helping relationship acknowledging the person's current emotional, mental, spiritual, and physical state. A helper suspends his or her own ideas or agenda in order to focus on the person's definition of the problems and the solutions.
- **Providing relevant information.** When persons need assistance with obtaining information or education on the problems and issues which affect them, human services helpers facilitate this, acting as resource persons and making linkages.
- **Making referrals.** Human services helpers know when it is appropriate to make referrals. Some situations for which human services helpers may want to be prepared to make referrals include: suicide, mental illness, alcohol or drug use, and relationship violence.
- **Being able to respond to a crisis.** Knowing the responsibilities of a human services helper when a crisis occurs and being able to respond themselves or refer to others is an important role of human services helpers. Another element of crisis intervention is being able to respond appropriately when someone is feeling or acting suicidal.
- **Problem solving.** Sometimes people are feeling overwhelmed or temporarily unable to figure out what to do about their current situation. Teaching problem solving skills helps persons to feel empowered and in control.
- **Being assertive.** Being able to recognize their own limitations as human services helpers and communicating these limitations increases the effectiveness of the helpers. These limitations may be because of time, emotional or physical energy, or psychological readiness.
- **Understanding a variety of models, populations, and issues.** Human services helpers need to understand case management, case and class advocacy, group work, family intervention, and community organization. Human services helpers often need knowledge about certain social and personal problems so that they can more effectively respond to persons who are fac-

ing these issues. For example, human services workers, students, and volunteers need information about persons who are experiencing grief, post-traumatic stress, or relationship violence; who are older or HIV positive; or have developmental disabilities, mental illnesses, or addictions.

Think About It #92

1. Identify the most important things that you learned about yourself during this process of becoming a more effective human services helper.
2. In what ways have these new areas of awareness made you a better human services helper?
3. In what ways have these new areas of awareness made you aware of your own limitations?
4. In what ways are you ready to assume the role of a human services helper? What are your strengths?
5. In what areas might you need more information or practice? How will you continue to learn?

REFERENCES

Achtenberg, J. (1990). *Woman as healer*. Boston: Shambhala.

Adlin, M. (1993). Health issues. In E. Sutton, A. R. Factor, B. A. Hawkins, T. Heller, & G. B. Seltzer (Eds.), *Older adults with developmental disabilities: Optimizing choice and change* (pp. 29–48). Baltimore: Paul Brookes.

Ainlay, S. C., Coleman, L. M., & Becker, G. (1986). Stigma reconsidered. In S. C. Ainlay, G. Becker, & L. M. Coleman (Eds.), *The dilemma of difference: A multidisciplinary view of stigma*. New York: Plenum Press.

Alcoholics Anonymous. (1991). *Twelve steps and twelve traditions*. New York: Alcoholics Anonymous World Services.

Allers, C. T. (1990). AIDS and the older adult. *The Gerontologist, 30*(3), 405–407.

Amaro, H. A., & Gornemann, I. (1991). Health care utilization for sexually transmitted diseases: Influence of patient and provider characteristics. In J. Waserheit, S. O. Aral, & K. K. Holmes (Eds.), *Research issues in human behavior and sexually transmitted diseases in the AIDS era*. Washington, DC: American Society of Microbiology.

American Association on Mental Retardation. (1992). *Mental retardation: Definitions, classification, and systems of supports* (9th ed.). Washington, DC: Author.

American Heritage College Dictionary. (1997). (3rd ed.). Boston: Houghton Mifflin.

American Hospital Association. (1992). Case management: An aid to quality and continuity of care. In S. M. Rose (Ed.), *Case management and social work practice* (pp. 149–159). New York: Longman.

American Psychiatric Association. (1988a). *Let's talk facts about mental illness: There are a lot of troubled people*. Washington, DC: Author.

American Psychiatric Association. (1988b). *Let's talk facts about schizophrenia*. Washington, DC: Author.

American Psychological Association. (1994). *Diagnostic and statistical manual of mental disorders* (4th ed.). Washington, DC: Author.

Anderson, D., & Polister, B. (1993). Psychotropic medication use among older adults with mental retardation. In E. Sutton, A. R. Factor, B. A. Hawkins, T. Heller, & G. B. Seltzer (Eds.), *Older adults with developmental disabilities: Optimizing choice and change* (pp. 61–76). Baltimore: Paul Brookes.

Anderson, J., Landry, C., & Kerby, J. (1991). *AIDS: Abstracts of the psychological behavioral literature*. Washington, DC: American Psychological Association.

Andrews, A. B. (1990). Interdisciplinary and interorganizational collaboration. In L. Ginsberg, S. Khinduka, J. A. Hall, F. Ross-Sheriff, & A. Hartman (Eds.), *Encyclopedia of Social Work* (18th ed., 1990 supplement, pp. 175–188). Silver Spring, MD: NASW.

Asch, A., & Mudrick, N. R. (1995). Disabilities. In R. L. Edwards & J. G. Hopps (Eds.), *Encyclopedia of Social Work* (19th ed., pp. 752–761). Washington, DC: NASW.

Baker, L. S. (1992). The perspective of families. In M. L. Stuber (Ed.), *Children with AIDS*. Washington, DC: American Psychiatric Press.

Barker, R. L. (1987). *The Social Work Dictionary*. Silver Spring, MD: NASW Press.

Barnett, A. W., Miller-Perrin, C. L., & Perrin, R. D. (1997). *Family violence across the lifespan*. Thousand Oaks, CA: Sage.

Baroff, G. S. (1991). *Developmental disabilities: Psychosocial aspects*. Austin, TX: Pro-Ed.

Bass, D. S. (1990). *Caring families: Supports and interventions*. Silver Spring, MD: NASW.

Belsky, J. K. (1984). *The psychology of aging: Theory, research, and practice*. Pacific Grove, CA: Brooks/Cole.

Benedek, E. P. (1989). Baseball, apple pie and violence: Is it American? In L. Dickstein & C. Nadelson (Eds.), *Family violence: Emerging issues of a national crisis* (pp. 1–13). Washington, DC: American Psychiatric Press.

Benjamin, A. (1969). *The helping interview* (2nd ed.) Boston: Houghton Mifflin.

Berkman, L. F. (1984). Assessing the physical health effects of social networks and social support. *Annual Review of Public Health, 5,* 413–432.

Birdwhistell, R. L. (1970). *Kinesics and context*. Philadelphia: University of Pennsylvania Press.

Blankenship, J. (1986). *In the center of the night*. Toronto: PaperJacks.

Blazer, D., & Palmore, E. (1976). Religion and aging in a longitudinal panel. *The Gerontologist, 16*(Pt. 1), *82–85.*

Bloom, M. L. (1973). Usefulness of the home visit for diagnosis and treatment. *Social Casework, 54,* 67–75.

Blume, S. (1988). *Alcohol/drug dependent women: New insights into their special problems, treatment, recovery*. Minneapolis, MN: Johnson Institute.

Boland, M. G., Czamiecki, L., & Haiken, H. J. (1992). Coordination care for children with HIV infection. In M. L. Stuber (Ed.), *Children with AIDS*. Washington, DC: American Psychiatric Press.

Borkman, T. S. (1990). Experiential, professional, and lay frames of reference. In T. J. Powell (Ed.), *Working with self help* (pp. 3–30). Silver Spring, MD: NASW.

Bower, B. (1997). Social links may encounter health risks. *Science News, 152,* 135.

Brammer, L. M. (1979). *The helping relationship: Process and skills*. Englewood Cliffs, NJ: Prentice Hall.

Brieland, D. (1982). Introduction. In M. MaHaffey & J. W. Hanks (Eds.), *Practical politics*. Silver Spring, MD: NASW.

Butler, R. (1969). Ageism: Another form of bigotry. *The Gerontologist, 9,* 243–246.

Cadwell, S. (1994). Twice removed: The stigma suffered by gay men with AIDS. In S. A. Cadwell, R. Burnham, & M. Forstein (Eds.), *Therapists on the front line: Psychotherapy with gay men in the age of AIDS* (pp. 3–24). Washington, DC: American Psychiatric Press.

Caine, L. (1988). *Being a widow*. New York: Penguin Books.

Caplan, G. (1964). *Principles of preventive psychiatry*. New York: Basic Books.

Cash, T., & Valentine, D. (1987). A decade of adult protective services: Case characteristics. *Journal of Gerontological Social Work, 10*(3/4), 47–60.

Centers for Disease Control. (1994). *HIV/AIDS surveillance report.* Atlanta, GA: Centers for Disease Control and Prevention.

Centers for Disease Control. (1995). *HIV/AIDS surveillance report.* Year-End Edition, 7(2). Atlanta, GA: Centers for Disease Control and Prevention.

Centers for Disease Control. (1997). *HIV/AIDS surveillance report.* Midyear Edition, 9(1). Atlanta, GA: Centers for Disease Control and Prevention.

CHADD (Children and Adults with Attention Deficit Disorders). (1993). *The disability named ADD: An overview of attention deficit disorders.* Plantation, FL: Author.

Clark, M., & Anderson, B. (1967). *Culture and aging.* Springfield, IL: Charles C Thomas.

Clifton, L. (1983). *Everett Anderson's goodbye.* New York: Holt.

Cohen, F. (1983). Stress, emotion, and illness. In L. Temoshok, C. Van Dyke, & L. S. Zegans (Eds.), *Emotions in health and illness: Theoretical and research foundations* (pp. 31–35). New York: Grune and Stratton.

Cohen, S., & Wills, T. A. (1985). Stress, social support, and the buffering hypothesis. *Psychological Bulletin, 98*(2), 310–357.

Colgrove, M., Bloomfield, H., & McWilliams, P. (1991). *Surviving, healing & growing: The workbook.* Los Angeles: Prelude Press.

Conway, P., & Valentine, D. (1988). Reproductive losses and grieving. In D. P. Valentine (Ed.), *Infertility and adoption: A guide for social work practice* (pp. 43–64). New York: Haworth.

Cormier, W. H., & Cormier, L. S. (1991). *Interviewing strategies for helpers: Fundamental skills and cognitive behavioral interventions.* Pacific Grove, CA: Brooks/Cole.

Costa, P. T., Jr. (1996). Work and personality: Use of the NEO-PI-R in industrial organisational psychology. *Applied Psychology: An International Review, 45*(3), 225–241.

Cottrell, L. (1942). The adjustment of the individual to his age and sex roles. *American Sociological Review, 7,* 617–620.

Cox, E. O., & Parsons, R. J. (1994). *Empowerment-oriented social work practice with the elderly.* Pacific Grove, CA: Brooks/Cole.

Cross, T., Bazron, B., Dennis, K., & Isaacs, M. (1989). *Towards a culturally competent system of care: A monograph for effective services for minority children who are severely emotionally disturbed.* Washington, DC: CASSP Technical Assistance Center, Georgetown University Child Development Center.

Cumming, E., & Henry, W. (1961). *Growing old: The process of disengagement.* New York: Basic Books.

Dardick, G. (1993). Access activism: On disabled people and their struggle for rights and respect. *Utne Reader, 56,* 98–110.

Dawson, D., Cynamon, M., & Fitti, J. (1987). AIDS knowledge and attitudes: Provisional data from the National Health Interview Survey. *Vital and Health Statistics for the National Center for Health.* Washington, DC: Department of Health and Human Services (No. 146, Nov. 19).

Developmental Disabilities Act of 1984. P.L. 98-527, 98 Stat.2662.

Developmental Disabilities Assistance and Bill of Rights Act Amendments of 1987. P.L. 100-146, 101 Stat. 840.

Developmental Disabilities Assistance and Bill of Rights Act Amendments of 1990. P.L. 101-496, 104 Stat. 1191.

DeWeaver, K. L. (1995). Developmental disabilities: Definitions and policies. In R. L. Edwards & J. G. Hopps (Eds.), *The Encyclopedia of Social Work* (19th ed, pp. 712–720). Washington, DC: NASW.

Dolnick, E. (1993). Deafness as culture. *Atlantic Monthly, 272*(3), 37–53.

Drew, C. J., Logan, D. R., & Hardman, M. L. (1988). *Mental retardation: A life cycle approach.* Columbus, OH: Merrill.

Dudley, J. R. (1987). Speaking for themselves: People who are labeled as mentally retarded. *Social Work, 32*(1), 80–82.

Dunst, C. J., Trivette, C. M., Davis, M., & Cornwall, J. (1988). Enabling and empowering families of children with health impairments. *Children's Health Care: Journal of the Association for the Care of Children's Health, 17*(2), 71–81.

Echterling, L. G., & Hartsough, D. M. (1980). Testing a model for the process of telephone crisis intervention. *American Journal of Community Psychology, 8*(6), 715–724.

Echterling, L. G., & Hartsough, D. M. (1989). Phases of helping in successful crisis telephone calls. *Journal of Community Psychology, 17* (July), 249–257.

Education for All Handicapped Children Act of 1975. P.L. 94-142, 89 Stat. 773.

Emlet, C. A. (1993). Service utilization among older people with AIDS: Implications for case management. *Journal of Case Management, 2*(4), 119–124.

Emlet, C. A. (1996). Case managing older people with AIDS: Bridging systems, recognizing diversity. *Journal of Gerontological Social Work, 27*(½), 55–71.

Epilepsy Association of America. (1989). *Recognizing the signs of childhood seizures.* Landover, MD: EFA.

Erikson, E. H. (1963). *Eight stages of man. Childhood and society.* New York: Norton.

Erikson, E. (1982). *The life cycle completed.* New York: Norton.

Evans, D. A., Funkenstein, H. H., Albert, M. S., Scherr, P. A., Cook, N. R., Chown, M. J., Hebert, L. E., Hennekens, C. H., & Taylor, J. O. (1989). Prevalence of Alzheimer's disease in a community population of older persons: Higher than previously reported. *Journal of the American Medical Association, 262*(18), 2551–2556.

Forsyth, B. (1995). A pandemic out of control: The epidemiology of AIDS. In S. Geballe, J. Gruendel, & W. Andiman (Eds.), *Forgotten children of the AIDS epidemic* (pp. 19–31). New Haven, CT: Yale University Press.

Fowler, J. W. (1981). *Stages of faith: The psychology of human development and the quest for meaning.* New York: Harper & Row.

Freud, S. (1953, reprinted from 1905). *Three essays in the theory of sexuality* (standard edition, vol. 7). London: Hogarth Press.

Friedan, B. (1981). *The second stage.* New York: Summit Books.

Friesan, B. J. (1989). National study of parents whose children have serious emotional disorders. In A. Algarin, R. Friedman, A. Duchnowski, D. Kutash, S. Silver, & M. Johnson (Eds.), *Second Annual Conference Proceedings—Children's Mental Health Services and Policy: Building a Research Base.* Tampa, FL: Research and Training Center for Children's Mental Health, University of South Florida.

Gastel, B. (1994). *Working with your older patient: A clinician's handbook.* Bethesda, MD: National Institute on Aging, National Institutes of Health.

Gelles, R. J., & Cornell, C. P. (1985). *Intimate violence in families.* Beverly Hills, CA: Sage.

George, L. K., Okun, M. A., & Landerman, R. (1985). Age as a moderator of the determinants of life satisfaction. *Research on Aging, 7*(2), 209–233.

Germain, C. B., & Patterson, S. L. (1988). Teaching about rural natural helpers as environmental resources. *Journal of Teaching in Social Work, 2*(1), 73–90.

Gibbons, K. (1989). *A virtuous woman.* New York: Vintage Books.

Golan, N. (1978). *Treatment in crisis situations.* New York: Free Press.

Golan, N. (1986). Crisis theory. In F. J. Turner (Ed.), *Social work treatment: Interlocking theoretical approaches,* (pp. 296–340). New York: Free Press.

Goodwin, F., & Brown, G. (1990). Summary and overview of risk factors in suicide. In L. Davidson & L. Markku (Eds.), *Risk factors for youth suicide.* New York: Hemisphere.

Gottlieb, B. H. (1985). Social networks and social support: An overview of research, practice, and policy implications. *Health Education Quarterly, 12*(1), 5–22.

Griffith, M. S. (1977). The influences of race on the psychotherapeutic relationship. *Psychiatry, 40*(1), 27–40.

Grossman, A. H. (1995). At risk, infected and invisible: Older gay men and HIV/AIDS. *Journal of the Association of Nurses in AIDS Care, 6*(6), 13–19.

Grossman, A. J. (Ed.). (1983). *Classification in mental retardation.* Washington, DC: American Association on Mental Deficiency.

Guttierrez, L. M. (1990). Working with women of color: An empowerment perspective. *Social Work, 35*(2), 149–153.

Hackl, K. L., Somlai, A. M., Kelly, J. A., & Kalichman, S. C. (1997). Women living with HIV/AIDS: The dual challenge of being a patient and caregiver. *Health and Social Work, 22*(1), 53–62.

Hallowell, N. (1994). *Driven to distraction: Recognizing and coping with attention deficit disorder from childhood through adulthood.* NY: Pantheon Books.

Hamilton, J. (1988). Child abuse and family violence. In N. Hutchings (Ed.), *The violent family: Victimization of women, children, and elders* (pp. 89–103). New York: Human Sciences Press.

Havighurst, R. J. (1963). Successful aging. In R. Williams, C. Tibbits, & W. Donahue (Eds.), *Process of aging* (Vol. 1). New York: Atherton Press.

Hepworth, D. H., & Larsen, J. A. (1990). *Direct social work practice.* Belmont, CA: Wadsworth.

Herek, G. M., & Glunt, E. K. (1988). An epidemic of stigma: Public reactions to AIDS. *American Psychologist, 43*(11), 886–891.

Heslin, P. (1993). When children are the only survivors—parents with AIDS. International Conference on AIDS, June 6–11, 9(2), 800 (abstract no. PO-D03-3496).

Hooyman, N., & Kiyak, H. A. (1996). *Social gerontology: A multidisciplinary perspective.* Boston: Allyn and Bacon.

Ingersoll, B. (1993). *Attention deficit disorder and learning disabilities.* New York: Doubleday.

Intagliata, J. (1992). Improving the quality of community care for the chronically mentally disabled: The role of case management. In S. M. Rose (Ed.), *Case management and social work practice* (pp. 25–55). New York: Longman.

Jett, A., & Branch, L. (1981). The Framingham disability study, II: Physical disability among the aging. *American Journal of Public Health, 71,* 1211–1216.

Johnson, M. (1993). Pushing for access: Loopholes in the ADA present continuing obstacles. *Utne Reader, 56,* 101–103.

Kadushin, A. (1983). *The social work interview.* New York: Columbia University Press.

Kane. R. A. (1985). Case management in health care settings. In M. Weil & J. M. Karls (Eds.), *Case management in human service practice: A systematic approach to mobilizing resources for clients* (pp. 170–203). San Francisco: Jossey-Bass.

Kane, R. A. (1988). Case management: Ethical pitfalls on the road to high quality managed care. *Quality Review Bulletin, 14,* 161–166.

Kane, R. A. (1990). The relevance of case management. In B. S. Fogel, A. Furino, & Gottlieb (Eds.), *Mental health policy for older Americans* (pp. 201–220). Washington, DC: American Psychiatric Press.

Kemper, P., & Murtaugh, C. M. (1991). Lifetime use of nursing home care. *New England Journal of Medicine, 324*(9), 595–600.

Kennedy, E., & Charles, S. (1990). *On becoming a counselor.* New York: Continuum.

Kisthardt, W. E., & Rapp, C. A. (1992). Bridging the gap between principles and practice: Implementing a strengths perspective in case management. In S. M. Rose (Ed.), *Case management and social work practice* (pp. 112–130). New York: Longman.

Koenig, H., & Blazer, D. (1992). Mood disorders and suicide. In J. E. Birren, R. B. Sloan, & G. Cohen (Eds.), *Handbook of mental health and aging.* San Diego, CA: Academic Press.

Kohlberg, L. (1981). *The philosophy of moral development.* New York: Harper & Row.

Kübler-Ross, E. (1975). *Death: The final stages of growth.* Englewood Cliffs, NJ: Prentice Hall.

Kushner, H. (1986). *When all you've ever wanted isn't enough.* New York: Simon & Schuster.

Laryea, M., & Gien, L. (1993). The impact of HIV-positive diagnosis on the individual, part 1: Stigma, rejection, and loneliness. *Clinical Nursing Residency, 2*(3), 245–266.

Levin, J. D. (1987). *Treatment of alcoholism and other addictions: A self-psychology approach.* Northvale, NJ: Jason Aronson.

Levine, C. (1990). AIDS and the changing concept of family. *The Milbank Quarterly, 68*(1), 33–58.

Levine, C. (Ed). (1993). *Orphans of the HIV epidemic.* New York: United Hospital Fund.

Levine, C. (1995). Today's challenges, tomorrow's dilemmas. In S. Geballe, J. Gruendel, & W. Andiman, *Forgotten children of the AIDS epidemic* (pp. 190–204). New Haven, CT: Yale University Press.

Lindemann, E. (1979). *Beyond grief: Studies in crisis intervention.* New York: Jason Aronson.

Linsk, N. L. (1994). HIV and the elderly. *Families in Society, 75,* 362–372.

Livneh, H. (1991). On the origins of negative attitudes towards people with disabilities. In R. P. Marinelli & E. Dell (Eds.), *Psychological and social impact of disabilities* (pp. 181–196). New York: Springer.

Longino, C. F. (1988). Who are the oldest Americans? *The Gerontologist, 28,* 515–523.

Lorde, A., (1997). *The cancer journals: Special edition.* San Francisco: aunt lute books.

Lowy, L. (1979). *Social work with the aging: The challenge and promise of the later years.* New York: Longman.

Lystad, M. (1986). *Violence in the home.* New York: Brunner-Mazel.

Maltsberger, J. (1986). *Suicide risk: The formulation of clinical judgment.* New York: New York University Press.

Markides, K. S., & Mindel, C. H. (1987). *Aging and ethnicity.* Newbury Park, CA: Sage.

McConachie, H. (1982). Fathers of mentally handicapped children. In N. Beril & J. McGuire (Eds.), *Fathers: Psychological perspective* (pp. 144–173). London: Junction Books.

McCubbin, H. I., McCubbin, M. A., Patterson, J., Cauble, S., Wilson, L., & Warwick, W. (1983). CHIP: Coping Health Inventory for Parents: An assessment of parental coping patterns in the care of the chronically ill child. *Journal of Marriage and the Family, 45*(2), 359–370.

McDermott, S., Valentine, D., Anderson, D., Gallup, D., & Thompson, S. (1995). Parents of adults with mental retardation living in-home and out-of-home: Caregiving burdens and gratifications. *American Journal of Orthopsychiatry, 67*(2), 323–329.

McDonald-Wikler, L. (1981). Chronic stresses of families of mentally retarded children. *Family Relations, 30,* 281–288.

McDonald-Wikler, L. (1987). Disabilities: Developmental. In A. Minahan, R. M. Becerra, C. J. Coulton, L. H. Ginsberg, J. G. Hopps, J. F. Longres, R. J. Patti, W. J. Reid, T. Tripodi,

& S. K. Khinduk (Eds.), *Encyclopedia of Social Work* (18th ed., pp. 422–434). Silver Spring, MD: NASW.

McKinlay, J. B., Skinner, K., Riley, J. W., & Zablotsky, D. (1993). On the relevance of social science concepts and perspectives. In M. W. Riley, M. G. Ory, & D. Zablotsky, (Eds.), *AIDS in an aging society.* New York: Springer.

Medvene, L. J. (1990). Family support organization: The functions of similarity. In T. Powell (Ed.), *Working with self help* (pp. 120–140). Silver Spring, MD: NASW.

Mellins, C. A., & Ehrhardt, A. A. (1994). Families affected by pediatric acquired immuno-deficiency syndrome: Sources of stress and coping. *Journal of Developmental and Behavioral Pediatrics, 15*(3), S54–60.

Meredith, N. (1987). Psychotherapy: Everybody's doin' it, but does it work? *Utne Reader,* March/April, 24–33.

Mesibov, G. B., & Bourgondien, M. E. V. (1992). Autism. In S. R. Hooper, G. W. Hynd, & R. E. Mattison (Eds.), *Developmental disorders: Diagnostic criteria and clinical assessment* (pp. 69–95). Hillsdale, NJ: Erlbaum.

Meyers, D. J. (1986). Fathers of handicapped children. In R. R. Fewell & P. F. Vadasy (Eds.), *Families of handicapped children,* (pp. 35–74). Austin, TX: PRO-ED.

Minkler, M., & Roe, K. M. (1993). *Grandmothers as caregivers: Raising children of the crack cocaine epidemic.* Newbury Park, CA: Sage.

Moustakas, C. E. (1961). *Loneliness.* New York: Prentice Hall.

Muschkin, C. G., & Ellis, M. (1993). Migration in search of family support: Elderly parents as caregivers for persons with AIDS. *Program Abstracts.* New Orleans, LA: Gerontological Society of America.

Nalty, L. P. (1997). *The social networks of women and men in residential treatment for chemical dependency.* University of South Carolina: Doctoral Thesis.

National Institute on Aging. (1996). *In search of the secrets of aging.* Bethesda, MD: National Institute on Aging, National Institutes of Health.

Netting, F. E., Kettner, P. M., & McMurtry, S. L. (1993). *Social work macro practice.* White Plains, NY: Longman.

Neugarten, B. L., & Hagestad, G. O. (1976). Age and the life course. In R. H. Binstock & E. Shanas (Eds.), *Handbook of aging in the social sciences.* New York: Roster and Reinhold Press.

Neugarten, B. L., Havighurst, R. J., & Tobin, S. S. (1968). In B. Neugarten, (Ed), *Personality and patterns of aging in middle age and aging* (pp. 173–177). Chicago: University of Chicago Press.

Norman, A. J. (1985). Applying theory to practice: The impact of organizational structure on programs and providers. In M.Weil, & J. M. Karls, *Case management in human service practice: A systematic approach to mobilizing resources for clients* (pp. 72–93). San Francisco: Jossey-Bass.

Nunnally, E., & Moy, C. (1989). *Communication basics for human service professionals.* Newbury Park, CA: Sage.

Ogu, C., & Wolfe, L. R. (1994). *Midlife and older women and HIV/AIDS.* Washington, DC: AARP.

Paradise, S. A. (1993). Older never married women: A cross-cultural investigation. *Journal of Women and Therapy, 14*(1–2), 129–139.

Payne, J. S., & Patton, J. R. (1981). *Mental retardation.* Columbus, OH: Merrill.

Pearlin, L. I. (1985). Social structure and processes of social support. In S. Cohen & S. L. Syme (Eds.), *Social support and health.* Orlando, FL: Academic Press.

Perlman, H. H. (1979). *Relationship: The heart of helping people.* Chicago: University of Chicago Press.

Piaget, J. (1970). *Science and education and the psychology of the child.* New York: Viking Press.

Poindexter, C. (1997). In the aftermath: Serial crisis intervention with persons with HIV. *Health and Social Work, 22*(2), 125–132.

Powell, T. H., & Ogle, P. A. (1985). *Brothers and sisters—a special part of exceptional families.* Baltimore: Paul Brookes.

Price, R. (1995). *A whole new life: An illness and a healing.* New York: Plume.

Puryear, D. A. (1979). *Helping people in crisis.* San Francisco: Jossey-Bass.

Rapoport, L. (1970). Crisis intervention as a mode of brief treatment. In R. Roberts & R. Nee (Eds.), *Theories of social casework.* Chicago: University of Chicago Press.

Rappaport, J. (1981). In praise of paradox: A social policy of empowerment over prevention. *American Journal of Community Psychology, 9,* 1–25.

Riley, M. W. (1993). AIDS and older people: The overlooked segment of the population. In M. W. Riley, M. Ory, & D. Zablotsky (Eds.), *AIDS in an aging society.* New York: Springer.

Roberts, A. R. (1990). *Crisis intervention handbook: Assessment, treatment and research.* Belmont, CA: Wadsworth.

Rockstein, M., & Sussman, M. (1979). *Biology of aging.* Belmont, CA: Wadsworth.

Rogers, C. (1951). *Client-centered therapy.* Boston: Houghton Mifflin.

Rogers, C. (1957). The necessary and sufficient conditions of therapeutic personality change. *Journal of Consulting Psychology, 22,* 95–103.

Rokusek, C. (1995). An introduction to the concept of interdisciplinary practice. In B. A. Thyer & N. P. Kropf (Eds.), *Developmental disabilities: A handbook for interdisciplinary practice* (pp. 1–12). Cambridge, MA: Brookline Books.

Rose, S. M. (1992). Case management: An advocacy/empowerment design. In Rose, S. M. (Ed.), *Case management and social work practice* (pp. 271–297). New York: Longman.

Rothman, J. (1994). *Practice with highly vulnerable clients: Case management and community-based service.* Upper Saddle River, NJ: Prentice Hall.

Rubin, A. (1992). Case management. In S. M. Rose, (Ed.), *Case management and social work practice* (pp. 5–20). New York: Longman.

Saleebey, D. (1996). The strengths perspective in social work practice: Extensions and cautions. *Social Work, 41*(3), 296–305.

Sallin, R. (Producer), & Meyer, N. (Director). (1982). *Star trek II: the wrath of Kahn* [film]. Paramount Pictures.

Saltzman, D. (1995). *The jester has lost his jingle.* Palos Verdes Estates, CA: Jester.

Sanders, C. (1989). *Grief: The mourning after.* New York: Wiley.

Sandler, I. N., & Barrera, M. (1984). Toward a multimethod approach to assessing the effects of social support. *American Journal of Community Psychology, 12*(1), 37–52.

Santos, K. D. (1995). Deafness. In R. L. Edwards & J. G. Hopps (Eds.), *Encyclopedia of Social Work* (19th ed., pp. 685–704). Washington, DC: NASW.

Satir, V., Banmen, J., Gerber, J., & Gomori, M. (1991). The transformation process. *The Satir model: Family therapy and beyond.* Palo Alto, CA: Science and Behavior Books.

Schaie, K. W. (1995). The road toward adult intellectual development. *Bollettino di Psicologia Applicata, 42*(214), 3–14.

Scheer, J., & Groce, N. (1988). Impairment as a human constant: Cross-cultural and historical perspectives on variation. *Journal of Social Issues. 44,* 173–188.

Schein, J. D., & Delk, M. T. (1974). *The deaf population of the United States.* Silver Spring, MD: National Association of the Deaf.

Schneider, J. (1984). *Stress, loss, and grief.* Baltimore: University Park Press.

Schur, E. M. (1983). *Labeling women deviant: Gender, stigma, and social control.* Philadelphia: Temple University Press.

Ship, J. A., Wolff, A., & Selik, R. M. (1991). Epidemiology of acquired immune deficiency syndrome in persons aged 50 and older. *Journal of Acquired Immune Deficiency Syndromes, 4,* 84–88.

Simeonsson, R. J. (1986). *Psychological and developmental assessment of special children.* Newton, MA: Allyn & Bacon.

Sloan, W., & Stevens, H. (1976). *A century of concern: A history of the American Association on Mental Deficiency 1876–1976.* Washington, DC: American Association on Mental Deficiency.

Small, J. (1990). *Becoming naturally therapeutic: A return to the true essence of helping.* New York: Bantam Books.

Sobsey, D. (1995). *Violence and abuse in the lives of people with disabilities: The end of silent acceptance.* Baltimore: Paul Brookes.

Sonnek, I. M. (1986). Grandparents and the extended family of handicapped children. In R. R. Fewer & P. F. Vadasy (Eds.), *Families of handicapped children* (pp. 99–119). Austin, TX: PRO-ED.

Southeast AIDS Education and Training Center. (1991). *Our patients' silent hopes.* Atlanta, GA: Emory University School of Medicine.

Stein, J. B., & Hodge, R. H. (1995). Substance abuse and HIV disease: A multidimensional challenge to caregivers. In V. J. Lynch, G. A. Lloyd, & M. F. Fimbres (Eds.), *The changing face of AIDS: Implications for social work practice* (pp. 199–212). Westport, CT: Auburn House.

Steinmetz, S. K. (1987). Family violence: Past, present, and future. In M. B. Sussman & S. K. Steinmetz (Eds.), *Handbook of marriage and the family* (pp. 725–765). New York: Plenum Press.

Strobe, W., & Strobe, M. (1987). *Bereavement and health: The psychological and physical consequences of partner loss.* New York: Cambridge University Press.

Tatara, T. (1993). Understanding the nature and scope of domestic elder abuse with the use of state aggregate data. *Journal of Elder Abuse and Neglect, 5*(4), 35–57.

Thompson, M. (Ed.). (1994). *Long road to freedom: The advocate history of the gay and lesbian movement.* New York: St. Martin's Press.

Thompson, R. A. (1997). Social support and the prevention of child maltreatment. In G. B. Melton & F. D. Barry (Eds.), *Protecting children from abuse and neglect: Foundations for a new national strategy.* New York: Guilford.

U.S. Bureau of the Census. (1991). 1980 and 1990 censuses of the population. *General population characteristics.* PC80-1-B1, Table 45.

U.S. Bureau of the Census. (1993). Population projections of the U.S., by age, sex, race, and Hispanic origin data: 1993 to 2050. *Current Population Reports.* Series P-25, No. 92-3.

Vadosy, P. F., Fewell, R. R., Mayer, D. J., & Shell, G. (1984). Siblings of handicapped children. *Family Relations, 33,* 155–167.

Valentine, D. (1988). Meeting the basic human needs of developmentally disabled children by empowering parents. *Empowering Families for Better Health.* Proceedings of the bi-regional conference for public health social workers in Regions IV and VI. May. Columbia: University of South Carolina.

Valentine, D. (1990). Double jeopardy: Child maltreatment and mental retardation. *Child and Adolescent Social Work, 7*(6), 487–499.

Valentine, D. P. (1993). Children with special needs: Sources of support and stress for families. *Journal of Social Work and Human Sexuality, 8*(2), 107–121.

Valentine, D., McDermott, S., & Anderson, D. (1998). Is race a factor in the experience of burden and gratification of parents caring for adult children with mental retardation? Unpublished manuscript.

Victoroff, V. (1983). *The suicidal patient: Recognition, intervention, management.* Oradell, NJ: Medical Economics Books.

Viorst, J. (1981). *If I were in charge of the world and other worries.* New York: Atheneum.

Viorst, J. (1986). *Necessary Losses.* New York: Simon and Schuster.

Viorst, J., (1995). *Alexander, who's not (do you hear me? I mean it!) going to move.* New York: Atheneum.

Walker, L. E. (1979). *The battered woman.* New York: Harper & Row.

Webb, N. B. (Ed.), (1993). *Helping bereaved children: A handbook for practitioners.* New York: Guilford Press.

Weil, M., (1985). Key components in providing efficient and effective services. In M. Weil & J. M. Karls (Eds.), *Case management in human service practice: A systematic approach to mobilizing resources for clients* (pp. 29–71). San Francisco: Jossey-Bass.

Weil, M., & Karls, J. M. (1985). Historical origins and recent developments. In M. Weil & J. M. Karls (Eds.), *Case management in human service practice: A systematic approach to mobilizing resources for clients* (pp. 1–28). San Francisco: Jossey-Bass.

Weiler, P. G. (1989). AIDS and dementia. *Generations, 13,* 16–18.

Weise, D., & Daro, D. (1995). *Current trends in child abuse reporting and fatalities: The results of the 1994 annual state survey.* Chicago: NCPCA.

Williams, G. (1980). Management and treatment of parental abuse and neglect of children: An overview. In G. Williams & J. Money (Eds.), *Traumatic abuse and neglect of children at home* (pp. 483–491). Baltimore: Johns Hopkins University Press.

A women's guide to coping with disability. (1994). Epilepsy. Lexington, MA: Resources for Rehabilitation.

Woodside, M., & McClam, T. (1998). *Generalist case management: A method of human services delivery.* Pacific Grove, CA: Brooks/Cole.

INDEX

TO THE OWNER OF THIS BOOK:

We hope that you have found *Essential Skills for Human Services* useful. So that this book can be improved in a future edition, would you take the time to complete this sheet and return it? Thank you.

School and address: ————————————————————————————

Department: ————————————————————————————————

Instructor's name: ———————————————————————————————

1. What I like most about this book is: ——————————————————

——

——

2. What I like least about this book is: ——————————————————

——

——

3. My general reaction to this book is: ———————————————————

——

4. The name of the course in which I used this book is: ————————————

——

5. Were all of the chapters of the book assigned for you to read? ———————

   If not, which ones weren't? —————————————————————————

6. In the space below, or on a separate sheet of paper, please write specific suggestions for improving this book and anything else you'd care to share about your experience in using the book.

——

——

——

——

——

Optional:

Your name: _____ Date: _____

May Brooks/Cole quote you, either in promotion for *Essential Skills for Human Services* or in future publishing ventures?

Yes: _____ No: _____

Sincerely,

Cynthia Cannon Poindexter
Deborah Valentine
Pat Conroy

FOLD HERE

- -

BUSINESS REPLY MAIL

FIRST CLASS   PERMIT NO. 358   PACIFIC GROVE, CA

POSTAGE WILL BE PAID BY ADDRESSEE

ATT: *Cynthia Cannon Poindexter*

Wadsworth Publishing Company
10 Davis Drive
Belmont, California 94002

FOLD HERE